Carbon Nanotube Reinforced Composites

PLASTICS DESIGN LIBRARY (PDL)
PDL HANDBOOK SERIES
Series Editor: Sina Ebnesajjad, PhD
President, FluoroConsultants Group, LLC
Chadds Ford, PA, USA
www.FluoroConsultants.com

The **PDL Handbook Series** is aimed at a wide range of engineers and other professionals working in the plastics industry, and related sectors using plastics and adhesives.

The PDL is a series of data books, reference works and practical guides covering plastics engineering, applications, processing, and manufacturing, and applied aspects of polymer science, elastomers and adhesives.

Recent titles in the series

Biron, Thermoplastics and Thermoplastic Composites, Second Edition (ISBN: 9781455778980)

Drobny, Ionizing Radiation and Polymers (ISBN: 9781455778812)

Ebnesajjad, Polyvinyl Fluoride (ISBN: 9781455778850)

Ebnesajjad, Plastic Films in Food Packaging (ISBN: 9781455731121)

Ebnesajjad, Handbook of Adhesives and Surface Preparation (ISBN: 9781437744613)

Ebnesajjad, Handbook of Biopolymers and Biodegradable Plastics (ISBN: 9781455774425)

Fink, Reactive Polymers, Second Edition (ISBN: 9781455731497)

Fischer, Handbook of Molded Part Shrinkage and Warpage, Second Edition (ISBN: 9781455725977)

Giles Jr., Wagner, Jr., Mount III, Extrusion, Second Edition (ISBN: 9781437734812)

Goodman & Dodiuk, Handbook of Thermoset Plastics, Third Edition (ISBN: 9781455731077)

Kutz, Applied Plastics Engineering Handbook (ISBN: 9781437735147)

Kutz, PEEK Biomaterials Handbook (ISBN: 9781437744637)

McKeen, The Effect of Long Term Thermal Exposure on Plastics and Elastomers (ISBN: 9780323221085)

McKeen, The Effect of Sterilization on Plastics and Elastomers, Third Edition (ISBN: 9781455725984)

McKeen, The Effect of UV Light and Weather on Plastics and Elastomers (ISBN: 9781455728510)

McKeen, Film Properties of Plastics and Elastomers, Third Edition (ISBN: 9781455725519)

McKeen, Permeability Properties of plastics and Elastomers, Third edition (ISBN: 9781437734690)

McKeen, The Effect of Creep and Other Time Related Factors on Plastics and Elastomers, Second Edition (ISBN: 9780815515852)

Modjarrad and Ebnesajjad, Handbook of Polymer Applications in Medicine and Medical Devices (ISBN: 9780323228053)

Niaounakis, Biopolymers Reuse, Recycling, and Disposal (ISBN: 9781455731459)

Sastri, Plastics with Medical Devices, Second Edition (ISBN: 9781455732012)

Sin, Rahmat and Rahman, Polylactic Acid (ISBN: 9781437744590)

Wagner, Multilayer Flexible Packaging (ISBN: 9780815520214)

Woishnis & Ebnesajjad, Chemical Resistance, Volumes 1 & 2, Chemical Resistance of Thermoplastics (ISBN: 9781455778966)

Woishnis & Ebnesajjad, Chemical Resistance, Volume 3, Chemical Resistance of Specialty Thermoplastics (ISBN: 9781455731107)

To submit a new book proposal for the series, please contact Sina Ebnesajjad, Series Editor
sina@FluoroConsultants.com

or

Matthew Deans, Senior Publisher
m.deans@elsevier.com

Carbon Nanotube Reinforced Composites

Marcio Loos

AMSTERDAM • BOSTON • HEIDELBERG • LONDON
NEW YORK • OXFORD • PARIS • SAN DIEGO
SAN FRANCISCO • SINGAPORE • SYDNEY • TOKYO
William Andrew is an imprint of Elsevier

William Andrew is an imprint of Elsevier
The Boulevard, Langford Lane, Kidlington, Oxford OX5 1GB, UK
225 Wyman Street, Waltham, MA 02451, USA

First edition 2015

ISBN: 978-1-4557-3195-4

British Library Cataloguing in Publication Data
A catalogue record for this book is available from the British Library

Library of Congress Cataloging-in-Publication Data
A catalog record for this book is availabe from the Library of Congress

For information on all William Andrew publications visit our web site at http://store.elsevier.com/

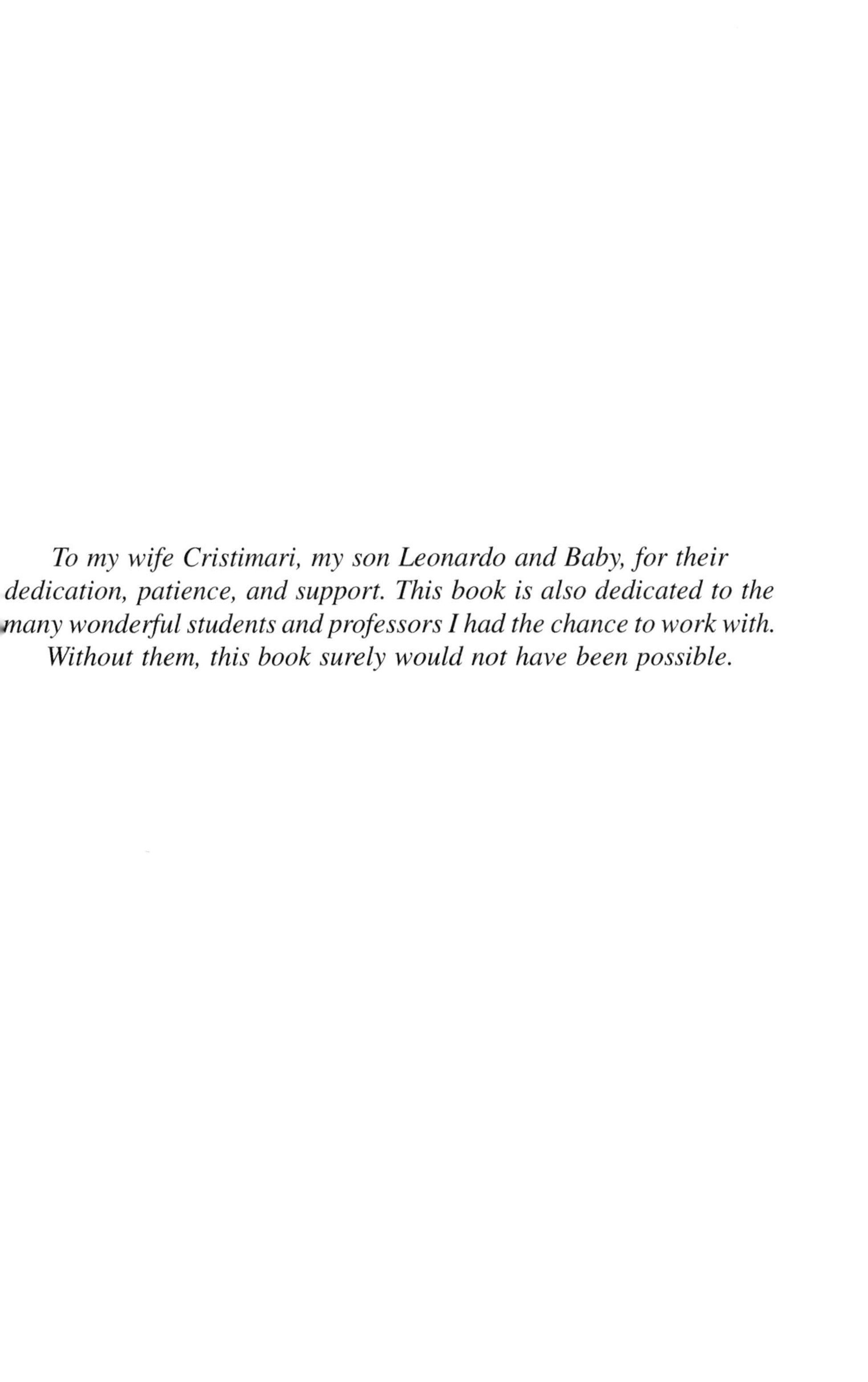

To my wife Cristimari, my son Leonardo and Baby, for their dedication, patience, and support. This book is also dedicated to the many wonderful students and professors I had the chance to work with. Without them, this book surely would not have been possible.

Contents

Foreword

It was the emergence of carbon nanotubes (CNTs) which really motivated funding agencies worldwide to support research in all fields of nanoscience and technology. Researchers immediately saw the great potential of this fascinating new, strong, electrical, and thermal conductive material as an excellent additive to reinforce polymers or to transfer them into a conductor when used as a filler.

However, in the last two decades, CNTs did not reach the expected success, partly due to overestimating its potential, partly due to an early involvement of multinational companies, trying to define own quality standards, which banned further optimization development toward the best-suited CNTs. It was partly not seen (suppressed) that CNTs can and must be tailored toward application. They can be sythesized in numerous shapes and structures (SWCNT, DWCNT, MWCNT, chirality, etc.), length, and diameter, all having influence on properties. They can even be optimized toward electrical conductivity. Finally, CNTs are graphite-based nanostructures, which need and can be produced in individual standards which best fit the properties needed for a specific application.

Some people think that with the emergence of graphene the CNTs' story ends. No, graphene is just another graphite-based nanostructure. It opens the door wide open for further applications, in which nanocomposites are only one field of application. While most of the findings on CNT/polymer composites are transferable to graphene composites, it is not only a book on CNTs.

Just at the right time this new textbook on CNTs is to be published for the scientific audience. The understanding of CNT growth has come to a level that further efforts in CNT development are promising and tailored CNTs will be produced. A profound understanding of chemical treatments for CNTs enables to optimize the tube/polymer interface with the result of best performance. The book gives a comprehensive overview on the state of art achieved in CNT/polymer research until today and does not omit critical aspects.

It is an excellent textbook for scientists who want to learn more about this exciting research field and for students to learn and be informed about CNTs and CNT/polymer composites in detail and generally about carbon-based nanoparticles. It is enriched in each chapter with questions, exercises, and examples, which best support learning and understanding.

Dr.-Ing. Karl Schulte (Professor Emeritus)
Hamburg University of Technology (TUHH)
Hamburg (Germany)

Preface

Why can't we write the entire 24 volumes of the Encyclopedia Britannica on the head of a pin?

This question, posed by Richard Feynman in 1959 during a lecture, can be considered one of the early milestones for nanoscience and nanotechnology (N&N). The problem of manipulating and controlling things on an atomic scale was then put into debate.

Since then many technological developments occurred: man walked on the moon, valves were replaced by tiny transistors, and electronic microscopes capable of increasing our ability to see detail by millions of times were invented.

In 1985 a new allotropic form of carbon, fullerene, a spherical molecule in which carbon atoms are bonded in an arrangement shaped like a soccer ball, was discovered by Richard Smalley, Robert Curl, and Harry Kroto. Sumio Iijima, 6 years later, published his article on carbon nanotubes (CNTs) and thereafter the interest of the scientific community and industries in the topic N&N has been extraordinary. The number of publications and patents covering CNT and N&N grows exponentially year after year. Nanotubes are 250 times stronger than steel, and also have the advantage of being 10 times lighter! CNTs are considered ideal for reinforcement of polymers. The addition of small amounts of CNT has the potential to impart thermal and electrical conductivities to insulating materials.

This book assumes that the reader is relatively new to the area of CNT-reinforced composites without extensive knowledge of the concepts of nanoscience, nanotechnology, composites, and nanotubes. Thus, the first chapters of the book aim to create a solid background in these topics, starting from basic themes such as the importance of size, and why the properties of materials change at the nanoscale. The following chapters will then apply the knowledge gained in the "basic concepts" part of the book creating an easy to follow flow of ideas and concepts. In addition the potential of CNTs to improve mechanical, electrical, and thermal properties of composites is presented with the relevant theoretical models. The expectations of using CNTs are compared to the results obtained in labs, and the reason for the discrepancy between theoretical and experimental results is presented. The processing of polymer composites containing CNTs, which is so important for new researchers on this area, is discussed. Finally a look forward on the potential of CNTs and application of CNTs in the manufacture of polymer composites is presented through two chapters entitled "Is it Worth the Effort to Reinforce Polymers with Carbon Nanotubes?" and "Reinforcement Efficiency of Carbon Nanotubes—Myth and Reality."

This book is written in simple language and is a great addition for undergraduate and graduate courses on the areas of physics, chemistry, and engineering. The book is also very useful for researchers working in the area of N&N, carbon nanotubes, and composites. The examples given in the book, such as applications of composites, are linked to our daily life making the text more attractive. Moreover many questions and exercises with answers are available as an appendix of the book.

Marcio R. Loos
Joinville—Brazil

CHAPTER

Nanoscience and Nanotechnology

1

CHAPTER OUTLINE

1.1 Introduction to the nanoscale

The term nano has etymological origins in the Greek, and means dwarf. This term indicates that physical dimensions are on the order of a billionth of a meter (10^{-9} m or nanometer). This range is colloquially called nanometric scale or simply nanoscale. By convention, dimensions between 1 and 100 nm are accepted as belonging to the nanoscale. Based on Table 1.1 we can understand the context of the nanoscale

Carbon Nanotube Reinforced Composites. http://dx.doi.org/10.1016/B978-1-4557-3195-4.00001-1

Table 1.1 Prefixes for the International System of Units (SI)

Factor	Prefix	Symbol
10^{-24}	yocto	y
10^{-21}	zepto	z
10^{-18}	atto	a
10^{-15}	femto	f
10^{-12}	pico	P
10^{-9}	nano	n
10^{-6}	micro	μ
10^{-3}	milli	m
10^{-2}	centi	c
10^{-1}	deci	d
10^{-1}	deka	da
10^{-2}	hecto	h
10^{-3}	kilo	k
10^{-6}	mega	M
10^{-9}	giga	G
10^{-12}	tera	T
10^{-15}	peta	P
10^{-18}	exa	E
10^{-21}	zetta	Z
10^{-24}	yotta	Y

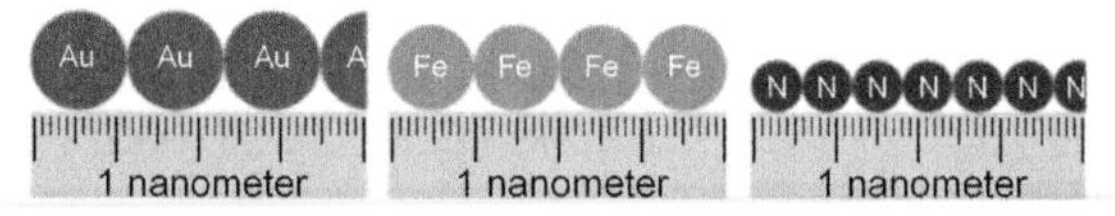

FIGURE 1.1 Different atoms aligned in a 1 nm long ruler: 3.5 gold atoms, 4 iron atoms, and 6.67 nitrogen atoms.

The atoms are considered as hard spheres and the covalent radius is assumed.

in relation to other scales of the international system (SI). Hydrogen atoms, for example, have a diameter of 0.074 nm. Thus, a cube with 1 nm edge could contain about 2500 atoms. The smallest integrated circuit currently known has a lateral dimension of 250 nm and contains 10^6 atoms in an atomic layer thickness. Considering the covalent radius (rigid sphere) of gold, iron, and nitrogen atoms as 0.144, 0.125, and 0.075 nm, respectively, a different amount for each type of atom may be aligned on a 1 nm long ruler (Figure 1.1).

Figure 1.2 compares the size of objects and natural organisms at different scales.

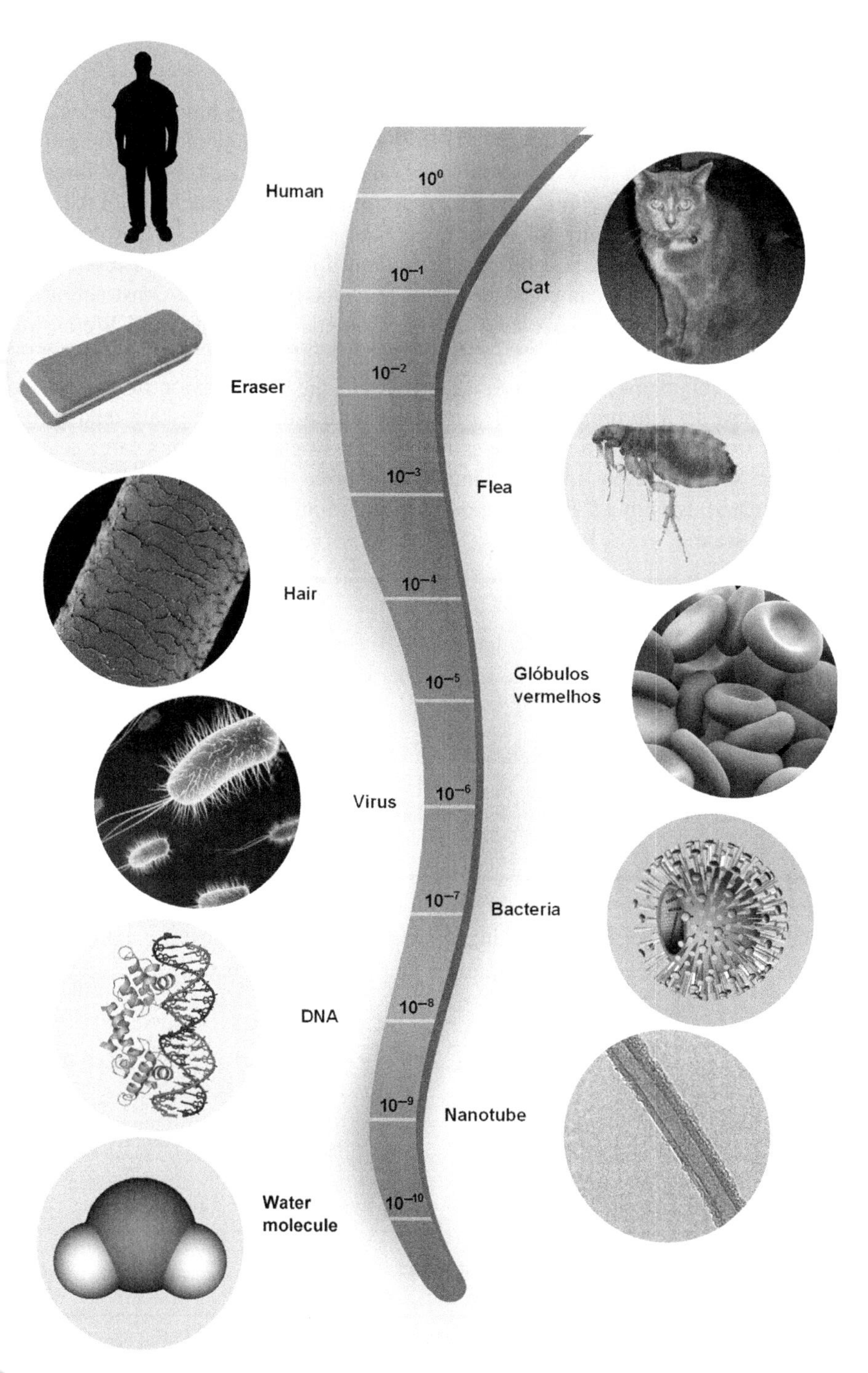

FIGURE 1.2 Size of different objects and natural organisms.

1.2 What makes the nanoscale important?

All materials you see around us, from a grain of sand to the largest galaxies, are formed by atoms. The difference between these materials emerges from the type of atom from which they are composed and the way they interact through bonds or chemical interactions. Only 92 types of atoms occur naturally in the universe. For comparison, only for didactic purposes, we can consider that the basic unit for building a house is a brick; i.e., several bricks form a house. Following this line of thought, we can consider the atom as the basic unit of construction of all that is around us. Thus, we can say that atoms are the basis of all life as we know it.

The size of a given solid has a large effect on the behavior of the atoms that compose it. The properties of a material on the nanometer scale tend to be different from the properties of the same material when viewed on a large scale.[a] There are several reasons for the changes observed in this range. The surface area of nanomaterials is much larger when compared with the same mass of material in a large scale.

EXAMPLE 1.1

A solid aluminum cube has a volume of 0.20 cm^3. Knowing that the density of aluminum is 2.7 g/cm^3, calculate the number of atoms contained in the cube.

Solution

Since the density equals mass divided by volume, the mass m of the cube is

$$m = \rho V = \left(\frac{2.7\ \text{g}}{\text{cm}^3}\right)(0.40\ \text{cm}^3) = 1.1\ \text{g}$$

To find the number of atoms N in the aluminum mass, we must use a proportion rule based on the fact that 1 mol of aluminum (27 g) contains 6.02×10^{23} atoms:

$$\frac{N_A}{27\ \text{g}} = \frac{N}{1.1\ \text{g}}$$

$$\frac{6.02 \times 10^{23}\ \text{atoms}}{27\ \text{g}} = \frac{N}{1.1\ \text{g}}$$

$$N = \frac{(1.1\ \text{g})(6.02 \times 10^{23}\ \text{atoms})}{27\ \text{g}} = 2.5 \times 10^{22}\ \text{atoms}$$

[a]The term *large scale* is used in this text to refer to macroscopic size, in the range of centimeters or meters.

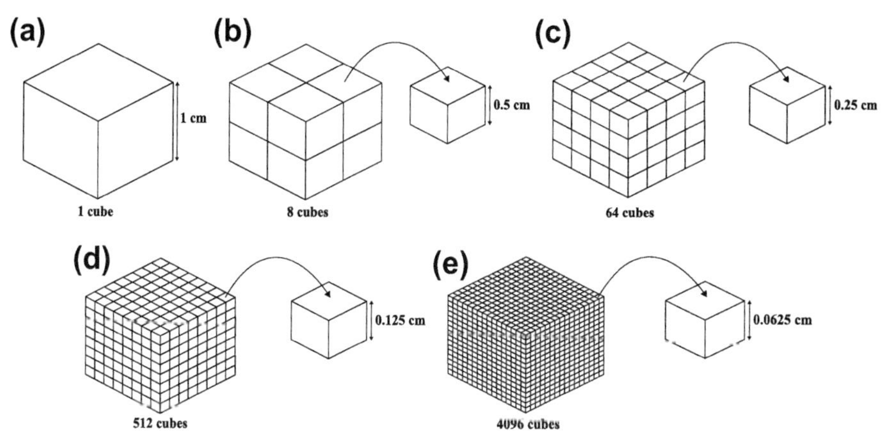

FIGURE 1.3 Changing of the specific surface area of a gold cube through its miniaturization. The edge of the cube varies from 1 cm to 0.0625 cm whereas the number of cubes varies from 1 to 4096.

Consider for example something that surely we would like to have: a gold cube with 1 cm edge (Figure 1.3(a)). In terms of the physical properties of the cube, we can affirm that the color is yellow (typical of gold), its melting point is 1063 °C, its specific surface is 0.31 cm^2/g, and its density is 19.3 g/cm^3, which reflects a mass of 19.3 g. Admit now that you can cut this cube into half-height, half-width, and half-length (Figure 1.3(b)). At the end of the process, we will have eight smaller gold pieces (cubes). Each piece will have the same color, density, and melting point. The weight of each piece is 19.3/8 g (neglecting losses during cutting) and the edge will now be 0.5 cm. However, the specific surface area will be greater and will have a value of 0.62 cm^2/g. If we repeat the same process (Figure 1.3(c)), we will have 64 smaller gold cubes. Again, color, density, and melting point do not vary. The mass of each piece is 19.3/64 g and the edge is 0.25 cm. Again the specific surface area is larger and its value is 1.24 cm^2/g. As you can verify, each time we repeat the process the specific surface area is twice of that obtained previously.

After the fourth cut (Figure 1.3(d)), there will be 512 smaller cubes with an edge of 0.125 cm and the specific surface area will be 2.49 cm^2/g. Yet, if we cut further the 512 cubes previously obtained (Figure 1.3(e)), we will get 4096 cubes with an edge of only 0.0625 cm and therefore the specific surface area will be 4.97 cm^2/g. After this process, we realize that there are some relationships between each current stage of the process and the previous stage. For example, the number of pieces N_p obtained from the starting cube to the end of each step n is described by $N_p = 8^n; n > 0$. In a similar fashion, the edge a of each piece is well described by $a = 0.5 \times 10^{-2}$ m. Finally, the specific surface area (S_{SA}) is given by $S_{SA} = 0.31 \times 2^n; n \geq 0$. Now imagine that you keep repeating the process until you cut the starting gold cube a total of 20 times. Using the above equations, we notice that we will have a total of 1.2×10^{18} cubes. The specific surface area increased dramatically to 325,000 cm^2/g or 32.5 m^2/g.

In addition, each of the pieces will have an edge of 9.5 nm. Certainly, we would not be able to see the new pieces of gold obtained by naked eye and would have to use an electron microscope. On this order of magnitude nanometric, something interesting starts to happen: the nanoscale effect starts to be detected. The melting point, which was 1063 °C, may decrease to 500 °C. The color that until then was constant now depends on the size of each piece. Pieces with different edges will have different colors ranging from blue to red in the spectrum. In addition, the fraction of atoms located on the surface of the material will be more reactive than the atoms located on its interior.

Thus, the reduction of dimensions makes the material more chemically reactive and can affect its electrical, magnetic, morphological, structural, thermal, optical, and mechanical properties. On this scale, such properties can no longer be described in terms of classical (Newtonian) mechanics, and quantum mechanics comes to be used for the explanation of various phenomena observed.

An important parameter to describe the phenomenon of variation in the properties of a material with size is the area/volume (A/V) ratio. For spherical particles with diameter d and radius R, we have

$$\frac{A}{V} = \frac{4\pi R^2}{\frac{4}{3}\pi R^3} = \frac{3}{R} = \frac{6}{d} \tag{1.1}$$

For particles in the form of a cylinder with diameter d, radius R, and height h, we have

$$\frac{A}{V} = \frac{2\pi R l}{\pi R^2 l} = \frac{2}{R} = \frac{4}{d} \tag{1.2}$$

As we can see, the A/V ratio varies with the inverse of the diameter ($1/d$) for different geometries. Thus, various properties vary linearly if plotted as a function of $1/d$. Table 1.2 shows how the area/volume ratio varies for cubes with edge from 1 m to 1 nm.

The diameter of spherical or cubic particles increases with the inverse of the third power of the number of atoms, i.e., $N^{1/3}$ [1]. In this context, the fraction of atoms found on the outer surface of a material (F_s) becomes relevant. Consider, for example, an agglomerate of cubic shape containing N atoms with radius r_0 at the edges of the cube. The total number of atoms forming the cube is $N = n^3$. The

Table 1.2 Variation of the Area, Volume, and Area/Volume Ratio for Cubes with Edges of Different Sizes

Dimension (m)	Area (m^2)	Volume (m^3)	Area/Volume
1	6	1	6
0.1	6×10^{-2}	6×10^{-3}	10
0.01	6×10^{-4}	6×10^{-6}	100
0.001	6×10^{-6}	6×10^{-9}	1×10^{3}
1×10^{-6}	6×10^{-12}	6×10^{-18}	1×10^{6}
1×10^{-9}	6×10^{-18}	6×10^{-27}	1×10^{9}

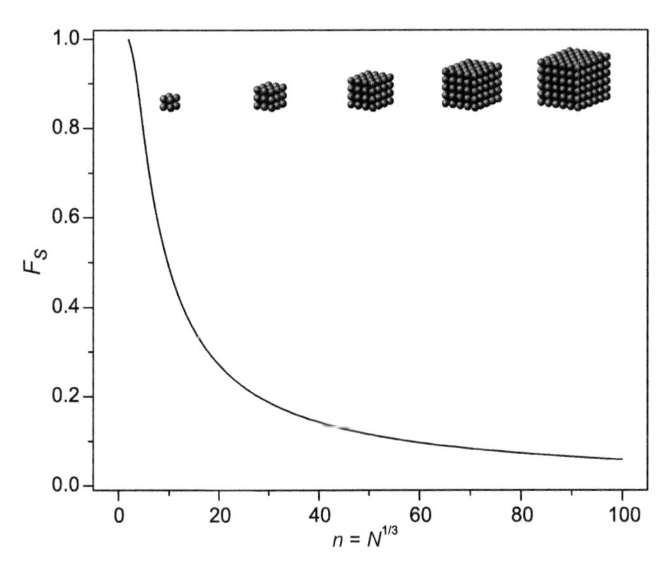

FIGURE 1.4 Variation of the fraction of atoms in the outer surface of a material as a function of the number of atoms.

Figure sourced from Nanoscopic Materials [1]

number of atoms at the surface is given by $6n^2$ (for the six faces) minus $12n$ (correcting double counting of the edges) and considering the eight corners:

$$F_s = \frac{6n^2 - 12n + 8}{n^3} = \frac{6}{N^{\frac{1}{3}}}\left(1 - \frac{2}{N^{\frac{1}{3}}} + \frac{8}{6N^{\frac{2}{3}}}\right) \approx \frac{6}{N^{\frac{1}{3}}} \tag{1.3}$$

Figure 1.4 shows a graph of F_s as a function of the number of atoms N. Note that for $N = 8$ ($n = 2$) all atoms forming the cube are found on the surface and therefore $F_s = 1$. As the number of atoms increases, the fraction of atoms encountered on the surface dramatically decreases: $F_s \rightarrow 0$ when $N \rightarrow \infty$.

For agglomerates in the form of a sphere of radius R containing atoms with radius r_0, the fraction of atoms found on the surface F_s is proportional to the volume of a layer with a $2r_0$ thickness divided by the total volume. In this case, the number of atoms can be approximated by $N = R^3/r_0^3$. Assuming a constant packing density along the sphere, F_s is given by [1]

$$\frac{A}{V} = \frac{4\pi R^2 2r_0}{\frac{4\pi R^3}{3}} = \frac{6r_0}{N^{\frac{1}{3}}} \tag{1.4}$$

Note that Eqn (1.4) is similar to Eqn (1.3). In fact, this equation is an approximation because the packing density of the atoms on the surface of the sphere is smaller than in its interior and in this case $F_s \sim 4/N^{1/3}$ [2].

Atoms and molecules near or on the surface or yet in the edges and corners have a smaller number of "neighbors" than atoms in the center of a cluster, and therefore are less linked (lower interaction). For this reason, the surface has higher energy and this affects other properties of the particle.

EXAMPLE 1.2

Calculate the mass of a copper atom (molecular weight of 63.55 g/mol).

Solution

As there are 6.02×10^{23} particles in 1 mol of any element, the mass per atom for a given element is

$$m_{atom} = \frac{\text{Molar mass}}{N_A}$$

Thus, the mass of a copper atom (Cu) is

$$m_{Cu} = \frac{63.55 \ \text{g/mol}}{6.02 \times 10^{23} \ \text{atom/mol}} = 1.06 \times 10^{-22} \ \text{g/atom}$$

A parameter of extreme importance to nanoparticles is their specific surface area S_{SA}, which is the ratio of the surface area by the volume of a material (S_A/V_m) divided by its density ρ_m.

$$S_{SA} = \frac{S_A}{V_m \rho_m} = \left[\frac{m^2}{\frac{g}{m^3} \cdot m^3}\right] = \left[m^2 \cdot g^{-1}\right] \tag{1.5}$$

In the following section, we will briefly present some of the observed changes in the properties of nanoscale materials, which can simply be called nanoparticles.

1.3 Properties of nanoparticles and effect of size

As we have seen above, the nanoscale materials tend to have unique characteristics different from the same material when considered on a large scale. Even properties such as melting point and dielectric constant, considered as specific properties, may change when the particles reach nanometric size. This change in the fundamental properties with the variation of particle size is known as the "size effect." Nanoparticles show a number of unique characteristics in their morphological, structural, thermal, electromagnetic, optical, and mechanical properties.

1.3.1 Morphological and structural properties

The elevated specific surface area of nanoparticles affects their reactivity and solubility. In processes such as sinterization—in which once compacted, the particles are heat treated—the heat and mass transfer among the nanoparticles and the surroundings plays an important role in determining the internal and surface properties of the final product. In some cases, the crystalline structure of the particles can change when at the nanoscale.

1.3.2 Thermal properties

As the atoms located on the surface of the particles are influenced by the nanoscale, the melting point of the nanoparticles decreases as compared to the same materials on a large scale. This is because at the nanoscale the motion of atoms tends to occur at lower temperatures. For example, the melting point of gold is 1063 °C and starts to decrease substantially to gold particles with diameters lower than 20 nm. Gold particles with diameters of 2 nm, for example, have a melting point up to 500 °C lower than gold particles on a large scale. Figure 1.5 shows how the melting point of agglomerated gold particles varies with the radius.

1.3.3 Electromagnetic properties

Nanoparticles are used as the base material for a large number of electronic devices. The size and electrical properties of nanoparticles are of great importance for improving the performance of these products [4]. The dielectric constant of particulates such as lead titanate ($PbTi0_3$) tends to increase substantially as the particles become smaller than 20 nm. Moreover, as a result of changes in electromagnetic

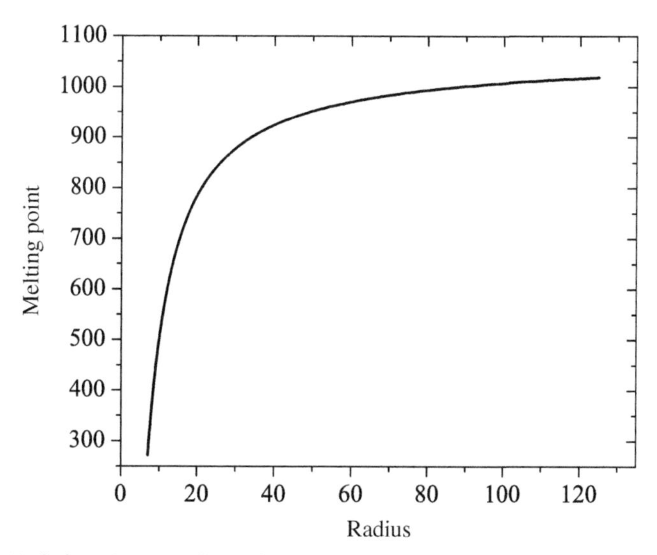

FIGURE 1.5 Variation of the melting point of agglomerated gold particles as a function of the radius.

Figure sourced from Atomic and molecular clusters [3]

properties, gold nanoparticles exhibit unique catalytic characteristics, unlike the gold which is stable at normal scale [5].

1.3.4 Optical properties

As the size of certain particles reaches the nanoscale, the absorption of light starts to take place at a specific wavelength (color) different from that observed when the same material was on a large scale [6]. Thus, the light absorption of certain nanoscale particles can be a function of the diameter and, of course, the type of material. For example, the maximum wavelength absorbed by gold particles with a diameter of 15 nm is 525 nm, while for particles with a diameter of 45 nm this value changes to 575 nm. Moreover, since nanoparticles are smaller than the wavelength of visible light, the light scattering over them can be considered negligible.

1.3.5 Mechanical properties

The hardness of crystalline materials generally increases with decreasing size of the crystals and thus the mechanical strength of these materials is high. Ceramic and metallic materials can exhibit this behavior when they are at the nanoscale.

Thus, based on the size effect on the properties of materials, we can actually claim that there is a new world on a nanoscale or, if you prefer, a nanoworld. Widely known materials exhibit completely different behaviors when they are totally at the nanoscale. These new properties observed at the nanoscale offer a wide range of new applications, new materials, and new technologies explored by nanoscience and nanotechnology (N&N).

1.4 N&N history

Generally, the starting point for the onset of N&N is credited to the American physicist and Nobel Prize laureate Richard Feynman (1918–1988). In 1959, Feynman gave a lecture entitled "There's Plenty of Room at the Bottom" at the annual meeting of the American Physical Society [7].

> *Why can't we write the entire 24 volumes of the Encyclopedia Britannica on the head of a pin?*

Making use of questions like that during his presentation, Feynman suggested that in the future we would be able to "manipulate and control things at the atomic scale" and "arrange the atoms the way we want," provided, of course, the laws of nature would not be violated. For teaching purposes, it would be equivalent to consider the atoms as parts of a game "Lego" and simply arrange them as we wish. It should be noted that at no time did Feynman use the prefix nano during his presentation. The transcription of Feynman's lecture can be found in the Appendix A at the end of this book.

Based on the ideas of Feynman, Kim Eric Drexler (1955 to present) began to develop the concept of molecular nanotechnology and published his book *Engines of Creation: the Coming Era of Nanotechnology* in 1986. In this work, Drexler suggested the concept of using molecular structures at the nanoscale to serve as machines to guide and activate the synthesis of large molecules. Drexler has proposed the use of billions of machines like this, similar to robots that would form the basis of molecular manufacturing technology, enabling to literally build anything, atom by atom and molecule by molecule. Since the publication of his book and after much work, this concept of "nanobots" is somehow still speculative. However, devices such as nanomotors have been recently developed [8]. The fact is that year after year the number of scientific papers on the theme "nano" is increasing exponentially.

One of the major milestones for the initiation of the nanotechnology revolution was the invention of the scanning tunneling microscope in 1981 by the physicists Gerd Binnig and Heinrich Rohrer, who were Nobel Prize laureates in Physics in 1986. The first evidence that atoms could be manipulated and accurately positioned using instruments was in 1989 when IBM scientists used a scanning tunneling microscope to manipulate 35 xenon atoms and form the letters IBM (Figure 1.6). Although still restricted to academic space, this technique aroused great interest in the fabrication of electronic devices on the atomic scale.

Up to now, we realized the importance of nanoscale and its effects on various properties. But what exactly do words like nanotechnology and nanoscience mean? Formulating a precise definition for these words is a difficult task. Even scientists in the field argue that the answer to this question depends on whom you are asking [9]. It is difficult to differentiate between science and technology. Science involves both theory and experiments. Technology involves the development, applications, and business implications. We can say that one feeds the other. Science forms the basis for technology. Through technology, better tools, equipment, materials, and specialties are created for the development of science. This relationship is depicted in Figure 1.7.

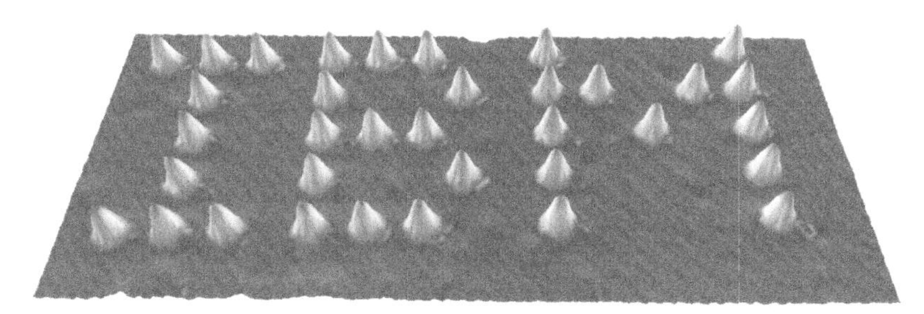

FIGURE 1.6 Xenon atoms positioned on a nickel substrate using a scanning tunneling microscope.

Figure reproduced with permission from IBM Research, Zurich

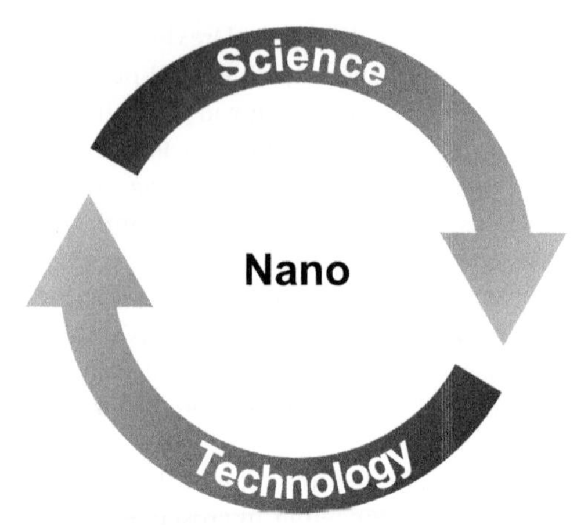

FIGURE 1.7 Relationship between science and technology.

Figure sourced from Introduction to nanoscience and nanotechnology [10]

The term nanoscience describes the scientific study of the properties of the materials at the nanoscale. Like science, this term does not describe a practical application. The term nanotechnology was defined by Norio Taniguchi (1912–1999) for the first time in a conference in 1974 [10]. At the time he used this term to describe the processing of a material with nanoscale precision. While many other definitions for nanotechnology exist, the National Nanotechnology Initiative (NNI) describes an area of research, development, and engineering as nanotechnology only if it involves all of the following [11]:

1. Research and technology development at the atomic, molecular, or macromolecular levels, in the length scale of approximately 1–100 nm range.
2. Creating and using structures, devices, and systems that have novel properties and functions because of their small and/or intermediate size.
3. Ability to control or manipulate on the atomic scale.

Therefore, nanotechnology can be defined as the ability to work on a molecular level, atom by atom, to create large structures with new fundamental properties and functions. Nanotechnology can also be described as the creation and precise manipulation of materials at the atomic scale [12]. Thus, nanotechnology is the application of nanoscience to control processes at the nanoscale, i.e., between 1 and 100 nm [13].

Nano is a big business! The American Foundation for Science predicts that the market for products and services related to nano can reach US$1 trillion in 2015 [14]. This makes the nano-industry one of the fastest growing industries in history, which is higher than that of the telecommunications industries and information technology combined. Nano is already a priority for large technology companies like HP (Hewlett Packard), NEC (NEC Corporation), and IBM

(International Business Machines Corporation) [15]. All indicate that the next industrial revolution will be based on N&N.

1.4.1 Types of nanotechnology

The fabrication of nanomaterials, or simply nanotechnology, is divided into two categories: top-down and bottom-up [16].

Top-Down [From top (great) to down (small)]: Mechanisms and structures are miniaturized to the nanoscale. This has been the most common application of nanotechnology to date. In particular, this miniaturization is predominant in the field of electronics. One of the processes most used in this category is lithography, where materials such as silicon, a semiconductor, are processed for electronic devices. One disadvantage of this process is the loss of material during the various stages.

Bottom-Up [From bottom (small) to top (large)]: In this category we start with nanometric structures, such as atoms or molecules, and through a process of self-assembly or assembly, mechanisms or devices greater than the initial ones are created. This category is considered the "true" nanotechnology and is beyond the limits of miniaturization, allowing control of matter with extreme precision. Some of the production methods used in this category are sol-gel, chemical vapor deposition, and laser pyrolysis. As we shall see, some of these methods are used for the manufacture of carbon nanotubes (CNTs). Some application examples of this category of nanotechnology are shown in Figure 1.8.

1.5 Nano in history

An extraordinary work with glass made by the Romans in the fifth century AD demonstrates one of the greatest examples of nanotechnology in the ancient world. The Lycurgus cup, which is part of the collection of the British Museum, shows King Lycurgus being dragged to the underworld by Ambrosia. Surprisingly, when lit outside the cup looks green (Figure 1.9(a)). When lit from inside the cup looks reddish and the King Lycurgus looks purple (Figure 1.9(b)).

The explanation for this phenomenon was obtained only in 1990 after scientists analyzed the cup using an atomic force microscope [18]. It was found that the dichroism (two colors) is observed due to the presence of nanoparticles, silver 66.2%, 31.2% gold, and 2.6% copper, up to 100 nm in size, dispersed in a glass matrix [19,20]. The red color observed is a result of absorption of light ($\sim$520 nm) by the gold particles. The purple color results due to the absorption by the larger particles while the green color is attributed to the light scattering by colloidal dispersions of silver particles with size >40 nm. The Lycurgus cup is recognized as one of the oldest synthetic nanocomposites [13].

Besides the Romans, medieval artisans also explored the effect of the addition of metal particles in glass to create stained glass windows [13]. Figure 1.10 shows an

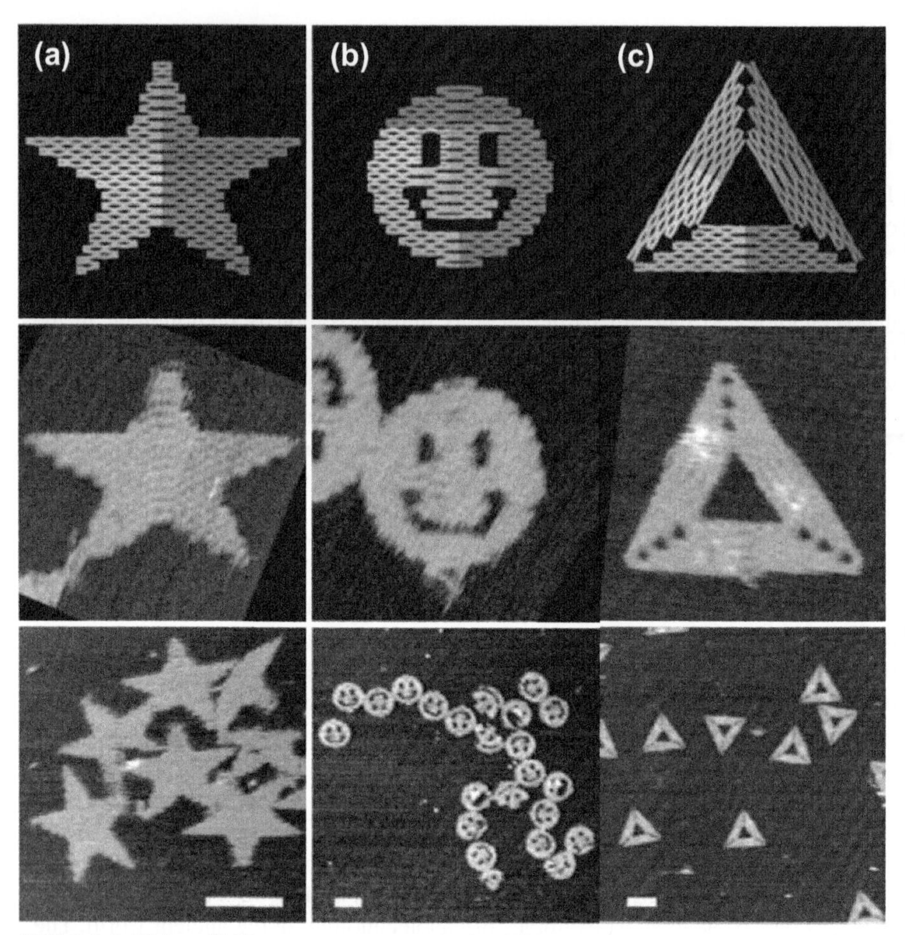

FIGURE 1.8 "Origami"-like structures created using DNA through the bottom-up fabrication.

(a) Star, (b) disc with three holes, and (c) triangle with trapezoidal domains.

Figure sourced from Folding DNA to create nanoscale shapes and patterns [17]

example of how gold and silver nanoparticles of different sizes were applied to stained glass windows.

Stained glass was also produced in Chinese ancient culture [23,24], and porcelain was also fabricated using gold nanoparticles with sizes in the range of 20–60 nm. The use of gold had been extended even to areas in medieval medicine, and work in this area was published by Hans Heicher in 1718 [25,26].

Alchemists had knowledge of colloidal dispersions and potential applications; however, they did not have science-based explanations for the observed phenomena and did not follow rigorous experimental methods. Figure 1.11 depicts alchemists experimenting around 1661.

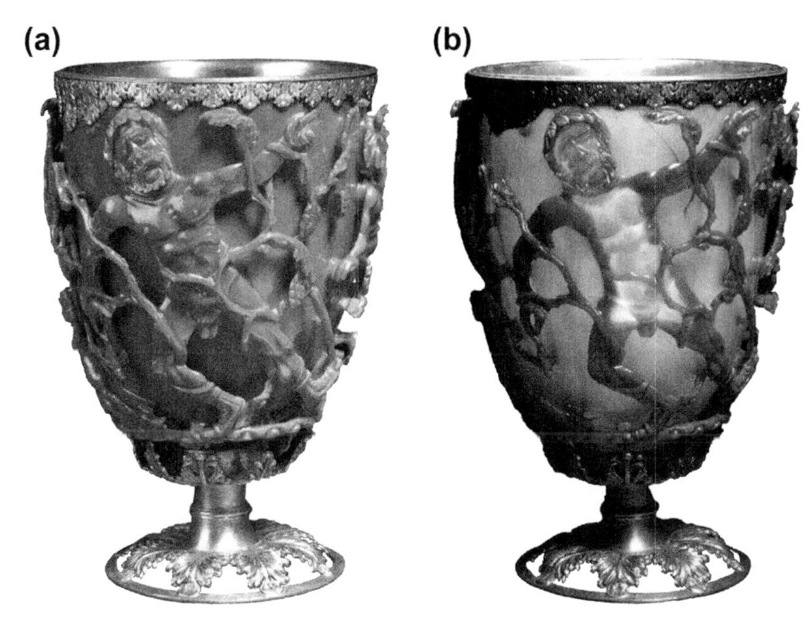

FIGURE 1.9 The Lycurgus cup.

The glass appears green when illuminated from the outside (a) and purple-red when illuminated from the inside (b).

Figure kindly provided by The British Museum [15]

In 1857, Michael Faraday was the first to study in a rigorous and scientific way the preparation and properties of colloidal suspensions of gold [28]. In this context, Table 1.3 presents a timeline of events relevant to N&N.

1.6 Moore's Law

Probably, the most important technological breakthrough in the second half of the twentieth century was the advent of the electronics of silicon. The microchip and its revolutionary applications are a result of the development of silicon technology. In the 1950s, televisions were black and white and with low-quality pictures. There were no digital clocks, internet, optical fiber, or cell phones. All these advances are a direct reflection of chip development. Furthermore, the current price of silicon contributes to the marketing of computers at prices much more affordable than before.

The quest for miniaturization of electronic components contributed to the consolidation of the nano era. The first transistor was devised by J. Bardeen, W. Shockley, and W. H. Brattain in 1947 (Figure 1.12) [31]. The invention earned the creators the Nobel Prize in 1956 (Figure 1.13.)

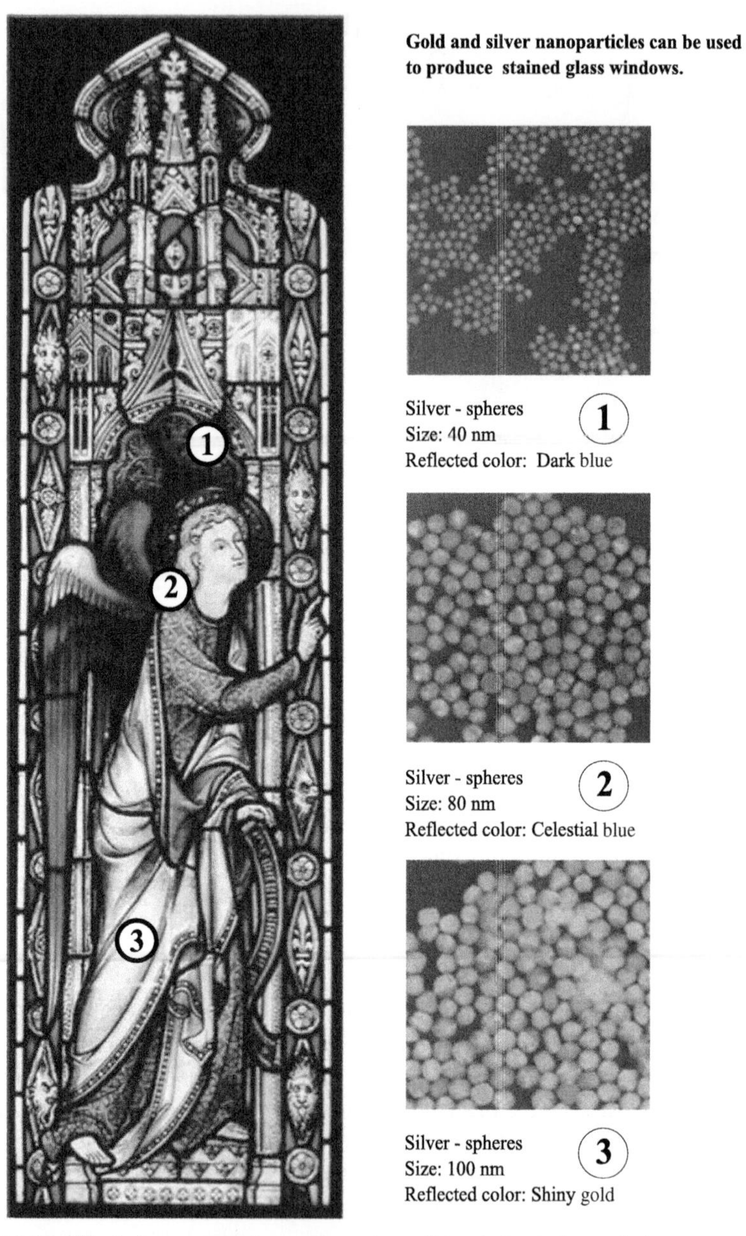

FIGURE 1.10 Effect of nanoparticles on the colors of the stained glass windows.

Figure reproduced with permission from Prof. Dr. Chad A. Mirkin, *Nanotechnology Northwestern University*

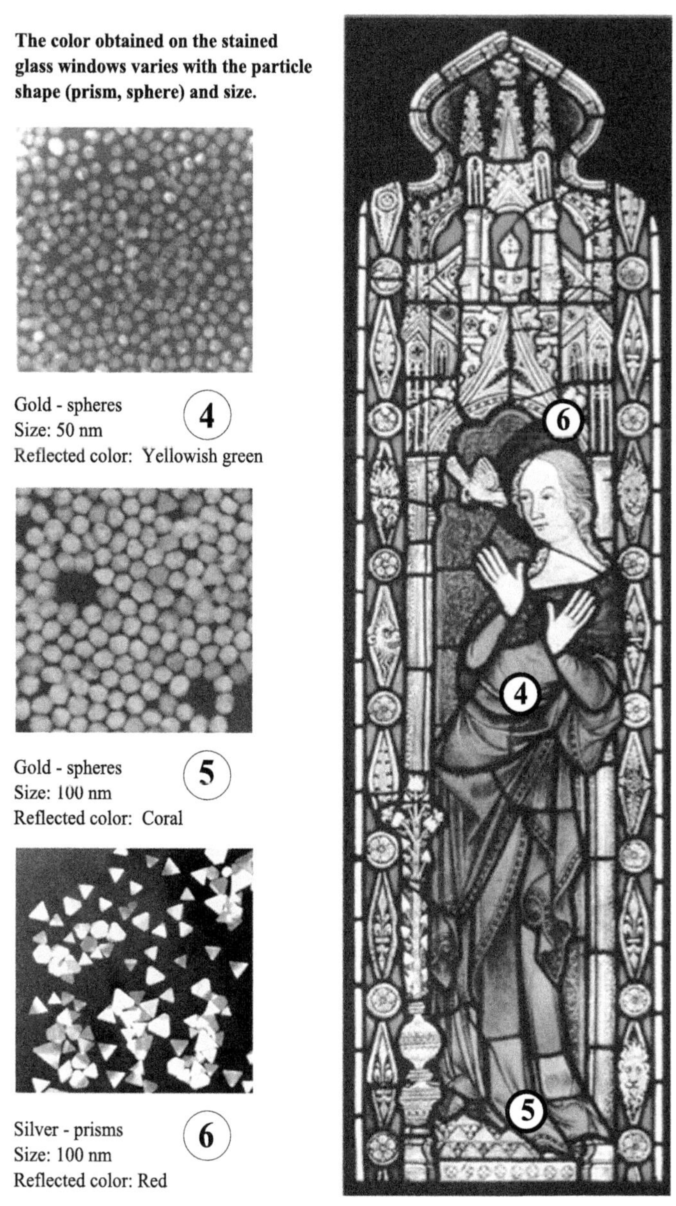

FIGURE 1.10 cont'd.

In 1951, the first transistor made of silicon was created, and in 1958, the first integrated circuit was produced. In 1965, Gordon E. Moore, who became one of the founders of Intel three years later, established Moore's Law [33]. Moore's Law stated that "considering the current rate of technological development, the

FIGURE 1.11 Nanotechnology in the past.

Picture "The Alchemist" by Pietro Longhi, 1661.

Figure sourced from Alchemy website [27]

complexity of an integrated circuit, seeking a minimum cost would double every 18 months." Figure 1.14 presents an adaptation of Moore's Law showing the evolution of computing power.

Currently Moore's Law is interpreted as "data density will double every 18 months." In 1965, a chip contained 30 transistors and in 1971, 2000 transistors (Figure 1.15.) Currently chips can contain up to about 40,000,000 transistors. According to the National Nanotechnology Initiative, transistors will be miniaturized to at least 9 nm, resulting in chips with billions of transistors.

Table 1.3 Timeline with Events Relevant to Nanoscience and Nanotechnology.

2600 YA	Kanada	Atomism
2600 YA	Anaximander	The Aeprion
2500 YA	Democritus	Atomism
4th Century	Lycurgus cup	Colored glass
1st Century Hero	Compressibility	
500–1450	Cathedrals	Stained glasses
1450–1600	Deruta pottery	Iridescence/metal clusters
1661	Robert Boyle	Minute masses
1805	John Dalton	Elements are atoms
1827	Photography	Silver nanoparticle
1857	Michael Faraday	Colored gold colloids
1908	Gustav Mie	Light scattering nanoparticles
1931	Ernst Ruska	Electron microscope
1938	Langmuir/Blodgett	LB films
1947	John Bardeen et al.	The first transistor
1951	Erwin Mueller	Ion-field electron microscope First to see atoms
1953	Watson and Crick	DNA
1958	Leo Esaki	Electron tunneling
1959	Richard Feynman	"There's plenty of room ..."
1960s	NASA	Ferrofluids
1960	Plank and Rosinski	Zeolites and catalysis
1965	Gordon E. Moore	Moore's Law
1970	John Pople	Atomistic modeling
1974	Norio Taniguchi	The term "nanotechnology"
1974	M. Ratner/Aviram	Molecular electronics
1977	Richard P. van Duyne	Surface enhanced Raman (SERS)
1980	Jacob Sagiv	Self-assembled monolayers
1981	G. Binnig/H. Rohrer	Scanning tunneling microscope
1985	Smalley/Croto/Curl	Buckminster fullerenes
1986	G. Binnig/C. Gerber	Atomic force microscope
1987	Averin/Likharev	Single-electron tunneling transfer
1988	Louis Brus	Quantum dots
1990	Donald Eigler et al. IBM with	Xe atoms
1991	Sumio Iijima	Multi-wall carbon nanotubes
1993	Iijima/Bethune	Single-wall carbon nanotubes
1996	Mirkin/Letsinger	SAM of DNA + gold colloids
2000	Hersam/Lyding	Feedback control lithography
2001	Mihail Roco	National Nanotechnology Initiative

Figure sourced from discovernano website [29]

FIGURE 1.12 First transistor invented in 1947.

Figure sourced from Alcatel-Lucent USA Inc. website [30]

FIGURE 1.13 Inventors of the transistor.

Figure sourced from Alcatel-Lucent USA Inc. website [32]

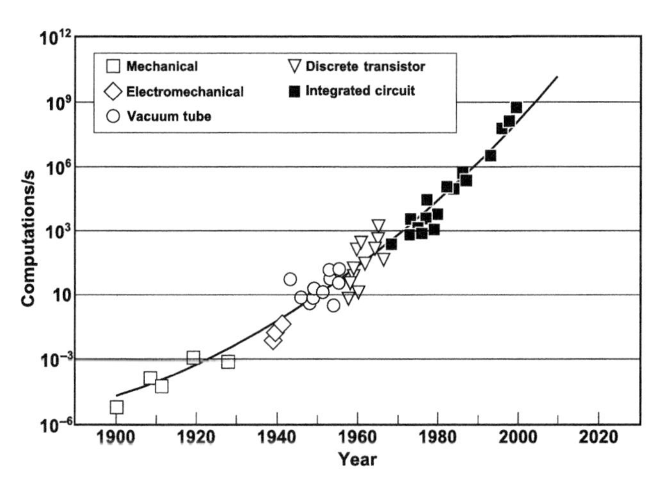

FIGURE 1.14 Adaptation of Moore's Law showing the evolution of computing power.

Figure reproduced with permission from IBM Research, Zurich

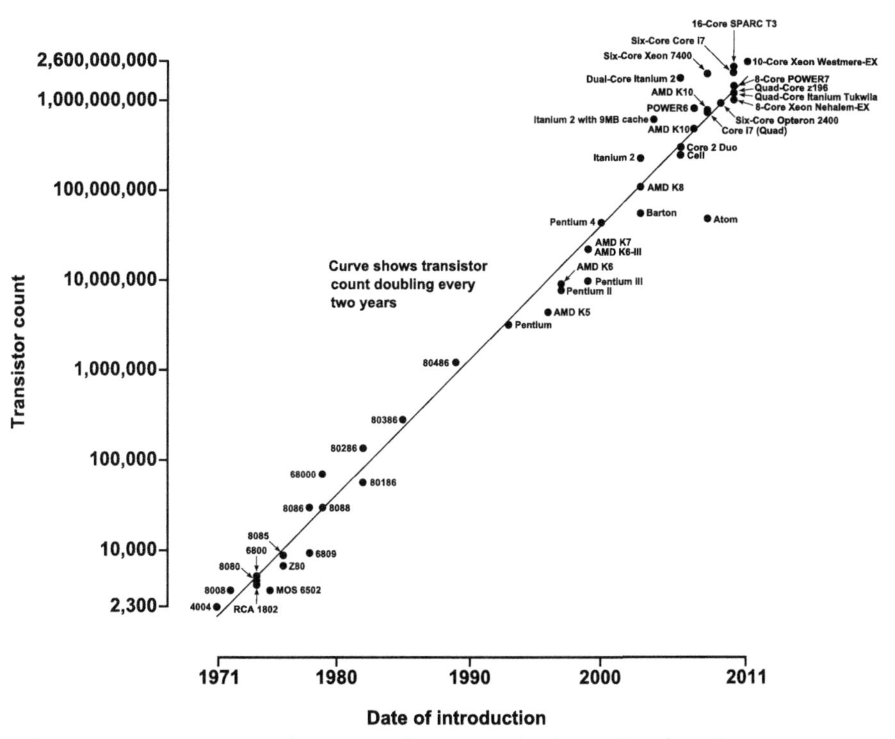

FIGURE 1.15 Number of transistors on an integrated circuit as a function of the year.

The number of transistors grows exponentially, doubling every two years.

Figure sourced from Wikipedia [34]

1.7 Applications of nanotechnology

The new nano world opens the door to more creative and, why not, unusual applications making use of the new features and phenomena observed at the nanoscale. Scientists around the world are constantly working on new ideas, mechanisms, and technologies that promise to effectively make the next industrial revolution a reality. The applications cover many areas, from the simplest such as food to the most complex such as space. Several products based on nanotechnology are already available on the market. In this section, we present some of the existing and envisioned applications in various areas.

1.7.1 Aviation and space

Using CNTs, researchers plan to build a "space elevator," a space transportation system that extends from the Earth's surface into space via a cable of length between 40,000 and 96,000 km [35]. In space, the cable will be attached to a counterweight (Figure 1.16). The idea of building a space elevator was presented in a 1960 article by Yuri N. Artsutanov. It is believed that the technology that will eventually make this visionary idea possible is the use of CNTs, one of the most resistant materials known to humans. The main purpose of the space elevator is to reduce the cost of shipping (launching) materials to the low earth orbit (LEO).

The current cost of sending materials to LEO ranges from $10,000 to $25,000 per kg [37]. In addition, the dream of starting human habitation on other planets would be somehow leveraged with respect to transportation. The prediction is that the space elevator will be ready by 2031 [38].

Another way to reduce the cost of space travel is the use of new materials in the manufacture of rockets and space shuttles. The major candidates for this purpose are composites prepared using high-performance polymers reinforced with nano-fillers. Ceramic matrix composites and metals are also used, but the application of polymers greatly reduces the mass of the aircraft and consequently the fuel consumption. The use of CNTs as a reinforcing agent for polymers also offers the possibility of increasing the structural strength and thermal stability of the composites.

1.7.2 Medicine

While most applications of nanotechnology in medicine are still under development, nano-silver crystals are already used as anti-microbial agents for the treatment of wounds [39]. Studies have shown that quantum dots (QD), particles with interesting optical and electronic properties, are able to selectively bind to cancer cells and mark them. In this case the QD act as fluorescent tags [40]. To transport the QD to diseased cells, researchers use a nanostructure that easily attaches to the plasma membrane of cells, known as dendrimer. In future, the use of dendrimers and QD coupled with other molecules such as folic acid could be used for the early diagnosis of cancer and also could be used as carriers for specific drugs against the disease itself,

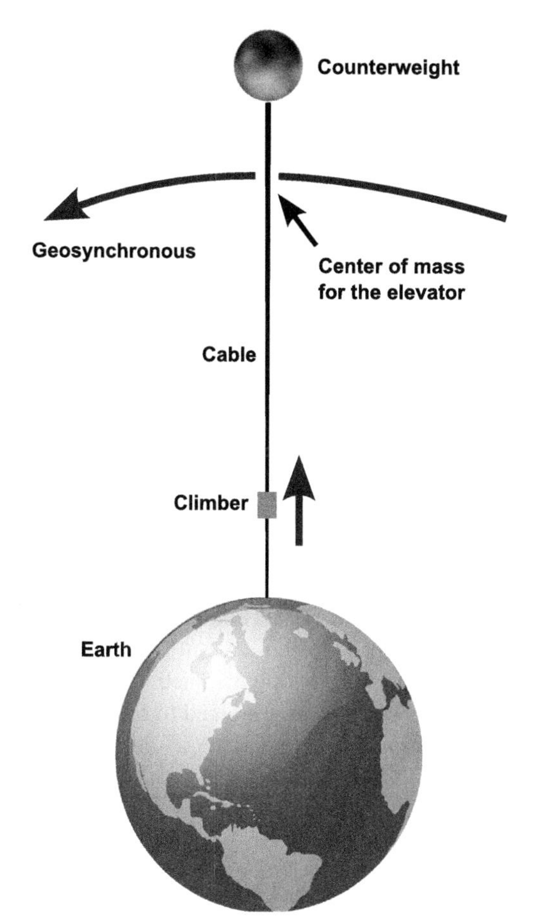

FIGURE 1.16 Representation of the space elevator designed to carry cargo from Earth's surface into space.

Figure sourced from spaceelevator.com [36]

revolutionizing the early diagnosis and treatment that currently exist for this disease. The use of nanoparticles to deliver chemotherapy drugs directly into tumor cells minimizes damage to healthy cells [41,42,43]. Following this same logic, the goal is the synthesis of nanocapsules that concentrate the energy of infrared light and X-rays to destroy cancer cells with minimal damage to the surrounding cells [44]. Figure 1.17 gives a good visual explanation of the use of nanocapsules. Yet, scientists are working on the preparation of nanoparticles that can bind to cells infected with various diseases and allow physicians to identify specific diseases through blood tests [46].

The healing process of a broken bone is long and tedious. The most effective method to repair broken bones is to hold the bone in the correct position during

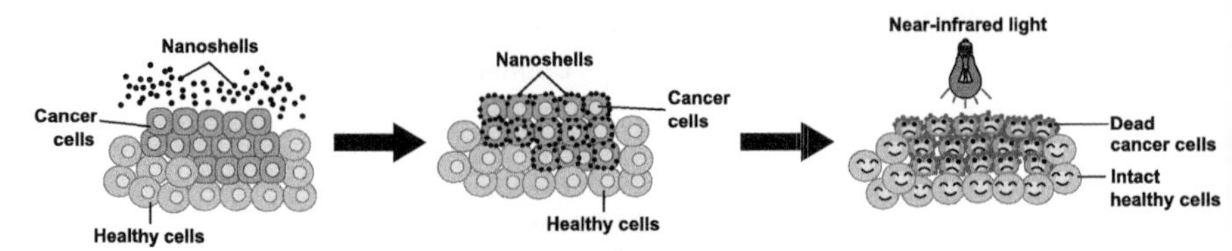

FIGURE 1.17 Principle of using nanocapsules for the treatment of cancer.

Figure kindly provided by the National Cancer Institute [45]

healing through the use of splints or plaster. The average time for healing a bone is six weeks. Scientists in Japan have found that CNTs can help to speed this process [47]. During the experiments, CNTs were placed in contact with fractured bones of rats, and the results show that CNTs help to regenerate bone tissue and reduce inflammation during healing. Microscopic images showed CNTs integration in the matrix of the bone itself, acting as a starting point for the growth of new bone tissue. According to researchers, healing is accelerated because CNTs act as a kind of "scaffolding" for bone regeneration. Although these results are incipient, the possibilities are vast. For example, CNTs could be used in the manufacture of biomaterials such as plates, screws, and implants used to repair damaged bone tissues.

One of the applications envisioned using nanotechnology is the construction of micro- and nano-robots that could be programmed to repair specific diseased cells, functioning similarly to antibodies in our natural healing process. Figure 1.18

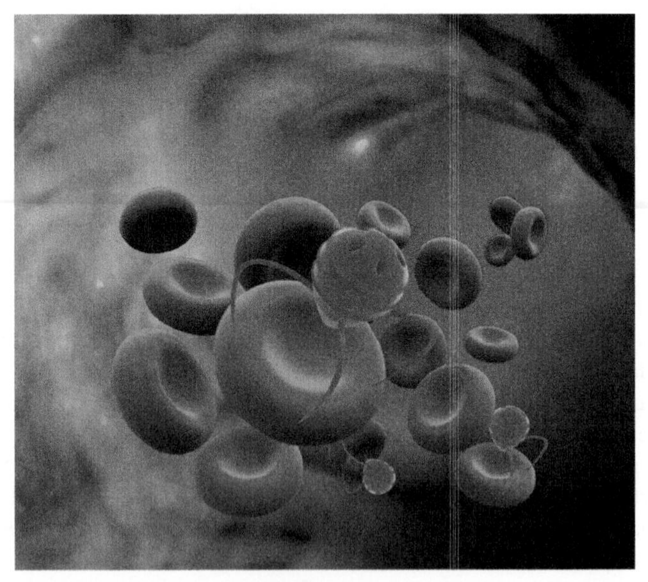

FIGURE 1.18 Representation of a micro-robot working similarly to antibodies in our healing process.

depicts a representation of how a micro-robot could be. The design is inspired by microorganisms found in deep sea.

1.7.3 Food

Some of the applications of nanotechnology are already being used in the food industry. Reinforced clay nanocomposites are used in bottles, packages, and films for obtaining impermeable barriers to gases such as oxygen, carbon dioxide, or moisture, reducing the possibility of food spoilage [48]. Silicates are also being applied for the same purpose [49] (Figure 1.19). Boxes used to store foods are being enhanced with silver nanoparticles. These nanoparticles eliminate bacteria from any food that was previously stored in boxes, minimizing health risks due to bacterial infection. This technology is already a reality in refrigerators manufactured in Europe.

The characteristics of foods and beverages can be changed through the use of nanomaterials [51]. Nanoparticles are currently used for delivery of vitamins and nutrients in foods and beverages without affecting taste or appearance [52]. These nanoparticles encapsulate nutrients and carry them through the stomach into the bloodstream. In some cases, this method allows a larger quantity of nutrients to be absorbed by the body. When encapsulated in nanoparticles, some nutrients that would be lost in the stomach are rather absorbed. Researchers also aim to develop nanocapsules containing nutrients and vitamins that shall be released when

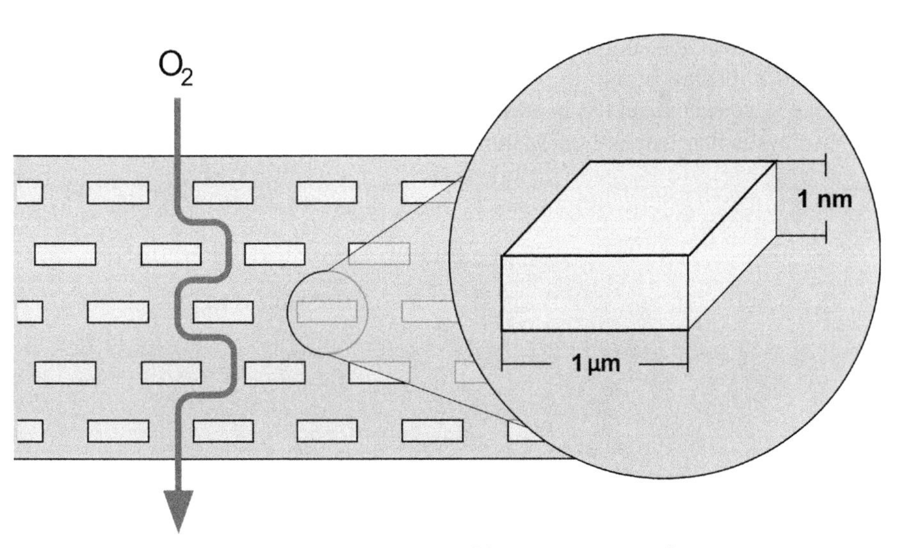

FIGURE 1.19 Plastic film (Durethan) containing silicate acts as a barrier to oxygen.

Before reaching their destination, the O_2 molecules must go the long way around the nanoparticles.

Figure sourced from nanoforum.org [50]

nanosensors detect a deficiency in the body. Basically, this research could result in a supply of vitamins in your body that releases the amount you need, when you need it.

Zinc oxide nanoparticles may be incorporated into plastic containers for blocking ultraviolet radiation and providing anti-bacterial protection. In addition, the maximum strength and stability of these materials is prolonged. Moreover, nanosensors will be used in plastic packaging to detect gases that are released when food starts to deteriorate. The color of the packaging itself will change to alert the user. Furthermore, nanosensors capable of detecting bacteria and contaminants are being developed and will be installed in food packaging industries and thus reducing the possibility of food contamination.

In agriculture, researchers are studying the possibility of manufacturing pesticides encapsulated in nanoparticles that will be released only in the stomach of insects, minimizing the contamination of plants. There are still researchers who envision the creation of a network of nanosensors and dispersers distributed in planting fields [50]. The nanosensors acknowledge when a plant needs nutrients or water and then dispersers release nutrients or water before the plant shows visible disability.

1.7.4 Electronics

One of the research efforts in this area involves the development of electrodes from nanowires that allow the fabrication of flexible displays thinner than the current ones. Prototypes have been fabricated (Figure 1.20) [54]. This technology involves using transistors made from transparent nanowires that are grown on glass or thin films of flexible plastic. The use of these nanowires allowed the creation of active-matrix displays similar to those in televisions and computer monitors. An active-matrix display is able to precisely direct the flow of electricity to produce the image since each pixel has its own control circuit [55]. A unique aspect of these new displays is that they are transparent until pixels are activated.

FIGURE 1.20 Flexible color display monitor developed by LG.Philips.

Figure sourced from voyle.net [53]

Researchers recently demonstrated that nanorings made of permalloy, in the form of rectangles, can store data practically instantly [53,56]. This finding may accelerate the commercialization of the so-called magnetic random access memory (MRAM). This type of technology uses magnetic fields instead of electric charges to store and access data randomly. In conventional dynamic random access memories (DRAMs), information is stored as electric charges. It must be constantly updated. Therefore, if the power is cut off, the information is lost. In the case of MRAM, even if there is no power, the magnetized memory layer remains magnetized and accessible. It is estimated that application of nanomagnetic rings could lead to the manufacture of data storage devices of up to 62 GB/cm^2 [57].

The fabrication of integrated circuits with components such as nanosized nanotransistors will increase the density of transistors in integrated circuits [58,59]. In addition, there is ongoing work on the development of self-aligned nanostructures for the manufacture of nanoscale integrated circuits [60].

All future applications of nanotechnology in electronics, mentioned above, will provide us with electronics such as cell phones, MP3s, laptops, etc., that are smaller, lighter, and modern.

1.7.5 Energy

Sectors related to energy such as fuel cells, solar cells, batteries, and fuel are now being enhanced by nanotechnology. A major option for power generation is the use of fuel cells. Platinum, a very expensive material, is the most widely used catalyst in fuel cells. The use of platinum nanoparticles reduces the amount of material needed due to a large increase in the surface area and consequently reactivity. Studies show that increase in the surface area and catalytic efficiency can be obtained by depositing platinum on alumina pores. Furthermore, catalysts will be incorporated into CNTs [61]. Another way to increase the efficiency of fuel cells is through the use of nano-projected hydrocarbon membranes [62].

The use of solar energy is also featured in today's world, but one of the biggest problems of this technology is the high cost and relatively low efficiency. The use of semiconductor nanoparticles allows the reduction of the manufacturing cost of solar cells and facilitates large-scale installations [63]. Furthermore, light-absorbing nanowires will be extended into flexible films for the production of low-cost flexible solar panels [64]. Nanoparticles dispersed in plastic films are already used to form solar cells that can be incorporated in mobile phones and notebooks [65]. Currently, backpacks and cell phone components with solar panels are already available commercially.

Once generated, energy must be stored in devices for use when needed. Energy sources like wind, for example, show increased production during the night: the time at which energy demand is minimal. It is crucial that this energy be stored.

Ultracapacitors using CNTs promise to be more efficient than batteries for use in hybrid cars (Figure 1.21). The coating of electrodes with nanoparticles will increase power and reduce battery recharge time [67]. Lithium-ion batteries using electrodes

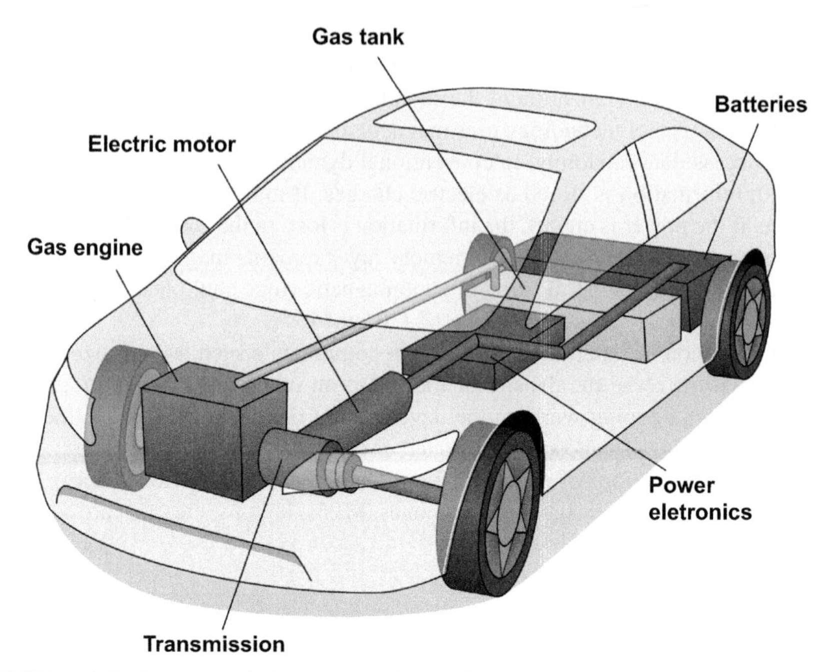

FIGURE 1.21 Typical transmission system of a hybrid car.

Hybrid cars are becoming more common, mainly due to their ability to reduce carbon dioxide (CO_2) emissions.

Figure sourced from car-question website [66]

in the form of nanoparticles can reach the safety standards required for the application of batteries in electric cars. The performance of these batteries is also high [68,69]. The use of less flammable electrode also reduces the possibility of accidents caused by fire.

There are studies that contemplate the use of nanotechnology to convert coal to gasoline and diesel fuel [70], reducing the cost of converting crude petroleum to fuel [71] and increasing diesel engine mileage and reducing air pollution [48]. The use of catalysts based on nanospheres can reduce the cost of production of biodiesel. There is still on-going research in modifying the bacteria that can make possible the production of methanol in one step and the modification of enzymes for converting cellulose into sugar, making the production of methanol via this method cost-effective.

1.7.6 Air pollution and water

Nanotechnology is being used to develop solutions for different problems related to air quality and water. Currently, some of the biggest problems in this area are the removal of gaseous air pollutants from air and the removal of salt, metals,

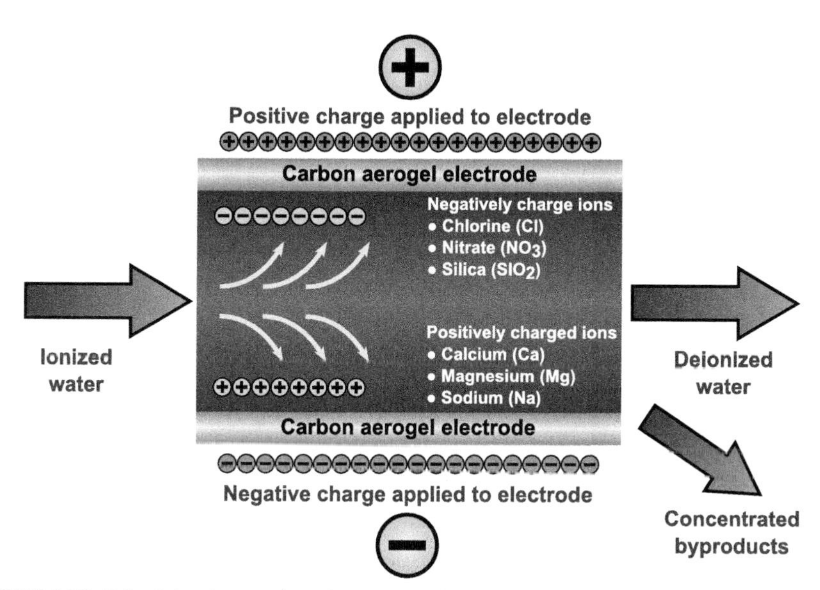

FIGURE 1.22 Principle of operation of a process for water desalination.

Figure sourced from cdtwater.com [73]

and industrial waste from water. Current research projects in this area include the following:

- Removal of volatile organic compounds from smoke emitted by industry through the use of membranes nanostructured or modified with CNTs and enzymes structured genetically [72].
- Development of electrodes composed of nanofibers for removal of metals and salt from water (Figure 1.22) [74,75].
- Reduction in amount of platinum used in catalytic converters. Currently platinum is used in automobiles for capturing gases such as carbon monoxide, which is highly toxic, and conversion of these pollutants into less harmful gases such as carbon dioxide [76].
- Conversion of carbon dioxide in CNTs, thereby helping reduce greenhouse gas emissions [77].
- Use of iron nanoparticles to clean carbon tetrachloride present in groundwater [73,78].
- Manufacture of filters with nanosized pores allowing for the removal of viruses [79].
- Nanoparticles that can absorb radioactive contaminants in wastewater.
- Nanoparticles that neutralize hydrophobic solvents present in water [80].
- Development of networks of nanowires that can absorb oil from spills.
- Manufacture of filters capable of retaining oil droplets using nanotubes and gold.

1.7.7 Textiles

The addition of nanoparticles or nanofibers to fabrics can improve its properties without increasing weight, thickness, or rigidity of these materials. Some of the new applications of nanotechnology under development in this area include the following:

- Use of nanowhiskers for the manufacture of stain-resistant and odor-proof fabrics [81].
- Addition of silver nanoparticles that kill bacteria in tissues, making clothes odor-proof [82]. Another option for the manufacture of stain-proof, anti-odor clothing already available is the use of bamboo nanoparticles (Figure 1.23).
- Nanopores to improve insulation in the soles of shoes in cold climates [84].
- Use of nanoparticles for the manufacture of stain-proof, self-cleaning clothes [85].

1.7.8 Sports

Nanotechnology is also applied in sports equipments. Some of the current applications of nanotechnology in this area are as follows:

- Use of nanoscale materials to fill voids in golf clubs [86,87].
- Use of nanocomposite films to extend the air inside tennis balls [88].
- Addition of CNTs in tennis rackets to increase rigidity and "power" [89,90].
- Increased energy transfer between golf balls and clubs through the use of nanostructured polymers [91].
- Improvement of bicycle handlebars made of carbon fiber by adding CNTs to fill the voids around the fibers. This process increases the resistance of handlebars and makes them lighter [92]. In fact, several bicycle components are already being reinforced with CNTs (Figure 1.24).

As we can see, there is a continuous rise in nanotechnology applications in our everyday lives. Many of these applications are found in the form of *polymer composites reinforced with CNTs*.

1.8 Nanoscience and nanotechnology: A look to the future

As we see in this chapter, scientists are now able to manipulate materials at the atomic scale, atom by atom. Although such outcomes are extraordinary, the use of such technology is far from becoming a reality on a commercial scale where the rate of production should be aimed at the highest possible performance and maximum profit. See, for example, in Figure 1.25 a nano-abacus obtained by manipulating atoms at the nanometer scale [95]. This abacus is functional and can be handled to perform calculations by using the probe of a scanning tunneling microscope, which moves atoms. Furthermore, the abacus can be operated at room temperature. This is a typical example of "bottom-up" nanotechnology.

Building nanoscale devices, atom by atom, using a microscope is not an attractive option for mass production, a requirement of the current economy. Nanosized

FIGURE 1.23 Anti-odor fabric made from bamboo nanoparticles.

Figure sourced from voyle.net [83]

devices will have a greater chance of application if manufactured in large quantities, preferably using nanotechnology.

The process of manufacturing single-walled carbon nanotubes (SWCNTs), which are in use currently, can be considered simple compared to the manipulation of atoms at the nanometer scale. A reasonable amount of nanotubes in gram scale can be obtained by synthesis. However, the situation is entirely different if the goal is to synthesize high purity SWCNTs, with the same diameter, orientation and quality.

According to Georde Whitesides from Harvard University, we are living in the era of nano-manufacturing, but the era of nanotechnology—finding practical applications for nanostructures—has not yet started [96,97]. One of the directors of the center for nanotechnology at NASA adds: "A lot of nanoscience, but little nanotechnology at this point" [96].

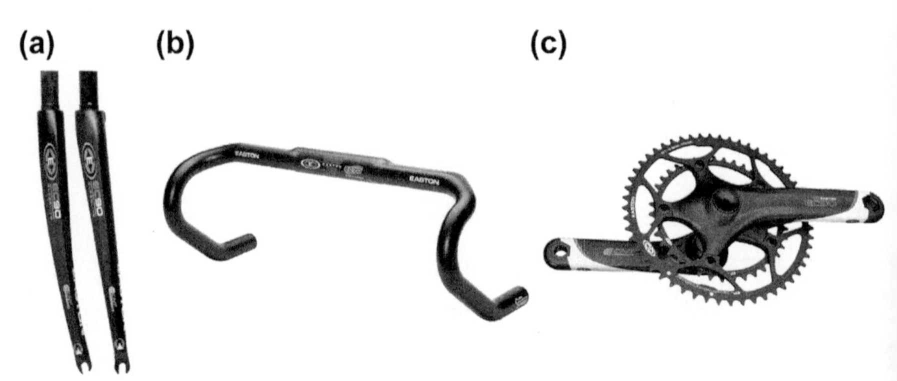

FIGURE 1.24 Different parts of a bicycle manufactured with CNTs: (a) fork, (b) handlebars, and (c) chain wheel.

Figure sourced from webersports.com [93]

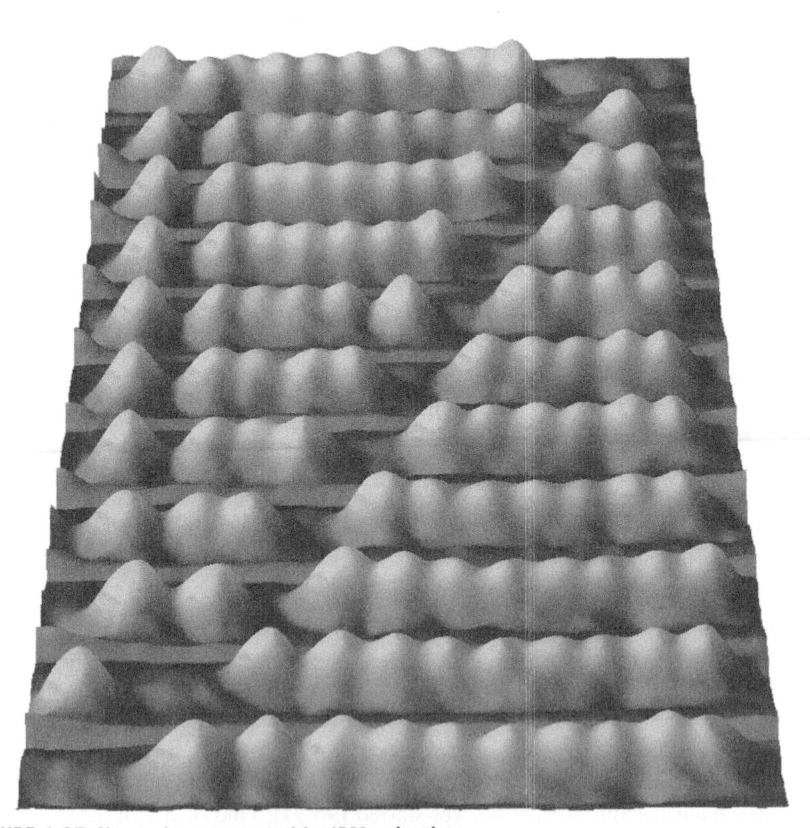

FIGURE 1.25 Nano-abacus created by IBM scientists.

The abacus can be operated by the probe of a scanning tunneling microscope.

Figure reproduced with permission from IBM Research, Zurich [94]

Two major challenges slowing the nanofabrication are:

1. Communication between the macro world and the nano world. This refers to the Heisenberg's uncertainty principle, which states that the state of a system is disturbed when measured. Therefore, a macroscopic detector would disturb a nanometric system from its ideal state [10].
2. The nanosurface. Devices of ultra-high purity manufactured on a nanometric scale need to be operated in environments of ultra-high vacuum. In other environments, contact with impurities would alter the properties. A similar principle is observed in transmission electron microscopes, where high vacuum is necessary to avoid the collision of the electron beam with impurities, such as the air itself.

N&N is one of the most studied topics in academia today and is slowly gaining traction in the industrial sector. It is estimated that products based on nanotechnology will contribute in $1 trillion to global gross domestic product and create over two million jobs by 2015 [98]. Figures like this highlight how nano may have mega or even giga consequences in our society.

It has been more than 50 years since Feynman delivered his talk. He was right! There's Plenty of Room at the Bottom!

To learn more...

We encourage readers to consult the following books:

1. Hornyak GL, Tibbals HF, Dutta J, Moore JJ. Introduction to nanoscience and nanotechnology. CRC Press; 2008, ISBN-13: 978-1420047790.
2. Roduner E, Cronin L. Nanoscopic materials: Size-dependent phenomena. Royal Society of Chemistry; 2006, ISBN-13: 978-0854048571.
3. Binns C. Introduction to nanoscience and nanotechnology (Wiley survival guides in engineering and science). Wiley; 2010, ISBN-13: 978-0471776475.
4. Boysen E, Muir NC, Dudley D, Peterson C. Nanotechnology for dummies. For Dummies; 2a edicão; 2011, ISBN-13: 978-0470891919.

References

[1] Roduner E. Nanoscopic materials: Size-dependent phenomena. 1st ed. Royal Society of Chemistry; August 29, 2006. ISBN-10: 085404857X, ISBN-13: 978−0854048571.

[2] Jortner J. Cluster size effects. Zeitschrift für Physik D Atoms. Mol Clust 1992;24(3): 247−75.

[3] Johnston Roy L. Atomic and molecular clusters. Series: Master's series in physics and astronomy. 1st ed. CRC Press; June 15, 2002. ISBN-10: 0748409319. ISBN-13: 978−0748409310.

[4] Matsui I. Nanoparticles for electronic device applications: A brief review. J Chem Eng Jpn 01/2005;38(8):535–46. http://dx.doi.org/10.1252/jcej.38.535.
[5] Haruta Masatake. Size- and support-dependency in the catalysis of gold. Catal Today April 25, 1997;36(1):153–66. ISSN: 0920-5861. http://dx.doi.org/10.1016/S0920-5861(96)00208-8.
[6] Kurokawa Y, Hosoya Y. Surface 1996;34(2):100–6 [in Japanese].
[7] Minoli D. Nanotechnology applications to telecommunications and networking. Wiley-Interscience. 1st ed. John Wiley and Sons, Inc.; October 28, 2005. ISBN-10: 0471716391, ISBN-13: 978–0471716396.
[8] Staff. Israelis bring nanoscale motors move closer to reality. Nanotechnol News; March 31, 2004.
[9] Collins PG, Avouris P. Nanotubes for electronics. Sci Am 2000;62(283):69. http://dx.doi.org/10.1038/scientificamerican1200-62.
[10] Hornyak GL, Tibbals HF, Dutta J, Moore JJ. Introduction to nanoscience and nanotechnology. Taylor and Francis Group, LLC; 2009. December 22, 2008. ISBN-10: 1420047795, ISBN-13: 978–1420047790.
[11] NSTC/NNI/NSET. National nanotechnology initiative: Research and development supporting the next industrial revolution. www.nano.gov; August 29, 2003.
[12] Forbes Nanotech Report. Online Journal. www.forbes-nanotech.net. Site não existe mais.
[13] Mansoori GA, Fauzi Soelaiman TA. Nanotechnology – an introduction for the standards community. J ASTM Int 2005;2(6):JAI13110.
[14] Ratner M, Ratner D. Nanotechnology: A gentle introduction to the next big idea. 1st ed. Prentice Hall; November 18, 2002. ISBN-10: 0131014005, ISBN-13: 978–0131014008.
[15] www.britishmuseum.org/explore/highlights/highlight objects/pe mla/t/the ly- curgus cup. aspx.
[16] Miller JC, Serrato RM, Represas-Cardenas JM, Kundahl G. The handbook of nanotechnology business, policy, and intellectual property law. 1st ed. October 19, 2004. ISBN-10: 0471666955, ISBN-13: 978–0471666950.
[17] Rothemund PWK. Folding dna to create nanoscale shapes and patterns. Nature 2006; 440:297–302. http://dx.doi.org/10.1038/nature04586.
[18] Barmer DJ, Freestone IC. An investigation of the origin of the colour of the Lycurgus cup by analytical transmission electron microscopy. Archaeometry 1990;32:33–45.
[19] Images of nanoparticles were kindly provided by Prof. Dr. Chad A. Mirkin. Department of chemistry and international institute for nanotechnology Northwestern University.
[20] Wagner FE, Haslbeck S, Stievano L, Calogero S, Pankhurst QA, Martinek K-P. Nature 2000;407:691–2. http://dx.doi.org/10.1038/35037661 [Before striking gold in gold-ruby glass].
[21] www.nytimes.com/imagepages/2005/02/21/science/20050222_NANO1_GRAPHIC.html.
[22] Pictures obtained from the stained glass museum. www.stainedglassmuseum.com/index.html.
[23] Ayers A. Ceramics of the world: From 4000 BC to the present. Abrams; 1992. ISBN-10: 0810931753, ISBN-13: 978–0810931756.
[24] Zhao B, Wang J, Li Z, Liu P, Chen D, Zhang Y. Mater Lett 2008;62:4380–2.
[25] Daniel M-C, Astruc D. Gold nanoparticles: assembly, supramolecular chemistry, quantum-size related properties, and applications towards biology, catalysis and nanotechnology. Chem Rev 2004;104:293–346.
[26] Reicher HH. Aurum potabile oder gold tinstur. Breslau (Leipzig); 1718.

[27] www.alchemywebsite.com/longhi.html.
[28] Faraday M. The Bakerian lecture: Experimental relations of gold (and other metals) to light. Philos Trans R Soc Lond 1857;147:145–81. http://dx.doi.org/10.1098/rstl.1857.0011.
[29] www.discovernano.northwestern.edu/whatis/history/.
[30] www.alcatel-lucent.com.
[31] Brittain WH. Laboratory notebook of December 24, 1947. Press Release. New York: Bell Telephone Laboratories; 1948.
[32] www.alcatel-lucent.com/wps/portal/!ut/p/kcxml/04_Sj9SPykssy0xPLMnMz0vM0Y_QjzKLd4w3MfQFSYGYRq6m-pEoYgbxjgiRIH1vfV-P_NxU_QD9gtzQiHJHR0UAAD_zXg!!/delta/base64xml/L0lJayEvUUd3QndJQSEvNElVRkNBISEvNl9BXzQ3QS9lbl93dw!!?LMSG_CABINET=Bell_Labs&LMSG_CONTENT_FILE=Photos_and_Videos/Photos/Photo_Detail_000001.xml.
[33] Moore GE. Cramming more components onto an integrated circuit. Electronics April 19, 1965;38.
[34] http://en.wikipedia.org/wiki/transistor count.
[35] www.spaceward.org.
[36] http://spaceelevator.com/.
[37] www.en.wikipedia.org; Topic: Non-rocket spacelaunch.
[38] www.lifport.com.
[39] www.nucryst.com.
[40] www.thisischile.cl/6491/4/702/chilenos-criam-revolucionario-nanocomposto-contra-o-cancer/article.aspx.
[41] www.cytimmune.com.
[42] www.nanotherapeutics.com.
[43] www.biodeliverysciences.com.
[44] www.nanobiotix.com.
[45] www.cancer.gov/cancertopics/understandingcancer/nanodevices/page18.
[46] www.oxonica.com.
[47] Usui Y, Aoki K, Narita N, Murakami N, Nakamura I, Nakamura K, et al. Carbon nanotubes with high bone-tissue compatibility and bone-formation acceleration effects. Small 2008;4(2):240–6. http://dx.doi.org/10.1002/smll.200700670.
[48] www.nanocor.com.
[49] www.research.bayer.com.
[50] www.nanoforum.org (nanotechnology in agriculture and food).
[51] www.aquanova.de.
[52] www.dummies.com/how-to/content/food-characteristics-changed-by-nanomaterials.html.
[53] www.voyle.net/nano.
[54] www.philips.com.
[55] www.motorola.com (nano-emissive display).
[56] Subramani A, Geerpuram D, Domanowski A, Baskaran V, Metlushko V. Vortex state in magnetic rings. Physica C 2004;404:241–5.
[57] www.freescale.com (mram).
[58] www.spectrum.ieee.org.
[59] www.intel.com (32 nm logic technology).
[60] www.hpl.hp.com.
[61] www.hoovers.com/pacific-fuel-cell.

[62] www.polyfuel.com.
[63] www.nanosolar.com.
[64] www.bloosolar.com.
[65] www.konarka.com.
[66] http://car-question.wikispaces.com/hybridcars.
[67] www.a123systems.com.
[68] www.nanoener.com.
[69] www.zmp.com.
[70] www.htigrp.com.
[71] www.refineryscience.com.
[72] www.americanelements.com.
[73] www.cdtwater.com/cdt technology summary.pdf.
[74] www.cdtwater.com.
[75] www.campbellap.com.
[76] www.nanostellar.com.
[77] www.nanotech-now.com.
[78] www.siremlab.com.
[79] www.argonide.com.
[80] www.ipp.nasa.gov.
[81] www.nano-tex.com.
[82] www.nanohorizons.com.
[83] www.voyle.net/nano.
[84] www.aerogel.com.
[85] www.basf.com.
[86] www.ncnanotechnology.com.
[87] www.accuflexgolf.com.
[88] www.inmate.com.
[89] www.babolat.com.
[90] www.wilson.com.
[91] www.ndmxgolf.com.
[92] www.eastonbike.com.
[93] www.webersports.com.
[94] www.research.ibm.com/atomic/nano/roomtemp.html.
[95] Cuberes MT, Schlitter RR, Gimzewski JK. Room-temperature repositioning of individual C60 molecules at Cu steps: Operation of a molecular counting device. Appl Phys Lett 1996;69:3016–8. http://dx.doi.org/10.1063/1.116824.
[96] Binks P. The challenges facing nanotechnology, interviewed by R. Williams, Ockham's razor. www.abc.net.au/rn/science/ockham/stories/s1304778.htm; 2005.
[97] Whitesides GM. The art of building small. In: Understanding nanotechnology. New York: Scientific American, Time-Warner Book Group; 2002.
[98] Cao G. Nanostructures and nanomaterials: Synthesis, properties and applications. 1st ed. Singapore: Imperial College Press; 2004. ISBN-10: 1860944809, ISBN-13: 978–1860944802, 1st ed.

CHAPTER

2 Composites

CHAPTER OUTLINE

Carbon Nanotube Reinforced Composites. http://dx.doi.org/10.1016/B978-1-4557-3195-4.00002-3

2.1 Conventional engineering materials

In our daily life, we are surrounded by products, utensils, and equipment. To manufacture these things, materials are required. These materials have to withstand loads, insulate or conduct heat and electricity, accept or reject magnetic flux, transmit or reflect light, be stable in hostile environments, and perform all these functions without damaging the environment or costing too much. Moreover, after choosing the right material, there is a need to choose the best process for manufacture. The correct choice of materials but an incorrect selection of process can cause a wide range of disasters. There is a need for compatibility between materials and processes.

Around 1890, only a few hundred materials were available in the market. There were no synthetic polymers available, like today, where we have more than 45,000 different choices. In fact, engineers and designers currently have more than 60,000 different types of materials available [1]. It is difficult to study all these materials and their properties individually, so a general classification is needed to enable simplification and characterization [2].

Engineering materials, depending on their main characteristics such as strength, stiffness, density, and melt temperature, may be conveniently divided into four main categories: (1) metal, (2) polymers, (3) ceramic, and (4) composites [2]. Some authors still divide the materials into additional groups such as semiconductors and biomaterials [3]. In this book, we will adopt the division based on four main categories. Table 2.1 shows the properties of some materials selected from each class.

2.1.1 Metals

Metallic materials are formed by atoms with metallic characteristics in which the valence shell electrons flow freely. Metals have been the dominant material for structural applications in the past. They are extremely good conductors of heat and electricity, are not transparent to visible light, and have high stiffness, mechanical strength, and thermal stability. Some common metals are aluminum, copper, iron, magnesium, zinc, nickel, and titanium. Alloys are more commonly used in structural applications than pure metals due to their superior properties. Alloys are materials with metallic properties that contain two or more chemical elements, of which at least one is metal. For example, steel is an alloy of iron and 0.008–2.11% carbon. The iron becomes tougher with the addition of carbon and even the addition of less than 1% of chromium makes it resistant to corrosion. Other common examples of alloys are the brass (an alloy of copper and zinc) and bronze (an alloy of copper and tin).

Table 2.1 Typical Properties of Engineering Materials

Material	ρ (g/cm^3)	E (GPa)	σ_m (MPa)	ε (%)	E/ρ	σ_m/ρ
Metals						
Aluminum 2124-T851	2.78	73	483	8	26	174
Steel 1020	7.87	200	450	36	25	57
Steel 4340	7.87	207	1280	12.2	26	163
Ceramics						
Aluminum oxide	3.98	380	282–551	–	95	71–138
Tungsten carbide	15.7	696	344	–	44	22
Silicon carbide	3.3	207–483	230–825	–	63–146	70–250
Polymers						
Epoxy	1.26	2.41	28–90	3–6	1.9	22–71
Nayon 6,6	1.14	1.6–3.8	95	15–80	1.4–3.3	83
Polycarbonate (PC)	1.20	2.38	63–72	110–150	2	52–60
Polyethylene terephthalate (PET)	1.35	2.8–4.1	48–72	30–300	2–3	36–53
Polypropylene (PP)	0.9	1.1–1.6	31–41	100–600	1.2–1.8	34–46
Polyvinyl chloride (PVC)	1.45	2.4–4,1	41–52	40–80	1.7–2.8	28-36
Polystyrene (PS)	1.05	2.3–3.3	36–52	1,2-2,5	2.2–3.1	34–50

Continued

Table 2.1 Typical Properties of Engineering Materials—cont'd

Material		ρ (g/cm^3)	E (GPa)	σ_m (MPa)	ε (%)	E/ρ	σ_m/ρ
Composites							
Aluminum 2124 + silicon carbide (25 vol%)		2.88	115	659	4.0	40	229
Epoxy + graphite (60 vol%)	Longitudinal	1.6	145	1240	0.9	91	775
	Transversal		10	41	0.4	6	26
Epoxy + glass fibers (60 vol%)	Longitudinal	2.1	45	1020	2.3	21	486
	Transversal		12	40	0.4	0.2	19
Epoxy + aramid (60 vol%)	Longitudinal	1.4	76	1380	1.8	1.3	986
	Transversal		5.5	30	0.5	4	21
Epoxy + boro (60 vol%)	Longitudinal	2.0	207	1320	0.6	104	660
	Transversal		19	72	0.4	10	36

ρ: Density; E: Young's modulus; ε: elongation at break; σ_m: Tensile strength.

The density of the metal is generally much greater than plastics and composites; exceptions are aluminum and magnesium. Metals may also be employed in applications requiring operating temperatures higher than those tolerable by polymers. In addition, metals exhibit a strong tendency not to form compounds with each other but have affinity for nonmetallic elements such as oxygen and sulfur, which form oxides and sulfides, respectively.

2.1.2 Polymers

Polymers are large molecules (macromolecules) composed of repeating structural units typically connected by covalent bonds. Because of their low density compared to other materials, easy processability, and corrosion resistance, polymers have become one of the most common engineering materials that are widely applied in the automotive and aerospace industries as well as various consumer goods. Polymers are divided into thermoplastics, thermosets, and elastomers. Some common polymers include polystyrene (PS), polypropylene (PP), polyethylene (PE), polyethylene terephthalate (PET), polyurethane (PU), and epoxy. Table 2.2 presents a selection of polymers and their chemical structures [4].

Despite all the advantages provided by polymeric materials, their application is limited due to their thermal stability, which is usually less than 150 °C. However, some high-performance polymers such as polyamides and poliazols, for example, exhibit superior thermal stability and mechanical properties [5—9].

2.1.3 Ceramics

By definition ceramic encompasses all materials that are inorganic, nonmetallic, and usually obtained after heat treatment at elevated temperatures. Ceramics can be classified into oxides (alumina and zirconia) and nonoxides (carbides, borides, nitrides, and silicides).

Ceramic materials have strong covalent bonds and have exceptional thermal stability and hardness in addition to low (or zero) thermal and electrical conductivities. This class of materials includes some of the most rigid materials existing in nature and they have high melting points. However, ceramics are extremely fragile and have virtually no ductility. The machining of ceramics is very difficult, and very expensive cutting tools such as diamond and carbide are necessary.

2.1.4 Composites

Composite materials have long been used to solve technological problems, but only in the 1960s, with the introduction of polymer-based composites, did these materials come to the attention of industries. Since then, composites are designed and manufactured to be applied in many different areas, taking the place of materials hitherto regarded as irreplaceable, such as steel and aluminum. Currently, the per capita use of composites is considered an indicator of technological development. While developed countries like the United States have high per capita use of

Table 2.2 Name and Structure of the Repeating Unit of Different Polymers

Common Name	Chemical Structure of the Repeating Unit
Polyethylene (PE)	$-[CH_2-CH_2]_n-$
Polypropylene (PP)	$-[CH(CH_3)-CH_2]_n-$
Polyethylene terephthalate (PET)	$-[(CH_2)_2-O-C(=O)-C_6H_4-C(=O)-O]_n-$
Polytetrafluoroethylene (PTFE)	$-[CF_2-CF_2]_n-$
Polycarbonate (PC)	$-[O-C_6H_4-C(CH_3)_2-C_6H_4-O-C(=O)]_n-$
Polyvinyl chloride (PVC)	$-[CH(Cl)-CH_2]_n-$
Poly(methyl methacrylate) (PMMA)	$-[C(CH_3)(C(=O)-O-CH_3)-CH_2]_n-$
Polyvinyl acetate (PVAc)	$-[CH(O-C(=O)-CH_3)-CH_2]_n-$

Table 2.2 Name and Structure of the Repeating Unit of Different Polymers—cont'd

Common Name	Chemical Structure of the Repeating Unit
Polycaprolactone (PCL)	$-[NH-C(=O)-(CH_2)_5]_n-$
Poly(*cis*-isoprene)	H; CH_3; n
Polyacrylonitrile (PAN)	$-[CH(CN)-CH_2]_n-$
Polyester	$-[O-R-O-C(=O)-R'-C(=O)]_n-$
Polystyrene (PS)	$-[CH(C_6H_5)-CH_2]_n-$
Polyimide	O, C, N, C, O, O, C, N, C, O; n
Polyvinylidene fluoride (PVDF)	$-[CF_2-CH_2]_n-$

Continued

Table 2.2 Name and Structure of the Repeating Unit of Different Polymers—cont'd

Common Name	Chemical Structure of the Repeating Unit
Polyvinyl alcohol (PVA)	$-[CH(OH)-CH_2]_n-$
Polydimethylsiloxane (PDMS)	$-[Si(CH_3)_2-O]_n-$
Poly neoprene (1,4-*cis*-polychloroprene)	$-[CH_2-C(Cl)=C(H)-CH_2]_n-$
Polyphenylene sulfide (PPS)	$-[C_6H_4-S]_n-$
Polyether ketone (PEK)	$-[O-C_6H_4-C(=O)-C_6H_4]_n-$
Polyether ether ketone (PEEK)	$-[O-C_6H_4-C(=O)-C_6H_4-O-C_6H_4]_n-$
Poly-paraphenylene terephthalamide (Kevlar®)	$-[HN-C_6H_4-NH-C(=O)-C_6H_4-C(=O)]_n-$
Poly(p-phenylene oxide) (PPO)	$-[C_6H_4-O]_n-$
Polybutylene terephthalate (PBT)	$-[(CH_2)_4-O-C(=O)-C_6H_4-C(=O)]_n-$

Table 2.2 Name and Structure of the Repeating Unit of Different Polymers—cont'd

Common Name	Chemical Structure of the Repeating Unit
Polyacetal (POM)	$-[CH(R)-O]_n-$
Polyethersulfone (PES)	$-[O-C_6H_4-SO_2-C_6H_4]_n-$
Polyurethane (PU)	$-[O-R-O-C(=O)-NH-R'-NH-C(=O)]_n-$

Sourced from Physical Properties of Polymers Handbook [4]

composites, (7.90 in 2005), China uses less than 0.5 kg of composites per capita annually. Table 2.3 shows the per capita use of composite materials in different countries in 1998 and 2005 [10].

Some of the major advantages of composite materials are their high mechanical properties and low mass. Replacing steel components with composites can mean a reduction of up to 80% in mass. As a result, today we can find composite materials in the automotive, aerospace, civil, marine, and sports areas.

2.2 The concept of composites

Broadly speaking, it is accepted that there are three criteria that must be satisfied before a material can be considered a composite. These criteria are as follows:

- The properties of the composite should be markedly different from those of the constituents;
- The constituents should be present in measurable (noticeable) proportions;
- The different materials (phases) should be separated by an interface.

There are several definitions for composite materials. In a broad sense, *a composite material is made by combining two or more materials to provide a unique combination of properties*. Composite materials may also be defined as *materials made from two or more components with different compositions, structures, and properties that are separated by an interface*. Yet another possible definition is that *the term composite material refers to any solid material composed of more than one component where they are in separate phases*. The last definition is broad and includes a wide variety of materials such as fiber-reinforced plastics, ordinary concrete,

Table 2.3 Use of Composite Materials in Different Countries in 1998 and 2005

Country	Per Capita (kg)	
	1998	2005
South Africa	1.06	1.25
Germany	3.20	4.10
Argentina	0.97	1.19
Australia	2.40	2.72
Brazil	0.64	0.81
Canada	2.90	3.70
Chile	0.48	0.55
China	0.22	0.40
Colombia	0.22	0.25
Spain	4.80	6.00
United States	6.50	7.90
France	4.70	6.20
Italy	4.40	5.80
Japan	5.29	6.49
Taiwan	4.30	6.65
Venezuela	0.65	0.71

Sourced from Composites [10]

concrete reinforced with steel, polymers reinforced with particles, etc. However, this book will focus on polymer matrix composites reinforced with carbon nanotubes (CNTs). Therefore, when the term "composite" or "composite material" is used the definition envisioned is as follows: *composites are materials composed of a solid matrix that surrounds and retains the reinforcement phase.*

Composite materials are composed of at least two phases: a continuous and generally less rigid called matrix phase and the discontinuous and generally more rigid so-called reinforcing phase (Figure 2.1).

The concept of composites was not invented by humans; this concept is found in nature. The number of examples is vast (Figure 2.2). Wood, for example, can be viewed as a composite consisting of a matrix of lignin reinforced by cellulose fibers. The shell of some invertebrates such as snails and oysters are examples of composites. In our own body, we find composite materials like bones and teeth.

The properties of composite materials depend on the properties of the constituents, geometry, and distribution of the constituent phases. One of the most important parameters is the volume fraction (or mass) of reinforcement. The distribution of the reinforcing phase determines the homogeneity or uniformity of the material. The less uniform the distribution of the reinforcement, the more heterogeneous the material and the greater the probability of failure in areas where stress concentrations occur. The geometry and orientation of the reinforcement affect the anisotropy of the system.

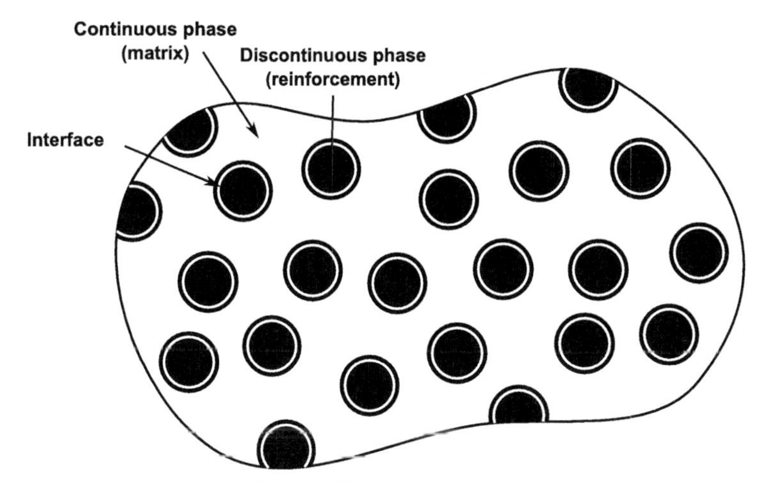

FIGURE 2.1 Phases of a composite material.

Figure sourced from Engineering mechanics of composite materials [11]

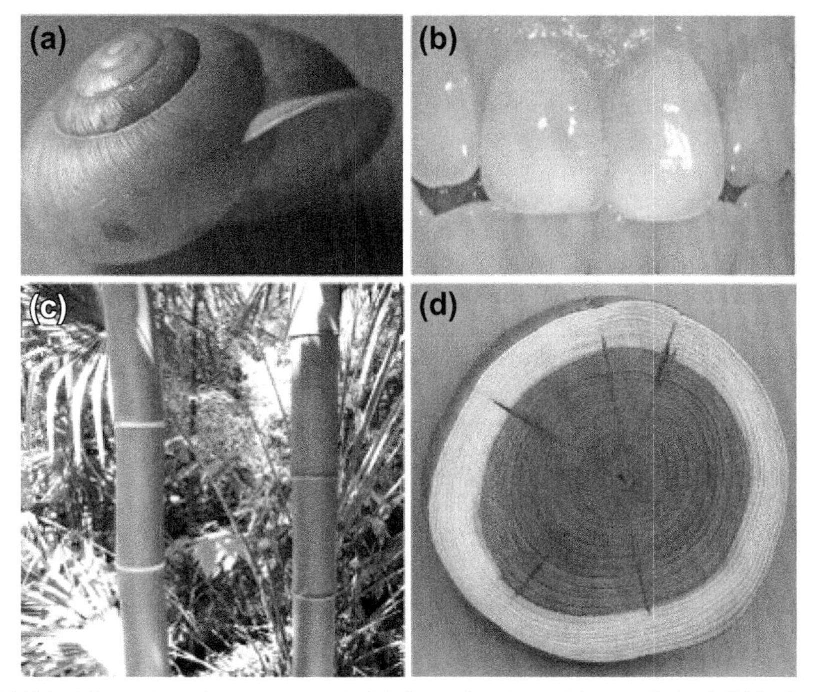

FIGURE 2.2 Examples of composite materials found in nature: (a) a snail shell, (b) teeth, (c) bamboo, and (d) timber.

The discovery that the combination of materials could produce a "new" material having properties superior to those of the same materials isolated first occurred thousands of years ago. Mud bricks reinforced with twigs were once used by the Israelites in Egypt. There are even biblical references to this fact. For example, in the Book of Exodus, there is a passage that says, "Do not give them straw for their bricks, do they seek their own straw." The straw, which the Egyptians denied the Hebrews, was used at that time as today, to reinforce mud bricks. The Mongolians manufactured composites bows by gluing together five pieces of wood (using glue made from bones and animal hooves) to form the core of the arc (center handle, two arms, and two ends) to which tendons of cattle were attached on the side of strain and horn pieces were glued to the compression side. Given its performance, this type of bow is considered one of the most powerful ever manufactured.

2.2.1 Functions of the reinforcement and matrix

As mentioned earlier, a composite material is formed by a matrix phase strengthened by a reinforcement phase. To understand the behavior of composites, it is necessary to understand the role played by the matrix and reinforcement phases. The main functions of the matrix and reinforcement phases are discussed below.

The matrix is the continuous phase of a composite material. It can be metallic, ceramic, or polymeric. Its main function is to shape the structure. As the continuous phase, the matrix surrounds and covers the reinforcement and is exposed directly to the environment. Thus the matrix is the first component of the composite to sustain any force imposed on the composite. Generally, the matrix phase is not as durable and rigid as the reinforcement phase and it is not expected to withstand forces or loads applied to the composite. However, the matrix must transfer the loads applied to the composite to the reinforcement phase. The effectiveness of load transfer is one of the most important requirements for obtaining materials with adequate performance. It is crucial that a synergistic effect occurs between matrix and reinforcement phases.

Depending on the material selected for the matrix, the composite performance characteristics such as ductility, impact resistance, etc., are influenced. A ductile matrix increases the toughness of the structure. Generally, thermoplastics are applied for this purpose.

In summary, the main functions of the matrix phase are as follows:

- Transfer loads applied to the composite to the reinforcement;
- Provide a good shape and surface quality to the composite;
- Protect the reinforcement against chemical attack or mechanical damage such as wear;
- Tie (retain) the reinforcing phase in position so that separated reinforcements such as fiber can act independently.

Regarding the reinforcement, its main function is to provide rigidity, strength, and other mechanical properties to the composite. In Figure 2.3 we see how the use of fibers can increase the rigidity of a resin by forming a composite. The

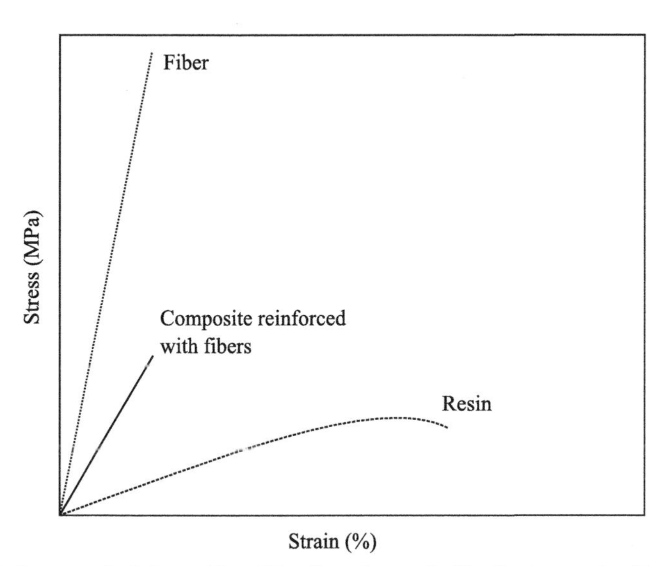

FIGURE 2.3 A composite is formed by adding fibers in a resin. The final property of the composite is a combination of the properties of the matrix and the fiber.

Figure sourced from gurit.com [12]

reinforcement phase may be composed of fibers, whiskers, or particles. Fibers can be continuous, long, or short. In addition, the fibers may be preferentially or randomly oriented. Similar to the matrix, reinforcements can be made from metals, ceramics, or polymers. When a composite is reinforced with fibers and these are preferentially oriented (unidirectional), composites exhibit anisotropic behavior. In this case it is expected, for example, that the mechanical properties of the composite are higher in the direction of the fibers. In the case of randomly oriented fibers, composites with isotropic properties (same in all directions) are obtained.

Thus, the main functions of the reinforcement phase in composites are:

- Provide strength, stiffness, and other structural properties to the composite;
- Provide insulation or electrical conductivity to the composite depending on the type of reinforcement used;
- Withstand the load applied to the composite and transferred by the matrix phase;
- Dominate the coefficient of thermal expansion;
- Prevent or slow the propagation of failures (Figure 2.4).

In a way, all the properties of a composite arise from the interaction or the presence of the matrix phase and reinforcement phase. However, some properties are dominated by the matrix phase and others by the reinforcement phase.

2.2.2 Classification of composites

There are several ways to classify composite materials. The classification can be based on the matrix phase, reinforcement phase, the form of arrangement of the

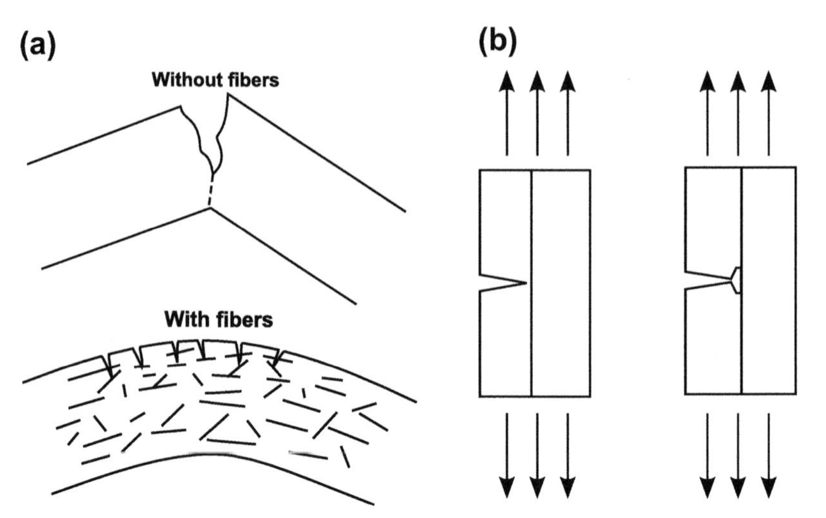

FIGURE 2.4 Representation of the failure propagation being stopped: (a) composite with short fibers and (b) composite with long fibers.

Figure sourced from Fiberglass & other composite materials [13]

reinforcement (geometry), and its architecture (Figure 2.5). Based on the matrix, as already discussed, composites can be classified as follows:

Metal Matrix Composites (MMC): These are widely found in the automotive industry. They are composed of a metallic matrix (aluminum, magnesium, steel, cobalt, copper) and a reinforcement (oxides, Carabidae, tungsten, molybdenum).

Ceramic Matrix Composites (CMC): These are mainly used in high-temperature environments. They are composed of a ceramic matrix phase and a reinforcement phase. The latter can be other ceramic material generally in the form of short fibers or whiskers made of silicon carbide or boron nitride.

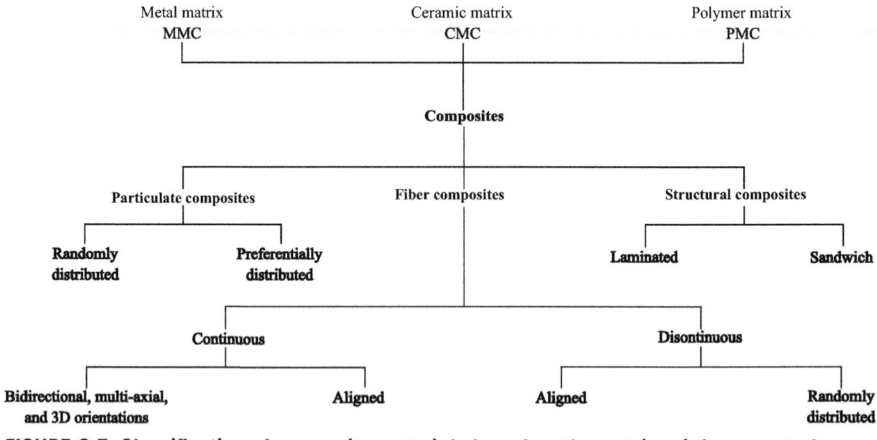

FIGURE 2.5 Classification of composite materials based on the matrix reinforcement phases.

Polymer matrix composites (PMC): These are the most common composites and are composed of a thermoset polymer (epoxy, vinyl ester, polyester, etc.), elastomers or thermoplastic (polycarbonate (PC), polyvinyl chloride (PVC), nylon, PS, etc.) matrix and a dispersed reinforcement phase consisting generally of glass, carbon, Kevlar®, or steel fibers.

Based on the reinforcement phase, composite materials can be classified into the following:

Particulate composites

These consist of a matrix reinforced by a dispersed phase in the form of particles. They can be further subdivided into the following:

1. Composites with randomly distributed particles (Figure 2.6(a));
2. Composites with preferential orientation distribution particles. In this case the dispersed phase consists of two-dimensional materials such as flakes parallel to each other (Figure 2.6(b)).

Fiber composites

These consist of a matrix reinforced by a dispersed phase in the form of fibers. They can be further subdivided into the following:

1. Composites with short fibers: they consist of a matrix reinforced by a dispersed phase in the form of discontinuous fibers. Depending on the orientation of the fibers, we have
   a. Composites with preferentially oriented fibers (Figure 2.7(a));
   b. Composites with randomly oriented fibers (Figure 2.7(b)).
2. Composites with long fibers: they consist of a matrix reinforced by a dispersed phase in the form of continuous fibers. Depending on the orientation of the fibers we have
   a. Unidirectional fiber orientation (Figure 2.8(a));
   b. Bidirectional fiber orientation (Figure 2.8(b)), multiaxial planar, random planar, and 3D.

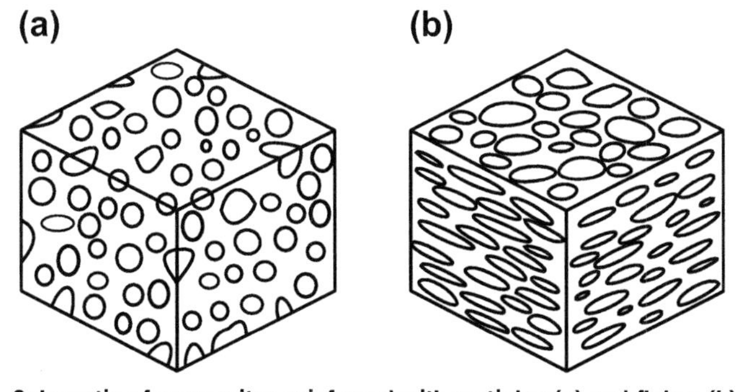

FIGURE 2.6 Schematic of composites reinforced with particles (a) and flakes (b).

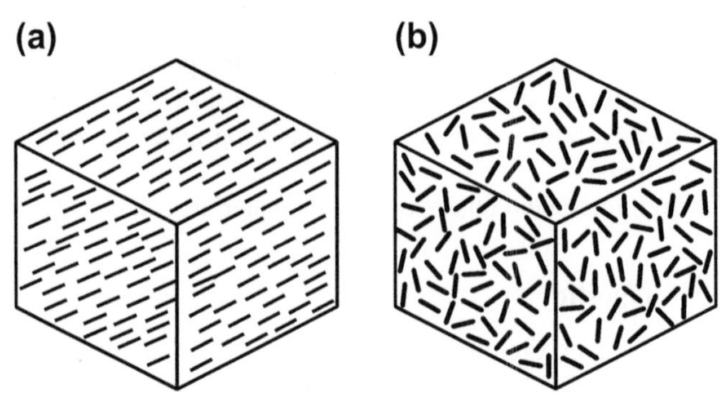

FIGURE 2.7 Schematic of composites reinforced with short fibers oriented preferentially (a) and randomly oriented (b).

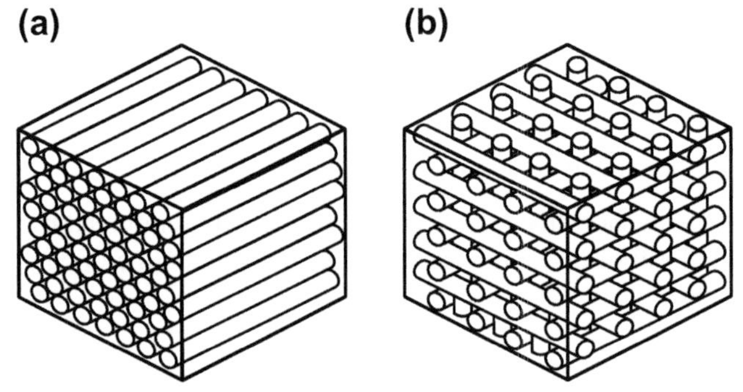

FIGURE 2.8 Schematic representation of composites reinforced with long fibers oriented in a unidirectional (a) and 3D fashion (b).

Structural composites

Typically composed of homogeneous materials and composites, simultaneously, in which the properties of the final material depend not only on the properties of the constituents, but also on the geometric design of the various structural elements. This class may be subdivided into the following:

1. Laminated composites: these are composites consisting of at least two layers connected together. The layers may be of different materials with different orientations, so as to obtain a final material with specific properties (Figure 2.9(a)).
2. Sandwich panels: these consist of two thin outboard laminates or faces, resilient, separated, and connected to a layer ("core") made of less dense material with lower strength and stiffness (Figure 2.9(b)). The faces bear most of the load in the plane and transversal bending stresses. The core serves to separate the surfaces and resist deformation (compression) perpendicular to the plane of the face, and to provide a certain degree of rigidity along the planes perpendicular to the faces.

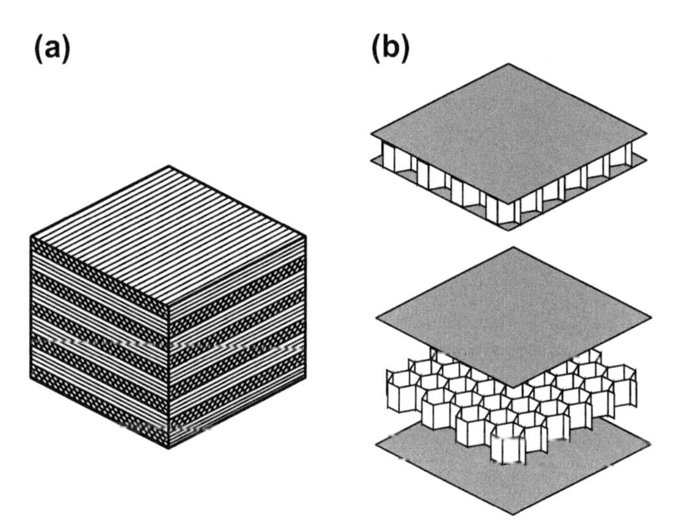

FIGURE 2.9 Schematic representation of structural composites. Laminated composites (a) and sandwich panels (b).

2.3 Raw material for manufacture of composites

In recent decades, many new types of composites were developed. Through careful selection of the matrix, reinforcement phase, and manufacturing process that unites these phases, engineers can manufacture composites that satisfy specific requirements. Composites can be designed to be resistant in one direction and weak in another direction where strength is not as important. In addition, properties such as resistance to heat, chemicals, and weather conditions can be selected by choosing the correct material for the matrix phase.

Most of the commercially produced composites use a polymer matrix often called resin. In this book we focus on composites of this class. As for the reinforcement phase, the use of fibers predominates. The volume fraction of fiber present in a composite can reach 60 vol%.

In the following section, we present some of the base materials used for the manufacture of composites.

2.3.1 Resins

The resins are divided into two major groups known as thermoplastics and thermosets. Thermoplastics become "soft" when heated, while they can be molded in a semifluid state and become rigid when cooled. Thermosets, on the other hand, are usually liquids or solids with low melting point in their original form. When used to produce end products, such resins are cured through the use of catalysts, heat, or a combination of both. Once cured, thermoset resins can no longer be converted to their original liquid form. Unlike thermoplastics, thermosets do not melt and flow and therefore cannot be remolded.

Although various resins are used in the composites industry, most of the structural parts are manufactured using polyester, epoxy, and vinyl ester resins.

2.3.1.1 Polyester

Polyester resins are the most widely used resins, particularly in the maritime and automotive industries. Most sailing boats, yachts, and boats built with composites make use of this resin system. Resins of this type are unsaturated and represent about 75% of total resin used in the composites industry. The unsaturated polyester resin is a thermoset, capable of being cured from the liquid or solid state when subjected to the right conditions. Saturated polyester, on the other hand, cannot be cured this way. It is common to refer to unsaturated polyester resins as "polyester resins" or simply "polyesters."

In organic chemistry, the reaction of an alcohol with an organic acid produces an ester and water. Through the use of special alcohols such as a glycol, in a reaction with dibasic acids, a polyester and water will be produced. This reaction along with the addition of components such as saturated dibasic acids and crosslinkable monomers (multifunctional) forms the basic process for manufacturing the polyester. As a result, there are a number of polyesters made from different acids, glycols, and monomers, all having different properties. Figure 2.10 shows the idealized chemical structure of a typical polyester.

Most polyester resins are clear viscous liquids consisting of a polyester solution in a monomer, which is typically styrene. The reason for this is that the addition of styrene in amounts up to 50% reduces the viscosity of the resin and facilitates its handling. Another vital function of styrene is to allow the resin to crosslink the chains of the polyester. Polyesters have a limited storage time due to their transformation into a gel when stored for long periods. This gelatinization process can be slowed down by the addition of inhibitors. The main advantages of this resin are its balance of properties (including mechanical, chemical, electrical), dimensional stability, cost, and ease of handling and processing.

2.3.1.2 Epoxy

The broad family of epoxy resins represents some of the highest performance resins that are available at this time. Epoxies generally outperform most other types of

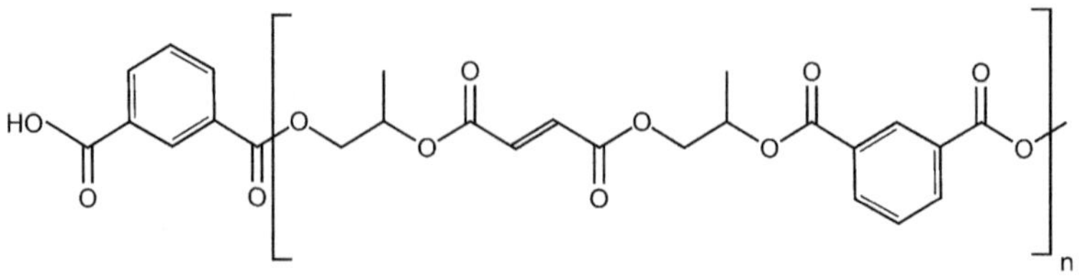

FIGURE 2.10 Chemical structure of a typical idealized polyester.

resins in terms of mechanical properties and resistance to environmental degradation, allowing almost exclusive application in aircraft components. As a laminating resin, its adhesive properties and high resistance to degradation by water make these resins ideal material for the manufacture of high-performance boats. The term "epoxy" refers to a chemical group consisting of an oxygen atom bonded to two carbon atoms already attached to some extent.

Usually identified by its amber or brown color, epoxy resins have a number of useful properties. Both the liquid resin and curing agent form easily processable systems with low viscosity. Epoxy resins are cured quickly and easily at any temperature between 5 °C and 150 °C depending on the choice of curing agent (hardener). However, in case of high viscosity, stages of postcuring may be required to obtain their final properties.

One of the biggest advantages of epoxies is their low shrinkage during curing, which minimizes internal stresses. High adhesive strength and mechanical properties are coupled to their high electrical insulation and chemical resistance.

Epoxy resins are formed from a molecular structure with long chains of reactive epoxy groups on the ends. The epoxy molecule also contains two aromatic rings in its center, which are capable of absorbing mechanical and thermal stresses better than linear groups and thus give the epoxy resin stiffness, toughness, and heat resistance. Figure 2.11 shows the chemical structure of diglycidyl ether of bisphenol A, a common type of epoxy resin.

2.3.1.3 Vinyl ester

Vinyl esters were developed to combine the simplicity and low cost of polyesters with thermal and mechanical properties of epoxies. The molecular structure of vinyl ester resins is similar to that of polyesters but differs primarily on the location of their reactive groups, which are positioned only at the ends of the chains. As the length of the chain is available to absorb impact loads, this makes vinyl ester resins more durable and resilient than polyesters. The vinyl ester molecule is also characterized by fewer ester groups. Such ester groups are susceptible to degradation via hydrolysis of water, which means that vinyl ester exhibits better resistance to water and many other chemicals than the polyester counterpart and as a consequence is often applied as transmission lines and tanks for storage of chemicals. Figure 2.12 shows the chemical structure of a typical vinyl ester.

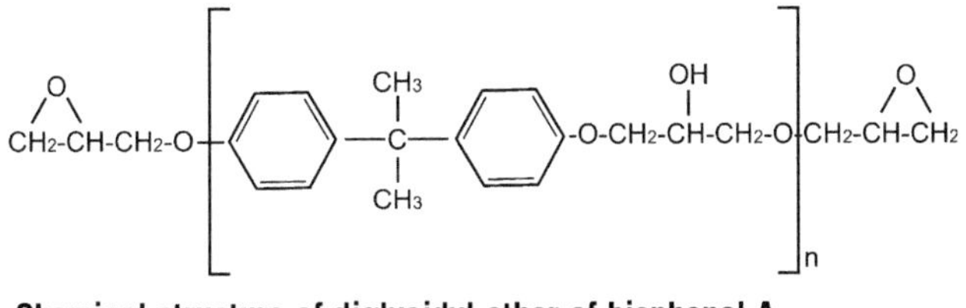

FIGURE 2.11 Chemical structure of diglycidyl ether of bisphenol A.

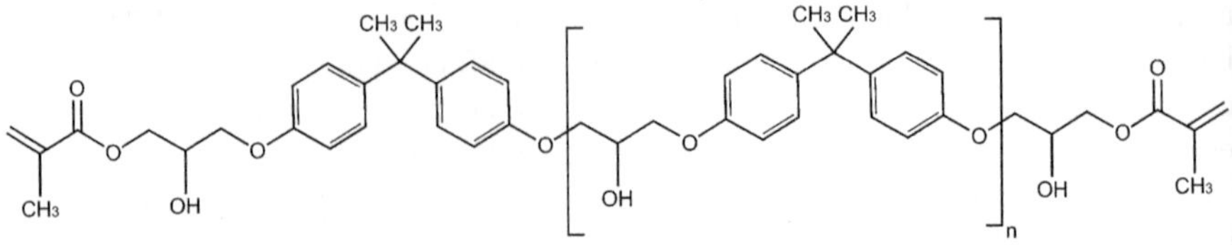

FIGURE 2.12 Chemical structure of a typical idealized ester based on epoxy.

Epoxy resins, polyesters, and vinyl esters, discussed so far, probably represent 90% of the thermosetting resins used in structural composites [7]. Table 2.4 summarizes the main advantages and disadvantages of each of these types of resin.

2.3.1.4 *Other resins used in composites*

Besides polyesters, epoxies, and vinyl esters, various other resins are used to manufacture composites due to their unique properties. As these other resins are less commonly used, they will be presented only briefly.

2.3.1.4.1 Phenolic resins

Used primarily in applications where fire resistance is required, phenolic resins retain their properties well at elevated temperatures. For materials cured at room temperature, corrosive acids are added, making the handling difficult. The cure by

Table 2.4 Advantages, Disadvantages, and Properties of Different Resins

Advantages	Disadvantages
Polyester	
Easy to handle Resin cheaper than epoxy and vinyl ester	Only moderate mechanical properties High styrene emission High shrinkage after cure Limited time frame for handling
Vinyl Ester	
High chemical resistance to the environment Mechanical properties superior to polyester	Need of postcure for superior mechanical properties High content of styrene More expensive than polyester High shrinkage after cure
Epoxy	
High mechanical and thermal properties Superior resistance to humidity Long handling periods possible Low shrinkage after cure	More expensive than vinyl ester Mixture of components is critical

a condensation reaction tends to lead to inclusion of voids (holes) and surface defects and the resins tend to be brittle and do not have high mechanical properties.

2.3.1.4.2 Ester Cyanates

These are mainly used in the aerospace industry. The excellent dielectric properties of this material make it applicable for manufacturing of protective dome cameras and antennas. Their thermal stability can reach 200 °C in humid environments.

2.3.1.4.3 Silicones

These are resins that include silicon structure in addition to carbon, oxygen, and hydrogen. They have good flame-retardant properties and thermal stability, and require high temperatures to cure. They are applied in the manufacture of missiles.

2.3.1.4.4 Polyurethanes

Polyurethanes are a family of polymers with a wide range of properties and applications based on the reaction of polyisocyanates with polyols. They are used as coatings, elastomers, foams, or adhesives. Polyurethanes present high chemical and mechanical resistance.

2.3.1.4.5 Bismaleimides

The bismaleimides are high-performance resins stable up to 230 °C in humid environments and 250 °C in dry environments. Typical applications include nacelles, missiles, fire barriers, and printed circuit boards.

2.3.1.4.6 Polyamides

This type of resin is used at elevated temperatures not supported by bismaleimides. Polyamides can be used up to 250 °C in humid environments and 300 °C in dry environments. Typical applications include missile and aircraft components.

In addition to thermosetting resins, thermoplastic resins are used in the manufacture of composite materials. These include nylon, PP, polyether ether ketone (PEEK), and polyphenylene sulfide (PPS).

2.3.2 Reinforcements

The reinforcements used in composite materials can be natural or synthetic. Various materials can be used to reinforce polymers. Some reinforcements are naturally produced as cellulose. However, most commercially available reinforcements are synthetic. Undoubtedly, the most widely used types of reinforcements are glass fibers, followed by carbon fibers and aramid. These types of fibers are discussed below.

2.3.2.1 Glass fibers

The glass fiber is a material obtained by making the molten glass flow through orifices and then it solidifies and has the flexibility to be used as fiber. Its diameter ranges from 5 to 24 μm. Different types of glass fibers are commercially available.

Manufactured based on alumino-boro-silicate, glass fibers of type "E" are considered to be the predominant reinforcement for PMC due to their high mechanical properties, low sensitivity to moisture, and superior electrical insulation. Fibers of type "S" have higher mechanical strength, thermal resistance, and elastic modulus.

Glass fibers usually have a good impact resistance, but their density is higher than that of carbon fibers and aramid. Composites reinforced with glass fibers are good electrical and thermal insulators. Moreover, such fibers are transparent to radio waves and are used in antennas and radars. Figure 2.13 shows the different forms of glass fibers available: roll, short fibers, and blankets.

2.3.2.2 Carbon fibers

Carbon fibers are produced by oxidation, carbonization, and controlled graphitization of carbon-rich organic precursors already in fiber form. The most common precursor is polyacrylonitrile (PAN) because it results in fibers with the best properties. However, fibers may also be made from cellulose and pitch. The variation in the controlled graphitization process produces fibers with high mechanical strength (~2600 °C) or high elastic modulus (~3000 °C) with other types among these. Once formed, the carbon fiber receives a surface treatment to improve the interaction with the matrix phase. Carbon fibers are more brittle (less elongation at break) than glass fibers and aramid. Like the glass fibers, the carbon fibers are commercially available in the form of continuous fibers, short fibers, and blankets (Figure 2.14).

2.3.2.3 Aramid fibers

Polymers are not only used as a matrix in composites, but can also serve as an excellent reinforcement. This is the case of aramid. The aramid fiber is a synthetic organic polymer produced by a spinning process of a solid fiber from a stock solution of polyamide. These yellowish filaments produced may have a variety of properties, but all have high mechanical strength and low density, providing a higher specific resistance. All classes of aramid fibers have good resistance to impact, and the

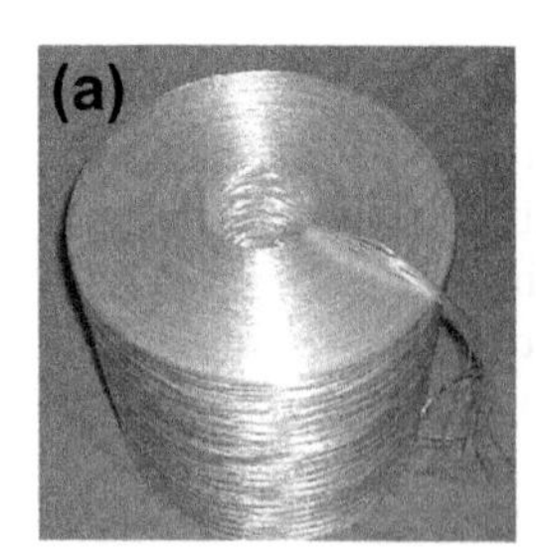

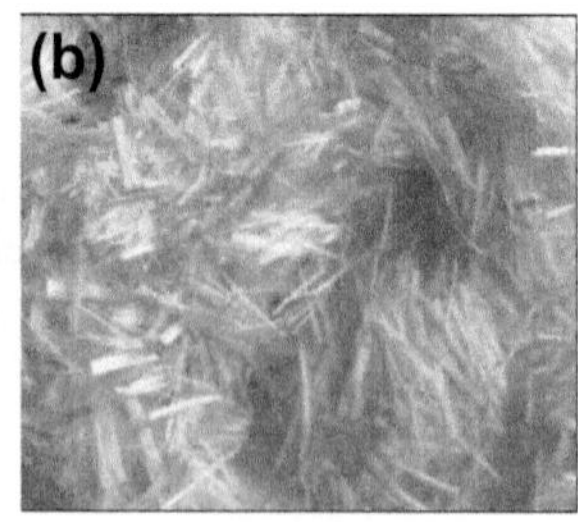

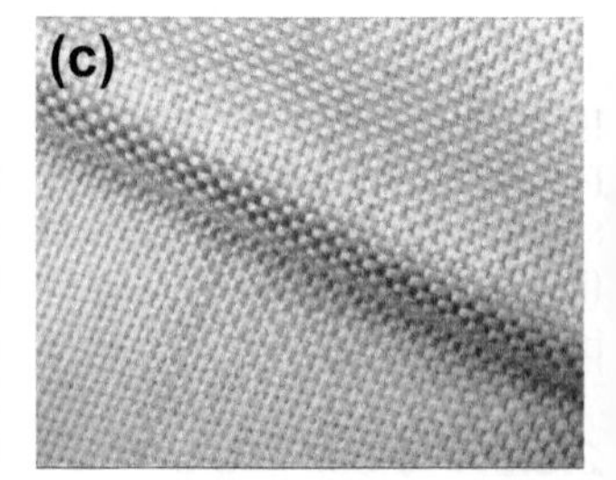

FIGURE 2.13 Glass fibers are marketed in the form of rolled continuous fibers (a), short fibers (b), and blankets (c).

Figure sourced from fibreglast.com [14]

FIGURE 2.14 Long carbon fibers (a), short fibers (b), and blankets (c).

Figure sourced from fibreglast.com [15]

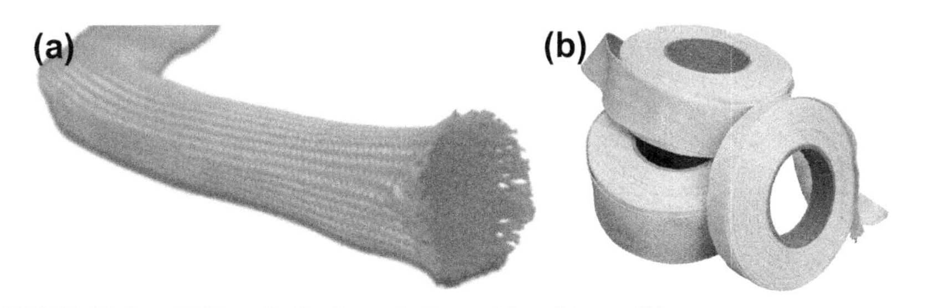

FIGURE 2.15 Aramid fibers in the form of sleeve (a) and tapes (b).

Figure sourced from fibreglast.com [16]

classes of lower moduli are widely used in ballistic applications. The resistance to compression, however, is similar to the glass fibers of the type "E." Figure 2.15 shows a sleeve and tapes made with aramid fiber.

Although commonly known by the trade name Kevlar® manufactured by DuPont, there are other suppliers of this type of fiber as the AkzoNobel company that uses the trade name Twaron®. Each supplier offers different classes of aramid with various combinations of stiffness and surface finish to meet different applications.

In addition to high strength, the fibers also offer good resistance to abrasion and thermal and chemical degradation. However, this type of fiber can slowly degrade when exposed to ultraviolet radiation.

A comparison of the mechanical properties of various types of fibers and a typical epoxy resin is shown in Figure 2.16 [7]. Notice how carbon fibers stand out with their high modulus. In addition, the properties of different types of glass, carbon, and aramid fibers are presented in Table 2.5.

2.3.2.4 Other fibers used in composites

There are a variety of other fibers that can be used for advanced structural composites but their use is less common. Some of these fibers are briefly discussed below.

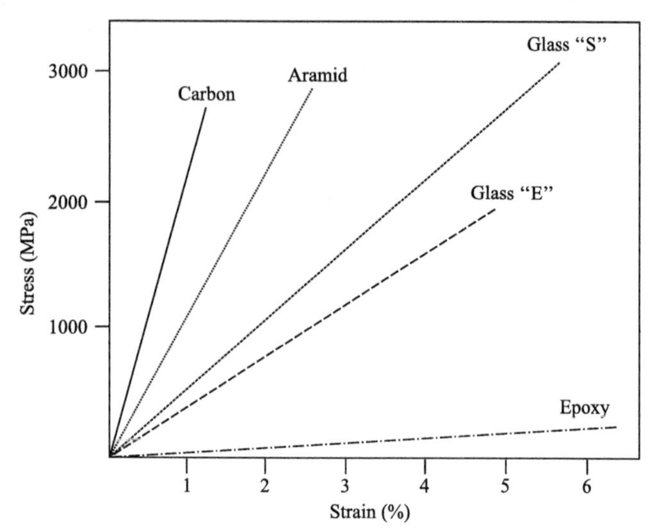

FIGURE 2.16 Comparison between stress–strain curves of various fibers and a typical epoxy resin.

Figure sourced from Fast and highly efficient one pot synthesis... [7]

2.3.2.4.1 Polyester fibers

Polyester fibers have low density compared to glass and carbon fibers, high toughness, and impact resistance, but low modulus. Their lack of stiffness restricts their inclusion in some composites, but it is usual in applications requiring low density, high impact, and abrasion resistance as well as low cost. It is also compatible with various types of resins.

2.3.2.4.2 Polyethylene

When in random orientation, molecules of PE of ultra-high molecular weight provide low mechanical properties. However, if dissolved and processed into fibers through spinning processing, the molecules become disentangled and aligne in the direction of the fiber. The molecular alignment gives a very high resistance to the fibers. These fibers have the highest specific strength of all the fibers described in this chapter. However, the modulus and maximum resistance of PE fibers are only slightly greater than those of the glass fibers of type "E" and smaller than those of carbon and aramid fibers. These factors, coupled with a high price and difficulty in creating a good interaction fiber/matrix, mean that PE fibers are not often used in the manufacture of composites.

2.3.2.4.3 Quartz

This is a version of the glass fiber rich in silicon with excellent mechanical properties and high resistance to elevated temperatures (>1000 °C). However, the manufacturing process and low production volumes imply a very high cost.

Table 2.5 Typical Properties of Carbon, Glass, and Aramid fibers

Fiber Type	ρ (g/cm^3)	E (GPa)	σ_m (Mpa)	ε (%)
Carbon Fibers—Polyacrylonitrile (PAN) Based				
T300	1.76	230	3.53	1.5
T400Hb	1.80	250	4.41	1.8
T700SC	1.80	230	4.90	2.1
T800HB	1.81	294	5.49	1.9
T1000GB	1.80	294	6.37	2.2
M35JB	1.75	343	4.70	1.4
M40JB	1.75	377	4.40	1.2
M50JB	1.88	475	4.12	0.9
M55JB	1.91	540	4.02	0.8
M60JB	1.93	588	3.82	0.7
M30SC	1.73	294	5.49	1.9
IM10	1.79	303	6.96	2.1
Carbon Fibers—Pinch Based				
K1352U	2.12	620	3.6	0.6
K1392U	2.15	760	3.7	0.5
K13C2U	2.20	900	3.8	0.4
K13D2U	2.20	935	3.7	0.4
Aramid Fibers				
Kevlar 29	1.44	70	2.92	3.6
Kevlar 49	1.44	112	3.00	2.4
Glass Fibers				
Type "E"	2.54	72	3.45	4.8
Type "S"	2.48	86	4.58	5.4

2.3.2.4.4 Ceramics

Ceramic fibers, generally in the form of very short "whiskers," are used primarily in areas that require strength at high temperatures. They are more often associated with nonpolymeric matrices such as metal alloys.

2.4 Advantages and disadvantages of composites

When we compare metals, ceramics, and polymers with composites, we see that the advantages and disadvantages of using a hybrid material depend on several factors such as the final application, manufacturing process, maintenance, and cost. Some

Table 2.6 Advantages/Properties and Disadvantages of Composite Materials

Advantages/Properties	Disadvantages
High strength/mass ratio	Costs
Light (low density)	Maintenance
Electrical properties	Manufacturing time/development
Chemical resistance to corrosion	Thermal stability
Color	Inspectability
Good fatigue resistance	Limited data bank of properties
Economy of material	
Reduced thermal expansion coefficient	
Reduced number of parts on a product	
Transparent to radars	
High stiffness and strength	

of the main advantages/disadvantages and properties of composites are summarized in Table 2.6.

2.5 Influence of fiber length in fiber composites

The reinforcement of polymer matrices with fibers may lead to materials with high specific strength and stiffness comparable to metal such as steel. The mechanism responsible for this improvement in properties is the transfer of load from the matrix to the fiber phase. Thus, the interfacial interaction between polymer and fiber is critical. Furthermore, the fiber length (l) also influences the load transfer and thus the performance of the composite. To have an efficient load transfer from the matrix phase to the fibers, the fiber length must exceed the so-called critical length (l_c) given by Eqn (2.1)

$$l_c = \frac{\sigma_f d}{2\tau_c} \tag{2.1}$$

where:

d is the diameter of the fiber;
σ_f is the ultimate strength of the fiber;
τ_c is the fiber-matrix bond strength.

For composites where $l < l_c$, no efficient load transfer from the matrix phase to the fibers takes place and consequently the reinforcement is small. Figure 2.17 shows the profile for transfer of tension in a composite material.

The classic approach to maximize the efficiency of the reinforcement is to increase the fiber length (Figure 2.17(a)). However, this results in processing difficulties. Conversely, by decreasing the diameter of the fiber, it is possible to

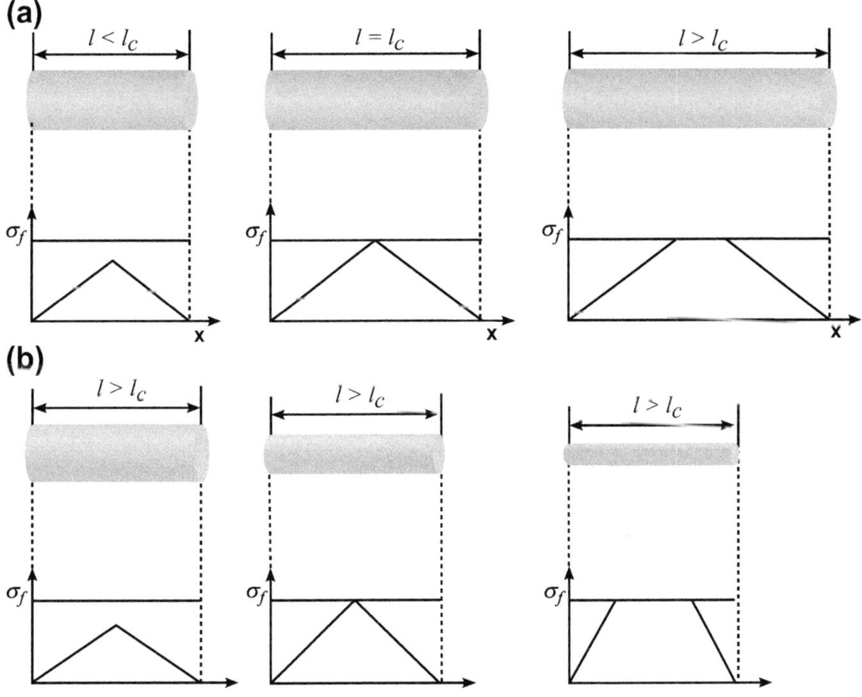

FIGURE 2.17 Profile of load transfer from the matrix to the fiber: (a) the composite resistance is increased by increasing the length of the fiber (classic composite) and (b) the resistance of the composite is increased by decreasing the diameter of the fiber (nanocomposites).

Figure sourced from The potential of carbon nanotubes in polymer composites [17]

increase the efficiency of reinforcement without sacrificing processability (Figure 2.17(b)). For this reason, nano-fillers such as CNTs and carbon nanofibers have attracted much interest from the scientific community and industry in the last decade. Due to their high surface area, as we shall see in the next chapter, CNTs can lead to further progress in which the counterbalance between performance and processability may not be a problem.

2.6 Applications of composites

Composites are used in a wide variety of components in the marine, automotive, aerospace, and civil construction areas, as well as in consumer goods. Much of the current composite technology was developed from aerospace applications. A typical example of the application of composites is the F-22A Raptor fighter aircraft (Figure 2.18). Approximately 24% of its structure (mass) consists of composites: bismaleimide and epoxy resins reinforced with carbon fibers.

FIGURE 2.18 Aircraft F-22A Raptor.

Part of its structure (24%) is made of composite materials.

Figure sourced from Airforce-Technology.com [18]

In the figure below, we see an example of how composite materials can replace metals such as aluminum. The diagram compares the materials used in the construction of the Boeing aircrafts from 1969 to 2010 [19].

The below diagram shows the evolution of materials used in the Boeing aircraft structure from 1969 to 2010:

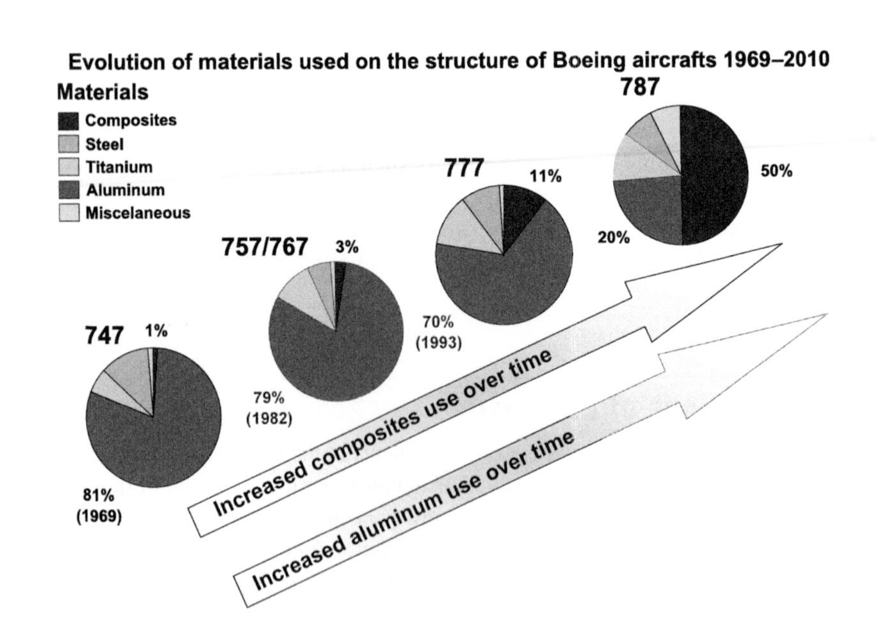

Probably, the greatest competition for the development of composites takes place in the aerospace industry. The two largest companies in the sector of manufacturing airplanes, Boeing and Airbus, have decided to expand the use of composites in their new aircrafts considerably. The world's largest aircraft, the A380 Super Jumbo, built by Airbus, contains 20% (by mass) of composites (Figure 2.19). On the other hand, the Boeing Company launched the 787 Dreamliner with 50% composites (Figure 2.20). The increased use of composites involves economies both by lower fuel consumption due to reduced mass as well as by the longer flight time of these aircrafts due to lower maintenance.

In 2005, the fastest convertible car in the world was released by Mercedes-Benz. The Mercedes CLK DTM AMG has a body made of PMC reinforced with carbon fibers (Figure 2.21).

Composites are also present in transportation with two wheels. With a mass of only 122 kg, Enertia, produced by Brammo Motorsports, is the first motorcycle in the world with zero carbon emissions (Figure 2.22.). Its frame is made from fiber-reinforced composite carbon, and the energy needed for the engine is supplied by a battery bank that is easily recharged in the power outlet.

Bulletproof vests are one of the most typical examples of the application of Kevlar® fibers (Figure 2.23). For thousands of years, people wore armor-like metal.

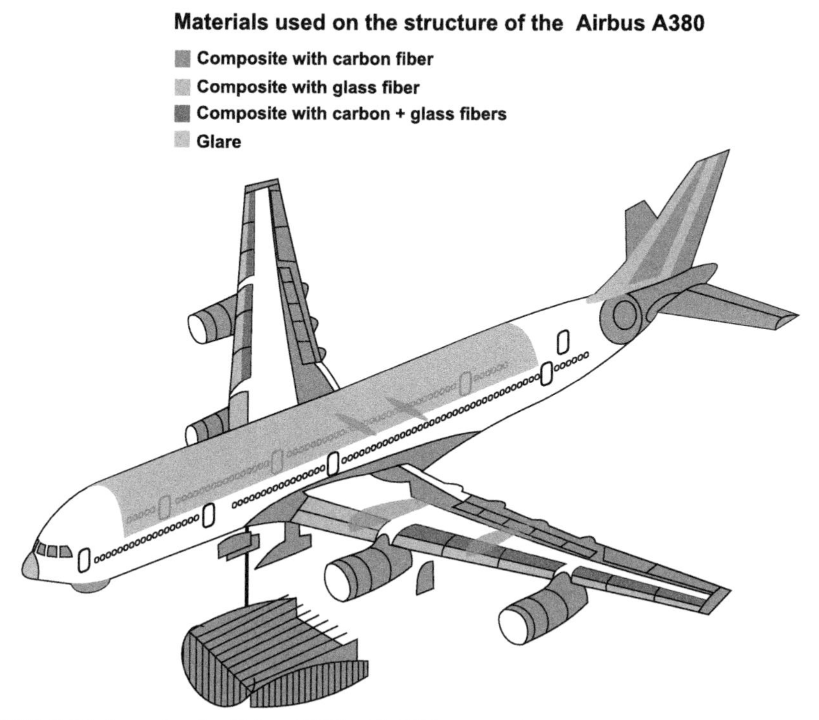

FIGURE 2.19 Super Jumbo Airbus A380 with 20% composites.

Figure sourced from airporttech website [20]

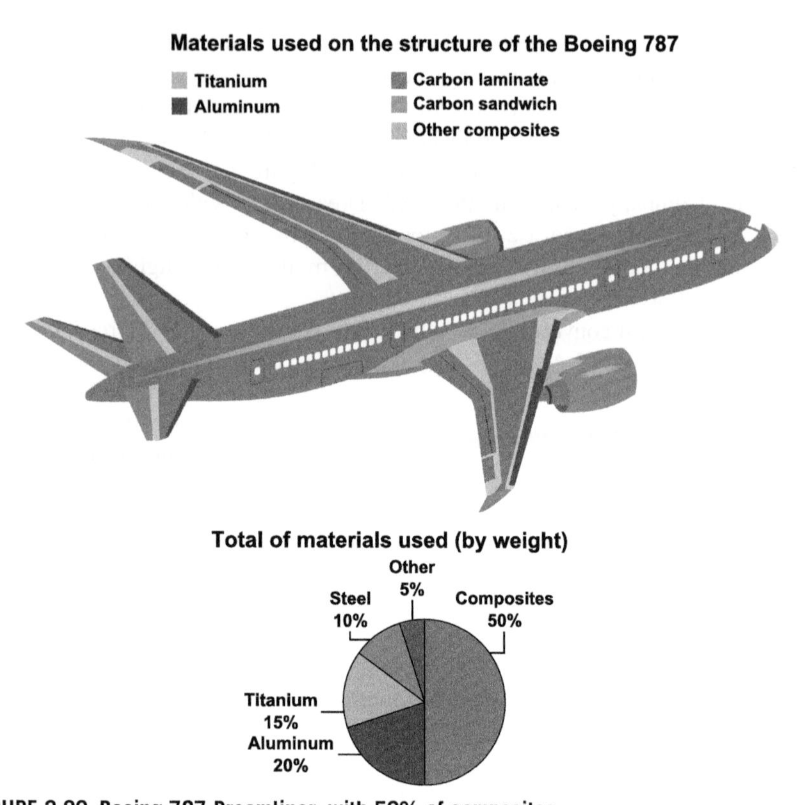

FIGURE 2.20 Boeing 787 Dreamliner, with 50% of composites.

Figure sourced from 787 Dreamliner [19]

FIGURE 2.21 Numerous body components of the Mercedes CLK Cabriolet are made of carbon fiber composite material, including the luggage compartment, doors, front wings, rear wheel arches flaring, the front and rear aprons, as well as the rear airfoil.

Figure sourced from Wikipedia [21]

FIGURE 2.22 Enertia is an electric bike with a frame reinforced with carbon fibers.

Figure sourced from Brammo Gallery [22]

FIGURE 2.23 Bulletproof vest formed by woven Kevlar® fibers.

Figure sourced from Engarde Body Armor [23]

However, with the development of firearms in the nineteenth century, the thickness of the metal used in armor had to be drastically increased to prevent the penetration of projectiles. This made the armor too heavy to be worn and inconvenient to use. It was only in the1960s that lightweight bulletproof vests were developed from woven fibers.

In the infrastructure sector, the use of composites is growing slowly. The main reason is that the mass reduction factor is less critical than in other industrial sectors, and in civil engineering, the cost of composites is much higher compared to traditional building materials such as concrete and steel. However, applications such as the repair of structures (pillars) or damaged poles are already a reality. In Figure 2.24 we can see how polymer composites containing carbon or glass fibers are used for the reinforcement of concrete beams on a bridge.

In 1997, the first bridge made of PMC reinforced with carbon and glass fibers was installed in Denmark (Figure 2.25). At 40 m long, the bridge was constructed in advance, so during the installation the traffic in that area suffered little perturbation.

In the sector of energy generation, composites are used in the manufacture of wind generators (Figure 2.26). The blades of the generators are generally made of epoxy resins reinforced with glass or carbon fibers. The high mechanical strength and specific elastic modulus of the composites added to its fatigue resistance make this material the best choice for the manufacture of wind turbine generators.

Polymeric materials and composites may also be used in research and biological applications [27,28]. Biodegradable polymers and bioactive ceramics (alumina, zirconia, hydroxyapatite, etc.) can be combined to form composite materials for tissue engineering (artificial skin). Moreover, metals and alloys, including stainless steel and gold, can be used as biomaterials. Polymers such as P, polyacetal (POM),

FIGURE 2.24 Reinforcement of concrete columns through the use of fiber composites.

Figure sourced from Schnell Contractors Inc. [24]

FIGURE 2.25 Plastic bridge reinforced with carbon and glass fibers in Denmark.

Figure sourced from Fiberline Inc. [25]

FIGURE 2.26 Wind generator with blades manufactured from composites.

Figure sourced from Environment-Green [26]

and corresponding composites are excellent candidates for biological applications when compared to other materials. For example, metals have low biocompatibility and exhibit less corrosion while the ceramic materials are brittle and dense. Figure 2.27 lists various applications of different biocomposite materials.

Other applications of composites include skies, bicycles, surfboards, bows, baseball bats, tennis rackets, jet skis, tires, helmets, seats, decks, pools, tanks, missiles, satellites, space shuttles, solar generators, bathrooms, windows, computers, dental implants, prostheses, bags, poles, etc.

As we have seen in this chapter, over time composite materials have taken the place of other engineering materials, often regarded as irreplaceable, such as steel, for example. The advent of composites reinforced with CNTs is expanding even more the possibilities for the future of composite materials.

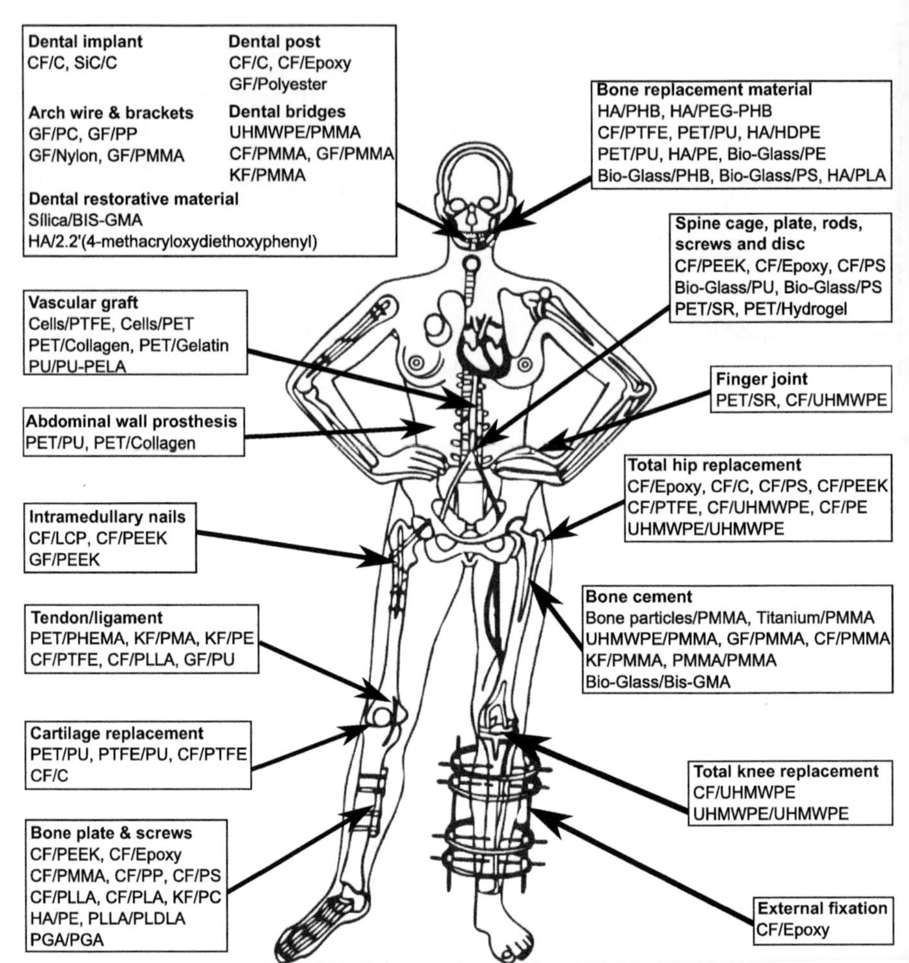

CF: carbon fibers, **C**: carbon, **GF**: glass fibers, **KF**: Kevlar fibers, **PMMA**: Polymethylmethacrylate, **PS**: polysulfone, **PP**: Polypropylene, **UHMWPE**: ultra-high-molecular weight polyethylene, **PLDLA**: poly(L-DL-lactide), **PLLA**: poly (L-lactic acid), **PGA**: polglycolic acid, **PC**: polycarbonate, **PEEK**: polyetheretherketone; **HA**: hydroxyapatite, **PMA**: polymethylacrylate, **BIS-GMA**: bis-phenol A glycidyl methacrylate, **PU**: polyurethane, **PTFE**: polytetrafluoroethylene, **PET**: polyethyleneterephthalate, **PEA**: poltethylacrylate, **SR**: silicone rubber, **PELA**: Block co-polymer of lactic acid and polyethylene glycol, **LCP**: liquid crystalline polymer, **PHB**: polyhydroxybutyrate, **PEG**: polyethyleneglycol, **PHEMA**: poly(20hydroxyethyl methacrylate)

FIGURE 2.27 Different applications of polymer composite biomaterials.

Figure sourced from Brammo Gallery [22] and Biomedical applications of polymer-composite materials [27]

To learn more...

We encourage readers to consult the following books:

1. Ma-zumdar S. Composites manufacturing: Materials, product, and process engineering. 1st ed. CRC Press; 2001, ISBN-13: 978-0849305856.
2. Sperling LH. Introduction to physical polymer science. 4th ed. Wiley-Interscience; 2005, ISBN-13: 978-0471706069.
3. Brent Strong A. Fundamentals of composites manufacturing: Materials, methods and applications. 2nd ed. Society of Manufacturing Engineers; 2007, ISBN-13: 978-0872638549.
4. Forbes Aird. HP Trade. Fiberglass & other composite materials: A guide to high performance non-metallic materials for race cars, street rods, body shops, boats, and aircraft. Revised ed. 2006, ISBN-13: 978-1557884985.

References

[1] Ashby MF, Shercliff H, Cebon D. Materials: Engineering, science, processing and design. 2nd ed. Butterworth–Heinemann; 2007. Publication Date: December 3, 2009. ISBN-10: 1856178951, ISBN-13: 978–1856178952.

[2] Mazumdar S. Composites manufacturing: Materials, product, and process engineering. 1st ed. CRC Press; 2001. Published: December 27, 2001. Content: 416 Pages.

[3] Callister Jr WD. Materials science and engineering: An introduction. 7th ed. Publication Date: December 30, 2009. ISBN-10: 0470419970, ISBN-13: 978–0470419977.

[4] Mark James E. Physical properties of polymers handbook. 2nd ed. Springer; 2002, ISBN 978-0-387-31235-4 (Print) 978-0-387-69002-5 (Online).

[5] Loos MR, Schulte K, Abetz V. A highly efficient one-pot method for the synthesis of carbon black/poly(4,4'-diphenylether-1,3,4-oxadiazoles) composites. Macromol Chem Phys 2011;212:1236–44.

[6] Loos MR, Schulte K, Abetz V. In situ synthesis of polyoxadiazoles (POD) and carbon black (CB) as an approach to POD/CB nanocomposites. Compos Part B: Eng 2011;42: 414–20.

[7] Loos MR, Schulte K, Abetz V. Fast and highly efficient one-pot synthesis of polyoxadiazole/carbon nanotubes nanocomposites in mild acid. Polym International 2010;60: 517–28.

[8] Loos MR, Schulte K, Abetz V. Dissolution of MWCNTs by using polyoxadiazoles, and highly effective reinforcement of their composite films. J Polym Sci Part A: Polym Chem 2010;48:5172–9.

[9] Loos MR, Gomes D. The effect of sulfonation level and molecular weight on the tensile properties of polyoxadiazoles. High Perform Polym 2009;21:697–708.

[10] Starr TF. Composites: A profile of the worldwide reinforced plastics industry. 2nd ed. Elsevier Advanced Technology; 1995. ISBN-10: 1856173542, ISBN-13: 978–1856173544.

[11] Daniel Isaac M. Engineering mechanics of composite materials. Oxford University Press; 1994. ISBN-10: 019515097X, ISBN-13: 978–0195150971.

[12] www.gurit.com/files/documents/guide-to-compositesv4pdf.pdf.

[13] Aird Forbes. Fiberglass & other composite materials: A guide to high performance nonmetallic materials for race cars, street rods, body shops, boats, and aircraft. HPTrade; 2006. ISBN-10: 1557884986 ASIN: B002HREKSW.
[14] www.fibreglast.com/category/fiberglass.
[15] www.fibreglast.com/category/carbon fiber.
[16] www.fibreglast.com/category/kevlar.
[17] Ciselli P. The potential of carbon nanotubes in polymer composites [Ph.D. thesis]. Eindhoven University of Technology; 2007.
[18] www.airforce-technology.com/projects/f22/f222.html.
[19] Cook Dave, Krause Randy. 787 Dreamliner: A new airplane for a new world. Presentation Singapore; October 17, 2007.
[20] www.airporttech.tc.faa.gov/safety/patterson2.asp.
[21] www.en.wikipedia.org/wiki/mercedes-benz clk-class.
[22] www.brammo.com/brammogallery enertia/.
[23] www.engardebodyarmor.com/concealable.htm dual.
[24] http://schnellcontractors.com/concrete-repair/carbon-fiber-strengthening/.
[25] www.fiberline.com/konstruktioner/profiles-and-decks-bridges/standard-concept-bridges.
[26] www.environment-green.com/wind power/wind generators.html.
[27] Ramakrishna S, Mayer J, Wintermantel E, Leong KW. Biomedical applications of polymer-composite materials: A review. Compos Sci Technol July 2001;61(9):1189–224 (36).
[28] Rezwan K, Chen QZ, Blaker JJ, Boccaccini AR. Biodegradable and bioactive porous polymer/inorganic composite scaffolds for bone tissue engineering. Biomaterials June 2006;27(18):3413–31.

CHAPTER

Allotropes of Carbon and Carbon Nanotubes

3

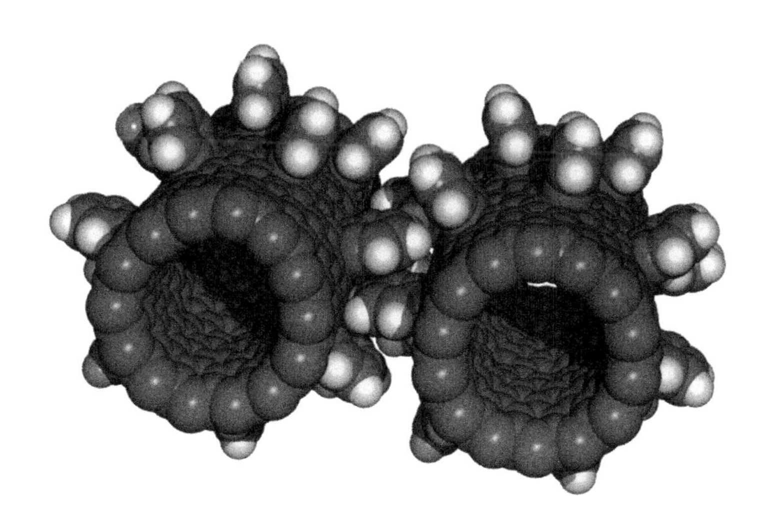

CHAPTER OUTLINE

Carbon Nanotube Reinforced Composites. http://dx.doi.org/10.1016/B978-1-4557-3195-4.00003-5

3.1 Allotropes of carbon

Carbon is the most important element found in living organisms. Life, as we know it, would not exist without carbon atoms. Carbon is the sixth most abundant element in the universe: about 0.5 ppm. One of the features that make carbon so important is its ability to form bonds with various elements in different ways. Most other elements form weak bonds between its atoms themselves. To get an idea, the energy of a single C–C bond is 348 kJ/mol. This energy is comparable with the energy of bonds formed between carbon and other elements such as oxygen, nitrogen, and hydrogen: C–O 360 kJ/mol, C–N 305 kJ/mol, and C–H 412 kJ/mol, respectively. Higher energies for atomic bonds involve molecules with greater structural stability.

Inorganic carbon is found in applications such as pencils, diamond, amorphous carbon filters, for hardening steel, lubricant, and polymeric materials. Organic carbon in the form of animals and plants is transformed into oil after millions of years, and this is a major source of energy for our civilization. Biological materials such as proteins, amino acids, lipids, and nucleic acid also have a carbon basis.

Diamond, an allotrope of carbon, is found in the cubic and hexagonal configurations. In a cubic diamond, each carbon atom is bonded to four other atoms equidistant via sp^3 bonds to form a tetrahedral arrangement (Figure 3.1(a)). The length of the C–C in diamond is 1544 Å, i.e., approximately 10% greater than the C–C in graphite.

Graphite is the most stable thermodynamic form of carbon, under normal conditions of temperature and pressure. The graphite structure is formed by layers of atoms in a hexagonal array (Figure 3.1(b)). The atoms of each layer are joined by covalent bonds, while the atoms in different layers are joined together by weak van der Waals bonds. These weak inter-planar links are responsible for the excellent lubricating properties of graphite, once planes can easily slide over each other.

There are no materials made of the same element that are as different from one another as in the case of graphite and diamond. Graphite is a soft material whereas diamond is one of the hardest. As a consequence, diamond is used as abrasive, and graphite as a lubricant. A curious fact is that, in space, graphite does not act as a good

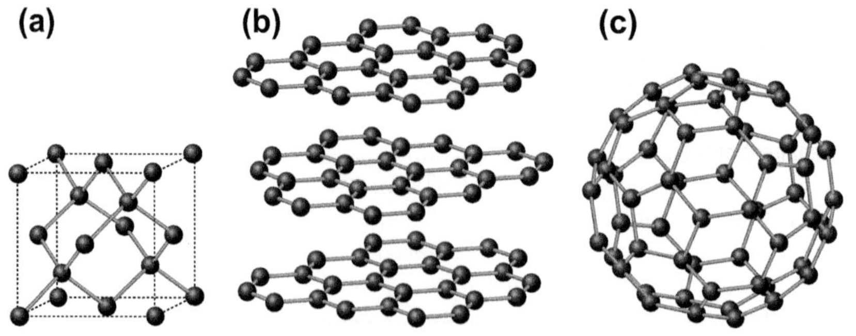

FIGURE 3.1 Allotropic structures of carbon (a) diamond, (b) graphite, and (c) fullerene.

lubricant as it does on Earth, possibly due to lack of moisture between layers. Graphite is electrically conductive and diamond is an insulator. Diamond is optically transparent, and graphite is optically dense. Diamond has an isotropic structure, while the structure of graphite is anisotropic. Diamond is a good conductor of heat but graphite, as a whole, has low thermal conductivity. Indeed, the conductivity of the graphite plane is comparable to or even greater than that of diamond. The low conductivity perpendicular to the plane causes the average conductivity of graphite (volume) to be low. Note that the graphene layers forming the graphite are oriented in an anisotropic fashion.

The term graphite was created by Abraham Gottlob Werner in 1789 due to its use in pencils and comes from the Greek word "graphein" meaning "to write."

Fullerenes, also called "buckyballs," are a third allotrope of carbon discovered in 1985 [1]. This molecular form consists of a hollow cluster containing 60 carbon atoms (or more) arranged in hexagons (six carbon atoms) and pentagons (five carbon atoms). These molecules are represented by C_{60}. There are also C_{70} molecules, C_{100}, etc. The average diameter of a fullerene is 1.1 nm. Figure 3.1(c) shows a C_{60} molecule composed of 12 pentagons and 20 hexagons; the format is similar to a soccer ball. Fullerenes are formed under conditions of high pressure and temperature as those found in space. Indeed, fullerenes have been found in a 4.6 billion-year-old meteorite that fell in Mexico [2].

These molecules may be more abundant than graphite and diamond. In pure form fullerenes are insulators, but the addition of impurities can make them conductors or semiconductors. The main method of synthesis of fullerenes is the arc discharge. When obtained in elongated form, fullerenes form carbon nanotubes (CNTs). In the following section we discuss the main characteristics of CNTs.

3.2 Carbon nanotubes

Nanotechnology has become the front line in the development of science and technology. CNTs are considered one of the most important building blocks of this new technology. Due to the combination of electronic, thermal, and mechanical properties, the scientific community's interest in CNTs' potential applications in composites, electronics, computers, hydrogen storage, efficiency medicine, sensors, and many other areas has grown rapidly. In fact, NASA is developing materials taking advantage of the high mechanical strength and modulus of CNTs for applications in space missions. It is believed that CNTs' reinforced composites may lead to production of materials with the greatest specific resistance (σ_{max}/ρ) ever manufactured by humans.

A CNT can be considered as a graphene sheet rolled to obtain a cylinder with a diameter of a few nanometers to a few micrometers length, and half fullerene at each end [3]. There are several types of CNTs, including single-walled CNTs (SWCNTs), double-walled CNTs (DWCNTs), and multi-walled CNTs (MWCNTs) (Figure 3.2). SWCNTs consist of a single cylinder [5,6]. DWCNTs consist of two concentric

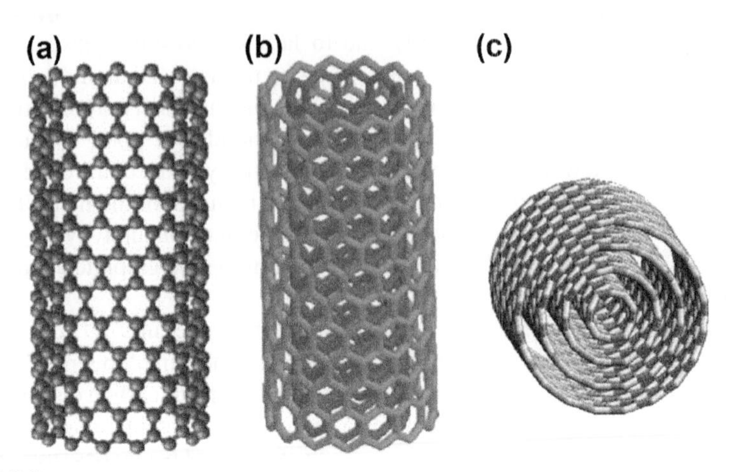

FIGURE 3.2

Schematic diagram of a (a) single-walled carbon nanotube, (b) double-walled carbon nanotube, and (c) multi-walled carbon nanotube.

Figure sourced from Nanowires and nanotubes [4]

cylinders [7], and MWCNTs are formed by a concentric arrangement of multiple CNTs separated by a distance of 0.034 nm from each other [8]. All layers of MWCNTs are held together only by weak van der Waals interactions, and the strength of this interaction is much smaller than C–C bonds in the plane of the graphene sheet. One mole of C–C bonds has an energy of 347 kJ/mol. The energy required to break a single C–C bond is 5.76×10^{19} J. Studies have shown that movement of the inner layers of MWCNTs occurs in a telescopic fashion due to low friction between the layers. The forces of static and dynamic friction between the layers have values $<2.3 \times 10^{14}$ N and $<1.5 \times 10^{14}$ N, respectively [9]. This means that the energy required to break a C–C bond is about 100 thousand times greater than the energy required to separate two layers of graphene on MWCNTs.

3.2.1 History of CNTs

Most published studies attribute the discovery of graphite hollow tubes with diameters in the nanometer range to the article by Iijima in 1991 [8]. The concept of SWCNTs was developed in 1993 in two independent papers published in the journal Nature, a work by Iijima [6] and another by Bethune [5]. However, an article published much earlier by Oberlin in 1976 [10] shows a picture illustrating a nanotube, possibly an SWCNT (Figure 3.3). The authors did not specify if this was the structure of a nanotube and due to problems with the magnification of microscopes of that time, the number of walls forming the nanotube cannot be determined. Thus, this structure is considered to be of an SWCNT or a DWCNT.

The article by Iijima in 1991 demonstrated the possibility of growing CNTs without the need for catalysts and boosted research on MWCNTs [8]. The possibility of growth of carbon filaments by the thermal decomposition of gases rich in

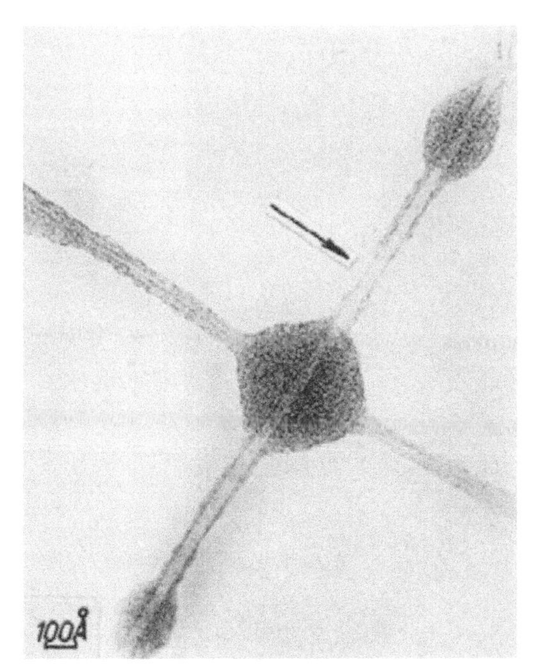

FIGURE 3.3 One of the first micrographs depicting a single-walled carbon nanotube or double-walled carbon nanotube.

Figure sourced from Filamentous growth of carbon through benzene decomposition [10]

hydrocarbons was first reported in 1889 [11]. Thus, the articles published by Schüzenberger [12] and Pelabon [13] can be considered the first evidence of the growth of CNTs.

With the invention of the transmission electron microscope (TEM) in 1939, significant progress was achieved in the characterization of carbon filaments. In 1952, Radushkevich [14] published what is believed to be the first image TEM of the tubular nature of carbon filaments with a diameter of nanometers (Figure 3.4).

According to several authors, Radushkevich should receive credit for the discovery that nano-carbon filaments may have a tubular structure, i.e., the discovery of CNTs [15]. However, the article by Iijima in 1991, showing high-resolution images of CNTs (Figure 3.5), caused an unprecedented shift in the science of carbon [8]. Twenty years after the article by Iijima, the number of published articles and patents related to CNTs is still growing exponentially (Figure 3.6).

3.2.2 Structure of CNTs

There are several ways to define the structure of CNTs. One possibility is to consider that CNTs are obtained by rolling a graphene sheet in a specific direction, keeping the circumference of the cross section [4]. The properties of the graphene are

FIGURE 3.4 Transmission electron microscopy of low resolution showing three hollow carbon fibers.

Magnification of 20,000×.

Figure sourced from About the structure of carbon formed by thermal decomposition of carbon monoxide on iron substrate [14]

governed by the carbon–carbon bonds with a bond length of 0.142 nm. As the microscopic structure of the CNTs is closely related to graphene, tubes are usually labeled in terms of the vectors associated with the lattice-shaped (honeycomb network) arrangement of carbon atoms forming the graphene sheet. In addition, the similarity to graphene enables the derivation of equations for the calculation of various properties of CNTs.

The atomic structure of the CNTs can be described in terms of helicity or chirality defined by the chiral vector, $\boldsymbol{C}_h$, and chiral angle, θ. In Figure 3.7, a sheet of graphene with a chiral vector and angle defined can be seen.

The chiral vector can be described in terms of chiral indexes (n, m) (also called Hamada indexes) and the unit vectors $\boldsymbol{a}_1$ and $\boldsymbol{a}_2$ (see Figure 3.7) by the equation

$$\boldsymbol{C}_h = n\boldsymbol{a}_1 + m\boldsymbol{a}_2 \tag{3.1}$$

where (n, m) are integers representing the number of steps along the carbon bonds of the zigzag hexagonal lattice. The module of the unit vectors $\boldsymbol{a}_1$ and $\boldsymbol{a}_2$ has value of

$$|\boldsymbol{a}_1| = |\boldsymbol{a}_2| = a = 0.246\,nm \tag{3.2}$$

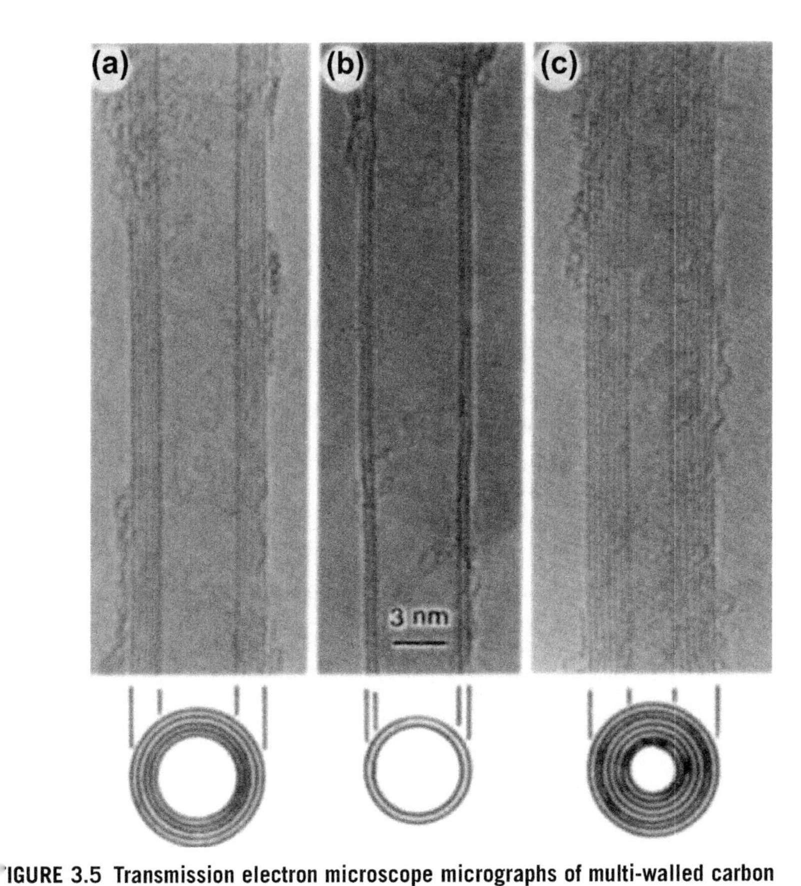

IGURE 3.5 Transmission electron microscope micrographs of multi-walled carbon anotubes consisting of (a) five, (b) two, and (c) seven concentric layers of graphene.

Figure sourced from Double wall carbon nanotubes with an inner diameter of 0.4 nm [7]

The C–C bond length (a_{C-C}) can be related to a unit vector by

$$a_{C-C} = \frac{a}{\sqrt{3}} = 0.142\,nm \tag{3.3}$$

The angle chiral θ determines the degree of "twisting" of the tube and is defined s the angle between the vectors $\boldsymbol{C}_h$ and $\boldsymbol{a}_1$, which varies in the range $0° \le |\theta| \le 30°$. n terms of integers (n, m), θ can be described by the following set of equations [4]

$$\sin\theta = \frac{\sqrt{3}m}{2\sqrt{n^2 + m^2 + nm}} \tag{3.4}$$

$$\cos\theta = \frac{2n + m}{2\sqrt{n^2 + m^2 + nm}} \tag{3.5}$$

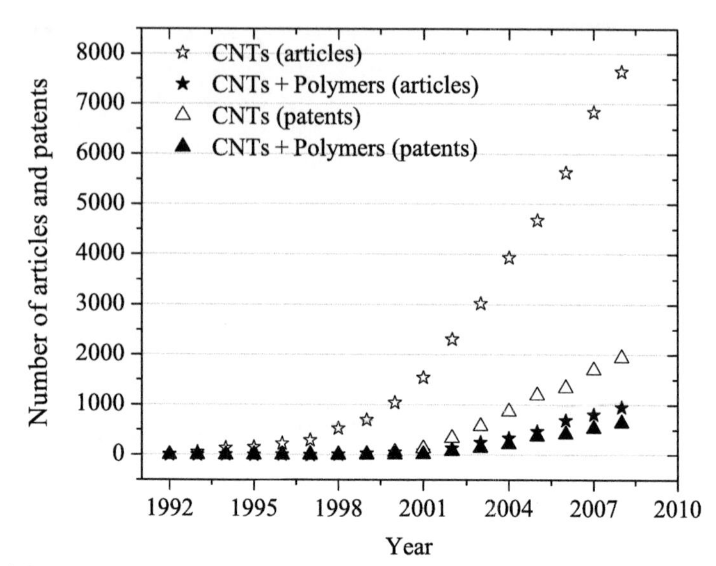

FIGURE 3.6 Number of patents and published articles on nanotubes and nanotube/polymer by year.

CNT, carbon nanotubes.

Figure sourced from Macromolecular theory and simulations [16]

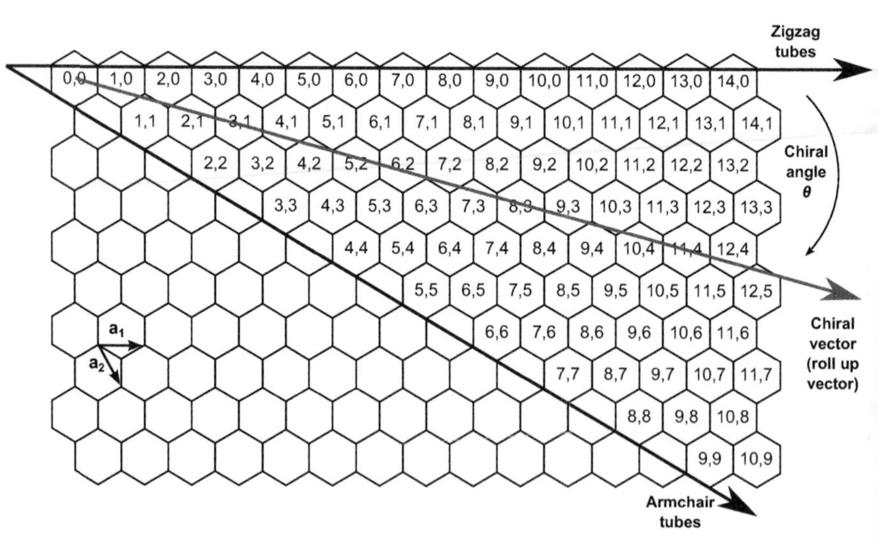

FIGURE 3.7 Schematic diagram showing how a sheet of graphene is "rolled" to form a carbon nanotube.

Figure reproduced with permission from Hornyak et al. [17], *"Introduction to nanoscience and nanotechnology"*

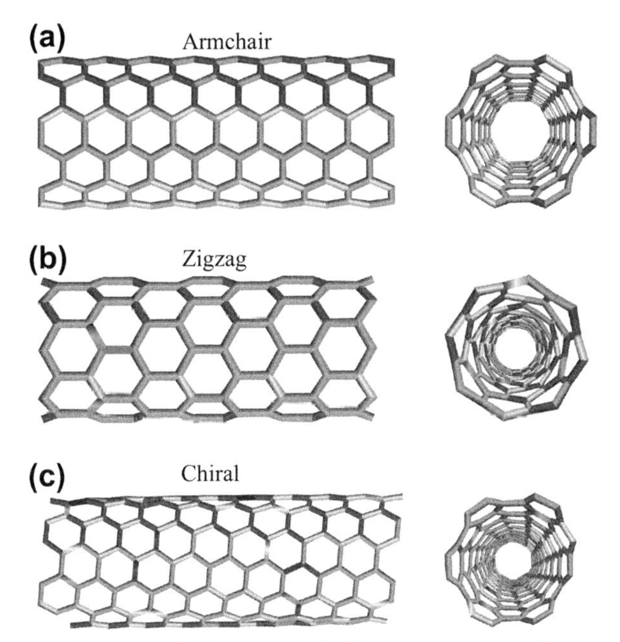

FIGURE 3.8 Schematic model of (a) an armchair, (b) zigzag, and (c) chiral nanotubes.

Figure sourced from Cobweb website [18]

$$\tan\theta = \frac{\sqrt{3}m}{2n+m} \tag{3.6}$$

Based on the geometry of carbon bonds around the circumference of the tube, there are two limiting cases, corresponding to the achiral tubes, known as the armchair ($\theta = 30°$) and zigzag ($\theta = 0°$) as shown in Figure 3.8. In addition, when $0° < |\theta| < 30°$ the nanotube is called chiral. In terms of the chiral vector, a nanotube is armchair for (n, m) and zigzag for $(n, 0)$. Also, as the interatomic spacing between the carbon atoms is known, the diameter of a nanotube (d) can be calculated as the ratio between the circumference of the tube L and π

$$d = \frac{L}{\pi} = \frac{\sqrt{n^2 + nm + m^2}a}{\pi} \tag{3.7}$$

or in terms of n, m, and b

$$d = \frac{|C_h|}{\pi} = \frac{\sqrt{3(n^2 + nm + m^2)}b}{\pi} \tag{3.8}$$

This equation can be used only for SWCNTs or for the outer layer of MWCNTs. Since $n = 0$ for zigzag nanotubes, it is easy to see from Eqn (3.8) that armchair nanotubes have a diameter greater than the zigzag type.

Due to the van der Waals attractions, nanotubes tend to aggregate and are usually found in the form of bundles. The spacing between nanotubes in a bundle is near that presented by graphene sheets: 0.34 nm. MWCNTs exhibit the same general spacing inter-tubes. However, the spacing between the different tubes in a bundle is dependent on its chirality. For armchair tubes, spacing is 0.338–0.341 nm, and zigzag tubes with chiral indices ($2n$, m) have a spacing of 0.339 nm.

EXAMPLE 3.1

Building an SWCNT.

Print an sheet of graphene on a transparency film like the type used for preparing slides for overhead projectors. You can print the graphene sheet in Appendix C.

Solution

To build a tube like the armchair type (5, 5), trace the chiral vector originating from (0, 0) and ending at (5, 5). Now roll the graphene sheet such that the atom (0, 0) is overlapped with the atom (5, 5) as shown in the Figure [17].

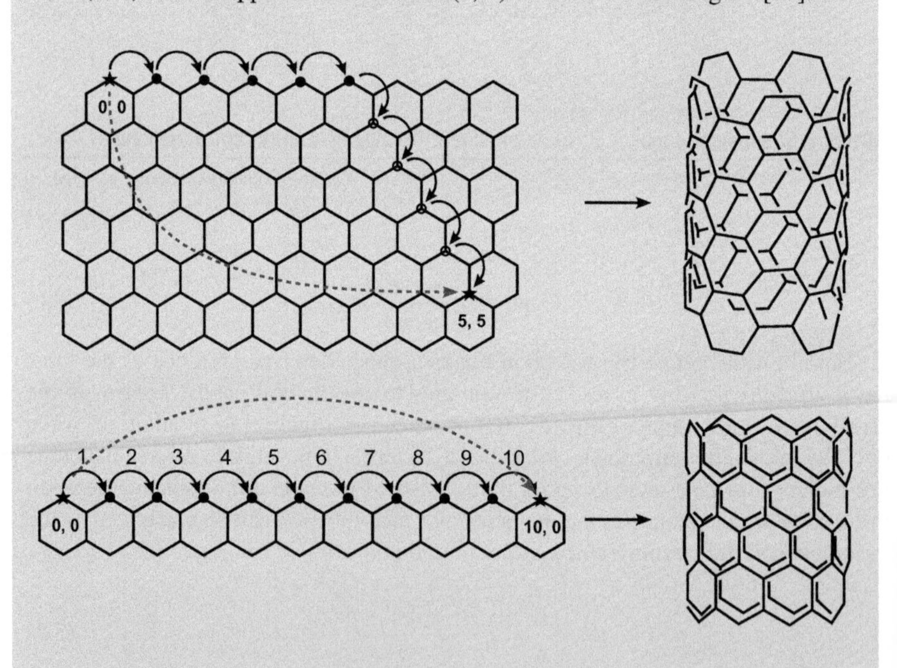

Note that when you roll the graphene sheet, the ends of the chiral vector coincide. Using the same method, build a zigzag nanotube (10, 0). Choose the index (n, m) suitable for a nanotube chiral and try to build it. In your opinion, what type of nanotube was easier to build? The tubes that you built, or conductors or semiconductors?

3.2.3 Mechanical properties of CNTs

Experimental and theoretical results show that CNTs are one of the strongest materials known to humans [19,20]. Nanotubes are formed by carbon bonds: the strongest type of bond in nature. Since many C atoms form a nanotube, the force required to break a CNT must be enormous. The covalent bonding that keeps the carbon atoms in position underlies the high mechanical strength of the structure of a CNT. It has been known for a long time that graphite has a modulus of 1.06 TPa in the plane and, due to sp^2 bonds, CNTs show similar properties [21,22]. As a result, several studies have focused on the feasibility of using CNTs as reinforcement to obtain lighter composites with ultra-high strength [23,24].

While the specific resistance value of the steel is 1.54×10^2 kN m/kg the specific resistance of nanotubes is in the order of 48.5×10^3 kN m/kg [25]. This combination of extremely high strength coupled with low density makes nanotube one of the most attractive materials for applications requiring low weight and high strength, such as airplanes, space shuttles, wind turbines, boats, etc.

EXAMPLE 3.2

Maximum resistance of a C–C bond.

The energy stored in a C–C bond is 348 kJ/mol and its length is 0.154 nm. The covalent radius of carbon is $r_c = 0.077$ nm. Assuming that the bond may elongate 2.5 times before breaking, what is the maximum bond resistance? What is the maximum mass that a C–C bond can withstand?

Solution

A force F_c is required to break the C–C bond. The connection is broken when the distance between atoms is 2.5 times higher than the bond distance, i.e., $d = (2.5a_{C\text{–}C} - a_{C\text{–}C}) = 0.231$ nm. The force required to break the bond is then given by

$$F_c = \frac{E_{C-C}}{d}$$

where $E_{C\text{–}C}$ is the energy stored in a C–C single bond and this is given by

$$E_{C-C} = \frac{E_C}{N_A} = \left(\frac{348\,\text{kJ}}{\text{mol}}\right)\left(\frac{\text{mol}}{6.022 \times 10^{23}}\right)\left(\frac{1000\,\text{J}}{1\,\text{kJ}}\right) = 5.78 \times 10^{-19}\ \text{J}$$

Thus the force required is

$$F_C = \frac{E_{C-C}}{d} = \frac{5.78 \times 10^{-19}\ \text{J}}{0.231\,nm}\left(\frac{1\,nm}{1 \times 10^{-9}}\right) = 2.50 \times 10^{-9}\ \text{N}$$

The maximum strength of the C–C bond ($\sigma_{C\text{–}C}$) is obtained by dividing the force by the area of the cylindrical connection ($A_{C\text{–}C}$)

$$\sigma_{C-C} = \frac{F_c}{A_{C-C}} = \frac{F_c}{\pi r^2{}_C} = \frac{2.50 \times 10^{-9}\ \text{N}}{3.14 \times (0.077 \times 10^{-9}\ \text{m})^2} = 134\ \text{GPa}$$

The total mass that the C–C bond can withstand is given by

$$m_{max} = \frac{F_c}{g} = \frac{2.50 \times 10^{-9}\ \text{N}}{9.81\ \text{m/s}^2} = 2.54 \times 10^{-10}\ \text{kg}$$

Despite the obvious experimental difficulties, several experimental studies on the mechanical properties of CNTs have been published. In 1997, the first direct measurement of the elasticity of CNTs was conducted by Wong [26]. The experiment was performed on CNTs produced by arc discharge using an atomic force microscope (AFM). A modulus of 1.28 ± 0.59 TPa was obtained. In addition, the first measurement of mechanical strength was performed, yielding a value of 14 GPa. Two years later, other researchers obtained a modulus of 810 ± 410 GPa for MWCNTs [27] and 1 TPa for SWCNTs [28]. Other techniques such as micro-Raman spectroscopy have been used to access the mechanical properties of CNTs [29]. Perhaps the most interesting is the experiment conducted by Yu in 2000, when a mechanical test of individual MWCNTs and SWCNTs clusters was performed by operating a load cell inside a scanning electron microscope (SEM) (Figure 3.9) [31].

For SWCNTs, a maximum strength ranging from 13 to 52 GPa and modulus between 320 and 1470 GPa were reported. As to MWCNTs, the resistance of the outer wall measured was 11–63 GPa, the modulus was 270–950 GPa, and elongation at break of 12% was obtained [32]. Note that the elongation at rupture of carbon fibers is generally less than 3%. For comparison, the maximum resistance of high-performance steel is 1.2 GPa [30]. Table 3.1 compares the mechanical properties of nanotubes to other materials.

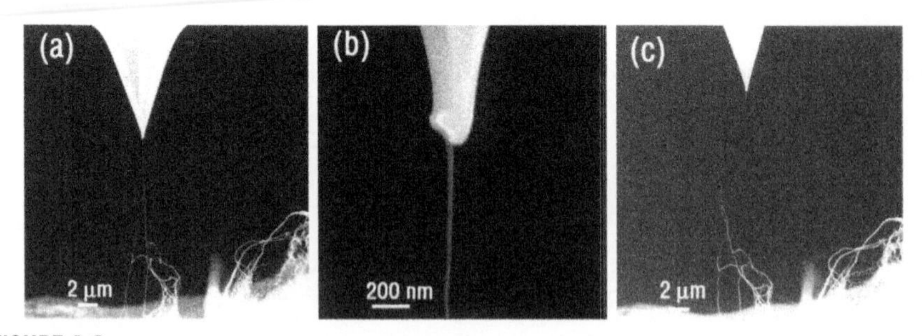

FIGURE 3.9

SEM images showing (a) carbon nanotubes (CNTs) bundles under tension applied by the probe of an atomic force microscope, (b) zoom showing the tip of the probe microscope and the CNTs, and (c) break of the bundle of CNTs.

Figure sourced from Strength and breaking mechanism of multiwalled carbon nanotubes under tensile load [30]

Table 3.1 Mechanical Properties of Materials.

Material	ρ (g/cm^3)	E (GPa)	σ_m (GPa)	ε (%)
SWCNTs	1.33	1054	150	12
MWCNTs	2.6	1200	150	12
Carbon fiber M60JB	1.93	588	3.82	0.7
Glass fiber type "S"	2.48	86	4.58	5.4
Kevlar 49	1.44	112	3.00	2.4
Aluminum 2219-T87	2.83	73	0.46	10
Steel 17 – 7 PH RH950	7.65	204	1.38	6
Epoxy	1.25	3.5	0.005	4

ρ: Density; E: Young's Modulus; ε: Elongation at Break; σ_m: Maximum Resistance.

EXAMPLE 3.3

Calculate the maximum resistance of a zigzag nanotube (12, 0). How many nanotubes would be required to support a 70 kg person?

Solution

As illustrated in the Example 3.3, there are 12 C–C bonds vertically aligned along the axis of the nanotube [17].

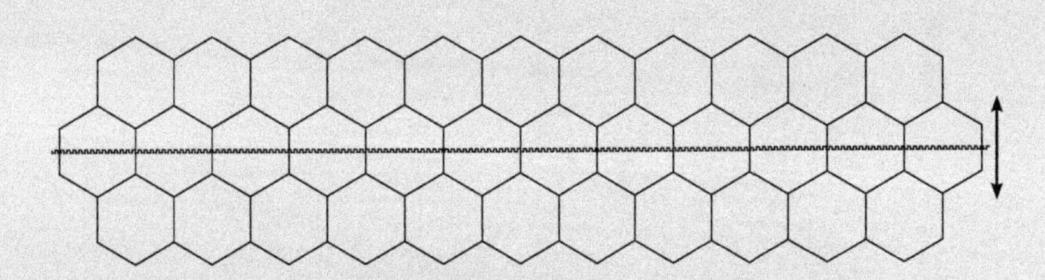

The diameter of the nanotube is given by Eqn (3.7)

$$d = \frac{\sqrt{n^2 + nm + m^2}a}{\pi} = \frac{\sqrt{0^2 + 12.0 + 12^2}0.246\,nm}{3.14} = 0.940\,nm$$

The force required for breaking 12 C–C bonds (F_{CNT}) is given by the force required to break a C–C bond (F_C) multiplied by the number of bonds

$$F_{CNT} = 12 \cdot F_C = 12 \cdot 2.50 \times 10^{-9}\ \text{N} = 3.0 \times 10^{-8}\ \text{N}$$

The cross-sectional area of the nanotube is

$$A_{\mathrm{CNT}} = \pi\left(\frac{d_{\mathrm{CNT}}}{2}\right)^2 = \pi\left(\frac{0.940 \times 10^{-9}\ \mathrm{m}}{2}\right)^2 = 2.21 \times 10^{-19}\ \mathrm{m}^2$$

The maximum resistance of the nanotube σ_{CNT} is given by

$$\sigma_{\mathrm{CNT}} = \frac{F_{\mathrm{CNT}}}{A_{\mathrm{CNT}}} = \frac{3.0 \times 10^{-8}\ \mathrm{N}}{2.21 \times 10^{-19}\ \mathrm{m}^2} = 136\ \mathrm{GPa}$$

The number of nanotubes (N_{CNT}) needed to raise a 70-kg person is then

$$N_{\mathrm{CNT}} = \frac{m \cdot g}{F_{\mathrm{CNT}}} = \frac{70\ \mathrm{kg} \cdot 9.81\ \mathrm{m/s}^2}{3.0 \times 10^{-8}\ \mathrm{N}} = 2.29 \times 10^{10}$$

If all of these nanotubes were united to form a cable, its diameter would be comparable to a hair!

The mechanical properties of CNTs have also been explored from a theoretical point of view [32–36] and agreement among experiment and simulations has been observed [37–39]. It is worth noting that SWCNTs are usually found in the form of bundles (Figure 3.10), and the slip among tubes affects the mechanical

FIGURE 3.10 An idealized bundle of single-walled carbon nanotubes (10,10).

In this bundle, individual nanotubes are held together by strong π–π bonds stacking.

Figure sourced from Functionalization of single-walled carbon nanotubes [40]

properties [3]. In the case of MWCNTs, the weak interaction between layers results in a poor load transfer from the outer layers to the internal layers. One reason that hinders comparisons of experimental results for the properties of CNTs is the difficulty of producing CNTs with similar characteristics (size, length, etc.). The probability of finding two identical nanotubes on a macroscopic sample is very small. Furthermore, the establishment of standards for the characterization of nanotubes is very slow, and only recently was a standard proposed and this contemplates only SWCNTs [41].

3.2.4 Transport properties of CNTs

3.2.4.1 Electrical conductivity

The electrical properties of CNTs are largely a consequence of its one-dimensional structure (1D) and the peculiar electronic structure of graphite [40]. Graphite is considered a semi-metal, but depending on the chiral vector related to integers (n, m), CNTs can be metallic (conductive) or semiconductors [42,43]. A nanotube is metallic either when $n = m$ or when $m - n$ or $n - m$ are multiples of 3, or m and n take values that satisfy the relation $|m - n| = 3i$ (where i is an integer) as (3, 0), (7, 1), (8, 5), etc. Therefore, all armchair nanotubes are metallic because $n - m = 0$. For all other cases, $|m - n| \neq 3i$, nanotubes are semiconductors with a band gap energy of around 0.5 eV. Figure 3.11 shows a nanotube unrolled and indicates which tubes are conductors and which are semiconductors in terms of n and m.

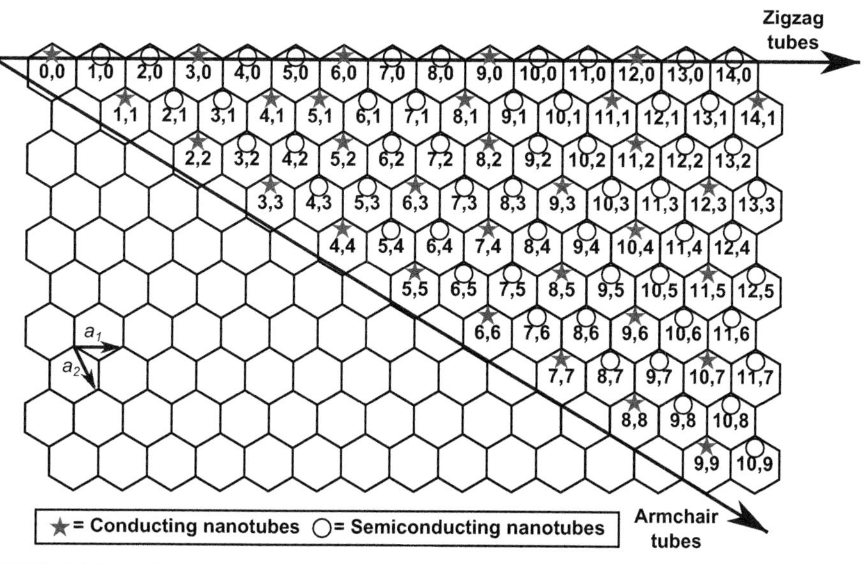

FIGURE 3.11 Semiconductor carbon nanotubes are represented by empty circles and conductors by stars.

Figure sourced from Introduction to nanoscience and nanotechnology [17]

Table 3.2 Electrical Conductivity (σ) of Materials

Material	σ (S/m)
CNT	10^6–10^7
Copper	6×10^7
Silver	6.3×10^7
Iron	1.00×10^7
Glass fiber	10^{-14}
Carbon fiber—Pitch	2–8.5×10^6
Carbon fiber—PAN	6.5–14×10^6
Epoxy	10^{-8}

The energy gap of semiconducting nanotubes varies with the diameter of the nanotube according to the equation

$$E_{gap} = \frac{2\gamma a_{C-C}}{d} \tag{3.9}$$

where γ, a_{C-C}, and d denote the superposition energy "tight-binding" (2.45 eV), the distance to the nearest C—C bond (1.42 Å), and the diameter of the nanotube, respectively.

On average, one-third of SWCNTs are metallic and two-thirds are semiconductors [44]. Chemical processes can be used to separate the metallic from semiconducting nanotubes, but the synthesis of only one of these types of nanotubes is still a challenge. In general, MWCNTs behave as one-dimensional conductors with high electrical conductivity [45]. Their metallic properties come from the multi-layer structure consisting of tubes with different electrical properties, where electronic coupling between layers occurs.

In 1D conductors, electrons move only in one direction and, in these circumstances the electron scattering is reduced and the electrical resistance of CNTs is low. Due to confinement, the electrical conduction is considered quantized, and as the free path of the electrons is of the order of nanometers, the conduction is ballistic. In addition, as the resistivity of nanotube is constant, CNTs are ideal for applications in high currents. In fact, CNTs can conduct the highest electrical current density ever measured for any material: 10^9 A/cm^2 [46,47]. Experimental results show that the electrical conductivity of metallic SWCNTs can reach 10^6 S/m and for semiconductors tubes, 10 S/m [44]. For bundles of SWCNTs, the conductivity ranges from 10^4 to 10^6 S/m [46,48,49]. MWCNTs have a conductivity of 2×10^7 S/m [50]. Table 3.2 compares the electrical conductivity of CNTs with different materials.

3.2.4.2 Thermal conductivity

Before the discovery of CNTs, diamond was considered the best heat conductor known. The thermal conductivity of CNTs can achieve double the conductivity of

Table 3.3 Thermal Conductivity (k) of Materials

Material	k (W/m K)
SWCNTs	>3000
Graphite	3000 in the plan/6 in the c-axis
Aluminum 2219-T87	120
Copper	400
Silver	420
Iron	80
Glass fiber	0.046–1.13
Carbon fiber K1352U	140
Carbon fiber K13D2U	800
Epoxy	0.12

diamond [51,52]. Thermal conductivity is a key factor in heat dissipation. In solids, the heat can be transported by phonons. Similarly, in CNTs the specific heat and thermal conductivity are determined by phonons.

Theoretical studies suggest that the thermal conductivity of individual SWCNTs can reach 6000 W/m K in the axial direction but is very low in the radial direction [53–55]. Values of up to 3000 W/m K have been experimentally obtained for MWCNTs [56]. For comparison, the thermal conductivity of various materials is shown in Table 3.3. In addition, CNTs are stable at temperatures of up to 2800 °C under vacuum.

3.3 Treatment of CNTs

One of the problems associated with the synthesis of nanotubes on a large scale is the purification. After synthesis, CNTs contains various impurities such as the catalyst particles used in the synthesis, graphite, and amorphous carbon. Such impurities are unwanted in the application of CNTs. Also, obtaining nanotubes as uniform as possible in size is desirable, but, in reality, the nanotubes obtained always have a distribution of diameters and length.

Some of the treatments used prior to application of CNTs are purification, oxidation, and functionalization.

3.3.1 Purification and oxidation of CNTs

Oxidative treatments are effective in removing carbonaceous impurities. One of the characteristics of oxidation is that not only are impurities oxidized but also the nanotube itself. However the damage suffered by CNTs is generally less than the damage caused to impurities since they have more defects and are more reactive. In addition, impurities tend to adhere to the metal catalyst particles, which act as a catalyst for oxidation [57].

One of the first methods used for oxidation was to heat nanotube samples to high temperatures (above 700 °C) in the presence of air. The result was the acquisition of open nanotubes, i.e., without half-fullerenes attached to each end. This indicates that the ends of the nanotubes are more reactive than its walls [58]. Nanotubes can be filled with metal oxides as nanoparticles or fullerenes. Gases may also be used (instead of air) for the purification of CNTs. Examples include CO_2, N_2O, NO, NO_2, O_3, and ClO_2 [59]. Results show that amorphous carbon is more susceptible to oxidation by gases than the CNTs.

Purification and oxidation may also be conducted in solution using extremely strong acids. The most common methods include the use of a single acid such as nitric acid (HNO_3), sulfuric acid (H_2SO_4), hydrochloric acid (HCl), or mixtures of acids such as "piranha" (a mixture of HNO_3 and H_2SO_4), sulfuric acid and potassium dichromate ($K_2Cr_2O_7 + H_2SO_4$), and sulfuric acid and potassium permanganate ($H_2SO_4 + KMnO$). Figure 3.12 shows CNTs before and after the purification process. Note how the amount of catalyst particles decreases after the reaction.

The oxidation process generates several functional end groups on the nanotubes and defects such as carboxylic groups (−COOH), hydroxyl (−OH), and carbonyl (−CO). In Figure 3.13 the main defects present in the nanotube structure and the interaction between CNTs and carboxylic groups are presented.

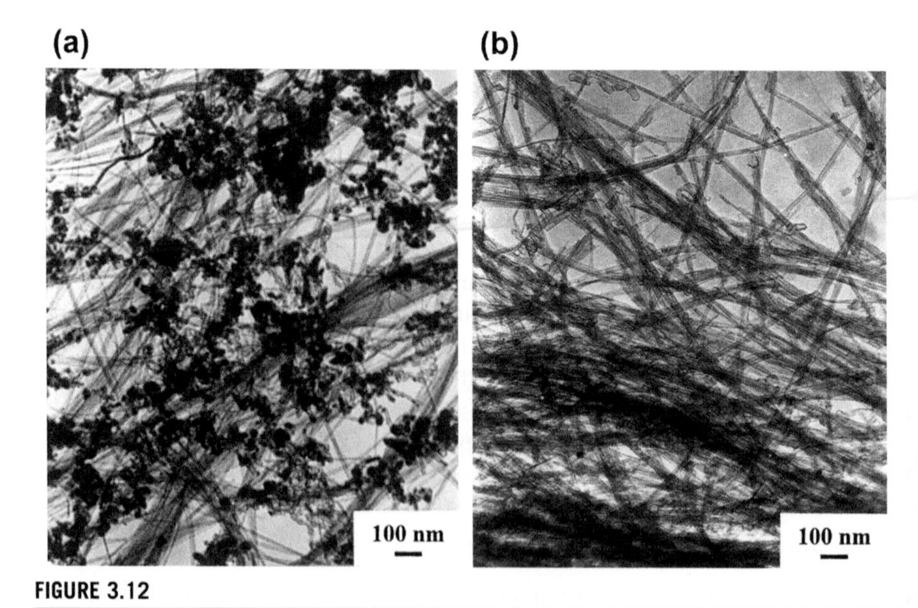

FIGURE 3.12

(a) Single-walled carbon nanotubes after being produced, containing impurities such as iron particles and amorphous carbon, and (b) carbon nanotubes after purification in hydrochloric acid (HCl).

Figure sourced from Multi-step purification of carbon nanotubes [60]

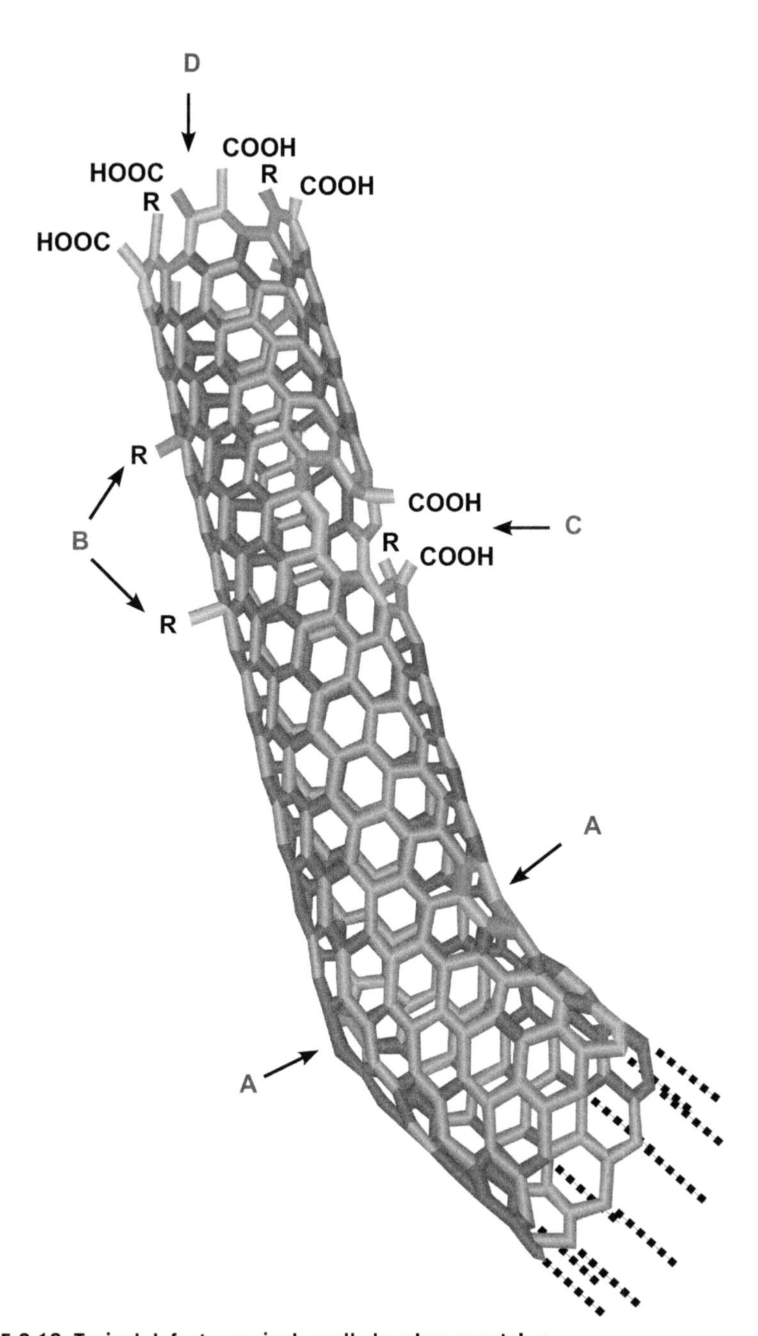

IGURE 3.13 Typical defects on single-walled carbon nanotubes.

a) Rings with five or seven carbon atoms instead of six, that makes the CNTs curved; b) defects sp^3 hybridized (R = H and OH); (c) damage to the structure of carbon nanotubes lue to oxidative conditions, causing the creation of a hole and links to groups –COOH; d) end of the open nanotube, linked to –COOH groups. Other chemical groups as NO_{2}-,)H, H, and = O are possible.

Figure sourced from Superaromaticity [42]

The degree of oxidation depends on the nature of the oxidizing agent and the reaction conditions (temperature, duration, etc.). The oxidation of SWCNTs initially occurs at the end of the tube and moves to the outermost layer of the bundle. In the case of MWCNTs, oxidation begins at the ends and gradually moves from the outermost layer of the nanotube to its interior, resulting in successive removal of graphene cylinders and thinner nanotubes. Oxidation of nanotubes can be characterized by several techniques such as Fourier transform infrared spectroscopy, UV/visible spectroscopies, Raman spectroscopy, and X-ray diffractometry.

EXAMPLE 3.4

Effect of thc oxidation temperature on the length of MWCNTs treated in a mixture of sulfuric and nitric acid (molar ratio 1:3) [61].

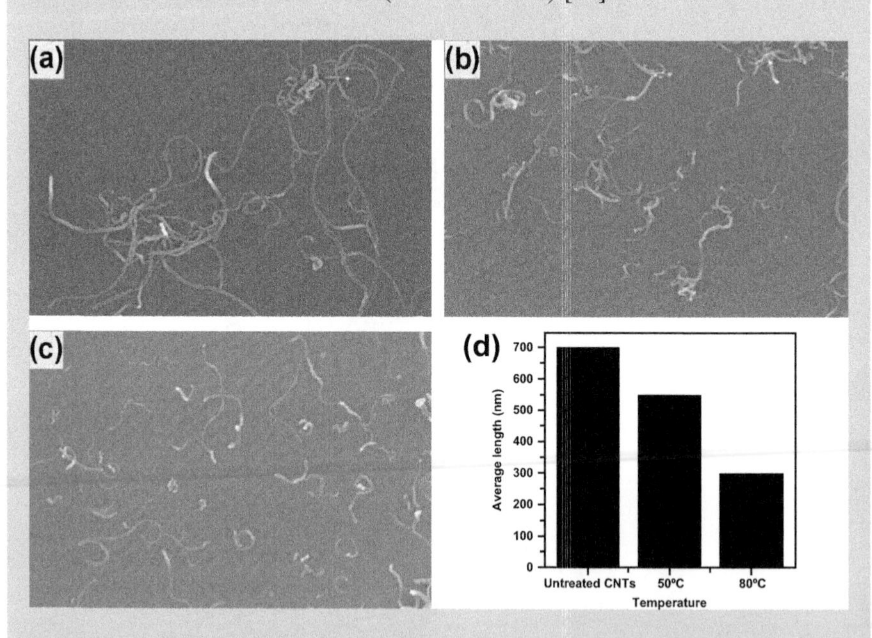

FIGURE: (a) Non-treated MWCNTs, (b) treated at 50 °C, (c) treated at 80 °C, and (d) the effect of temperature on the average length of the CNTs. The average length of the nanotubes decreased exponentially with the rising of temperature during oxidation.

Figure sourced from Covalent surface chemistry of single-walled carbon nanotubes [62]

The process of purification and oxidation of nanotubes can improve the interactions between CNTs and solvents and polymer matrices. However, to improve control of chemical bonding between CNTs and other systems, it is necessary to further the functionalization of the nanotubes. Functionalized nanotubes should present properties that differ from those of unmodified nanotubes.

EXAMPLE 3.5

The effect of oxidation time on the length of MWCNTs treated in a mixture of sulfuric and nitric acid (molar ratio 1:3) [62].

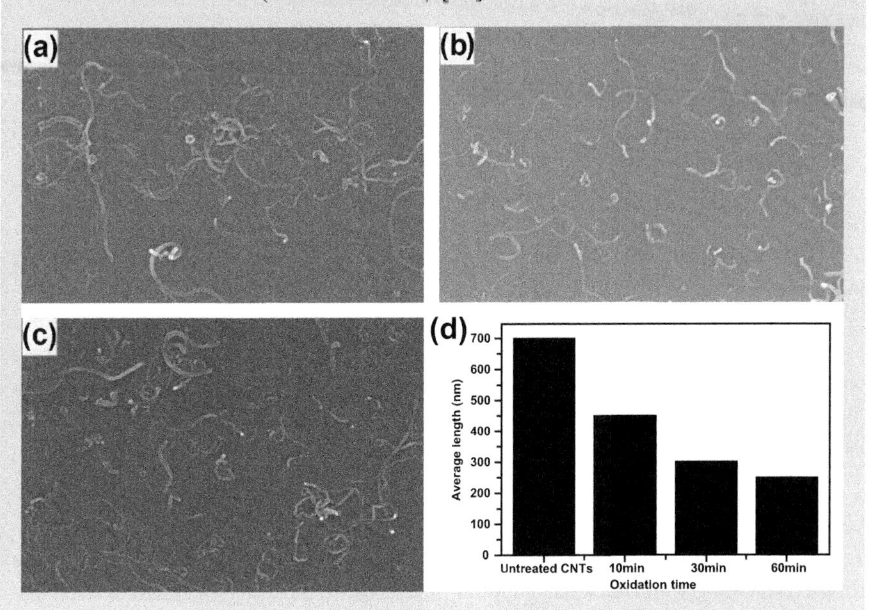

FIGURE: (a) MWCNTs treated for 10 min, (b) 30 min, and (c) 60 min, and (d) effect of treatment duration on the average length of the CNTs. The average length of the nanotubes dramatically decreased initially and after 30 min the reduction rate declines.

Figure sourced from Covalent surface chemistry of single-walled carbon nanotubes [62]

3.3.2 Chemical functionalization of CNTs

In general, the functionalization of CNTs is divided into non-covalent and covalent. The non-covalent functionalization is based on the intermolecular interaction between nanotubes and molecules while the covalent modification is based on the breaking of bonds and creation of new covalent bonds. More details are presented next.

3.3.2.1 Non-covalent functionalization

The non-covalent functionalization is based on the ability of the extended π electron system of the sidewall of the nanotube in interacting with molecules via $\pi-\pi$ interactions [62]. Other approaches take advantage only from absorptions via van der Waals interactions between the adsorbate and the nanotubes. The advantage of this technique comes from the preservation of the properties of CNTs, while its disadvantage is the weak forces between coupled molecules and nanotubes, which can decrease the load transfer in the composite [63]. In Figure 3.14 we see simulation results showing the process of a nanotube (10,10) being wrapped by a polyethylene polymer chain [65].

The use of surfactants is also a non-covalent method widely used for modification of CNTs. Studies show that the surfactant molecules will interact with the hydrophobic surface of the nanotubes, thus giving them a hydrophilic character that makes them soluble in various solvents, including water [64,66–68]. However, the mechanism of adsorption of the surfactant molecules on the walls of the nanotubes is not completely understood, and continues to be studied. There are currently three models commonly used to describe adsorption (Figure 3.15): encapsulation of the CNTs within the micelle adsorption, hemimicellar, and random adsorption of molecules on the surface of CNTs.

Block copolymers are an alternative to the dispersion of CNTs and stabilization of suspensions containing CNTs. In Figure 3.16 we see how a block copolymer consisting of a lyophilic block and another lyophobic interacts with CNTs. The lyophobic part adsorbs on the surface of the nanotubes while the lyophilic is dissolved by the solution, preventing the agglomeration of CNTs.

3.3.2.2 Covalent functionalization

The defect functionalization of CNTs is based on the conversion of carboxylic groups (–COOH) and other oxygenated places formed during the oxidative purification process [63]. The carboxylic groups located primarily on the ends of the nanotubes can be coupled to different chemical groups.

The sidewall functionalization is based on covalent attachment of functional groups on the nanotube sidewalls. The sidewall functionalization is associated with the change from sp^2 to sp^3 hybridization and the simultaneous loss of conjugation. This type of functionalization usually requires very reactive reagent, and a high degree of chemical functionalization is achievable [63].

Figure 3.17 illustrates a nanotube (10,10) with 5% of their carbon atoms functionalized with different chemical groups (carboxylic groups COOH, amide group $CONH_2$, alkyl C_6H_{11} group, or phenyl group C_6H_5) [64]. The change in the structure of the carbon atom to which the groups are connected is illustrated on the right side of the figure and hybridization changes from sp^2 to sp^3.

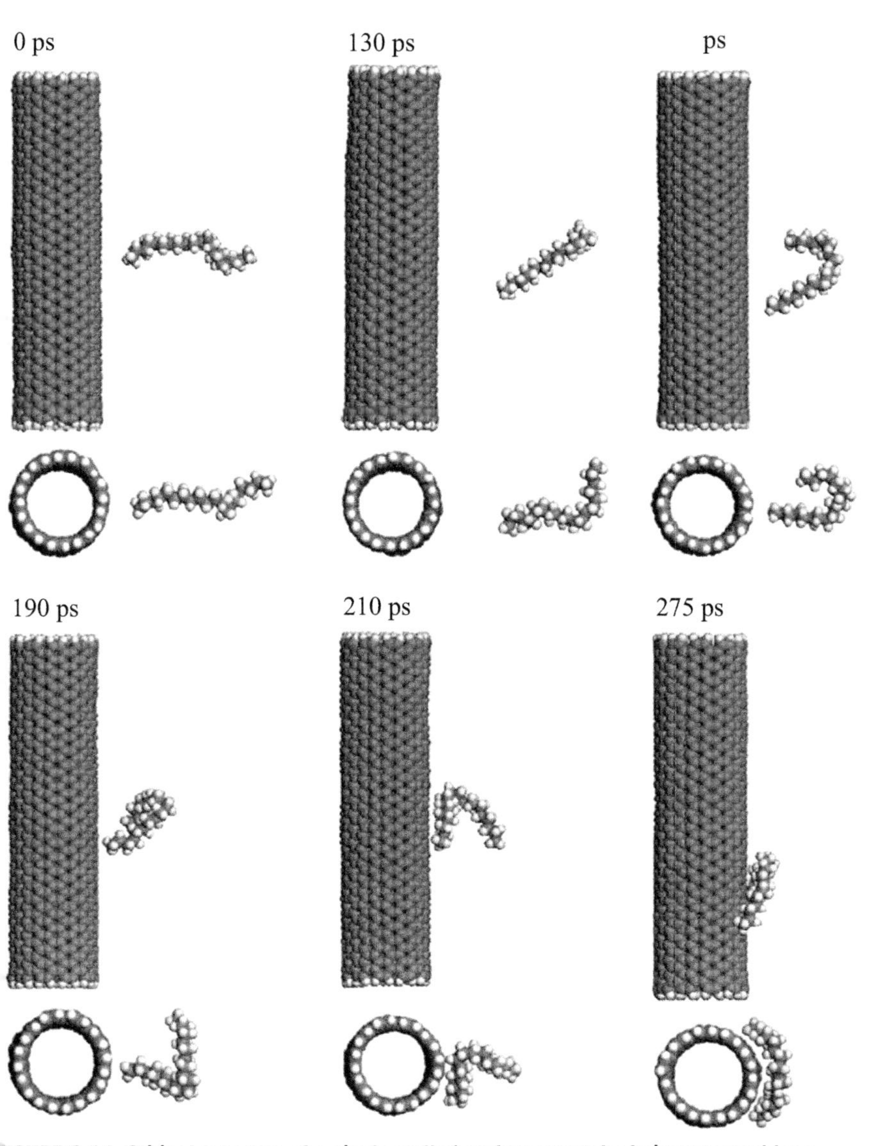

IGURE 3.14 Critical moments of a single-walled carbon nanotube being wrapped by a olyethylene polymer chain.

)uring simulation, one end of the macromolecule is placed close to the wall of the carbon ıanotubes (CNTs) and the other is moved away. The result shows that the chain of PE twists" and moves in the direction of CNTs until equilibrium is reached.

Figure sourced from Purification of single-wall carbon nanotubes by microfiltration [64]

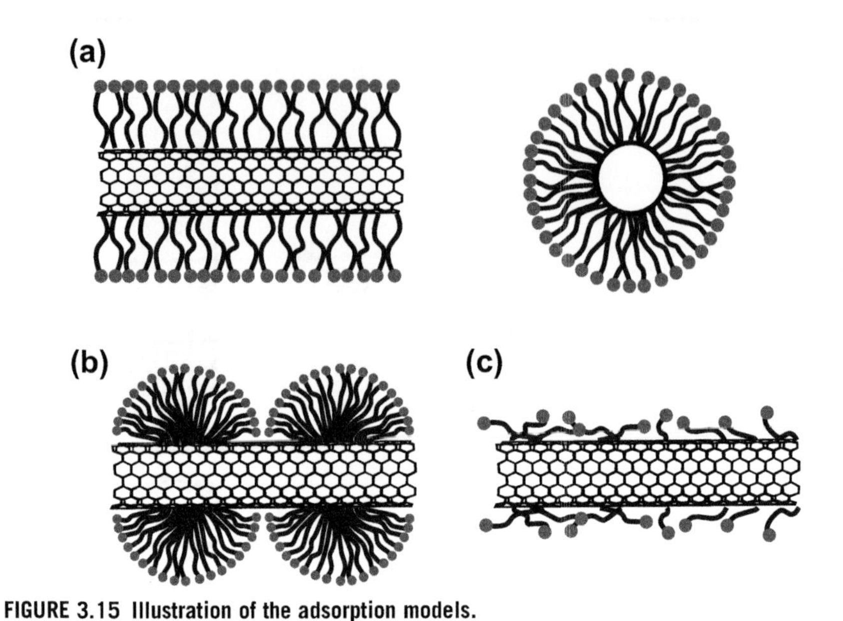

FIGURE 3.15 Illustration of the adsorption models.

(a) Cylindrical micelle, side view and front; (b) hemimicellar; and (c) random adsorption.

Figure sourced from Carbon nanotube self-assembly with lipids and detergent [69]

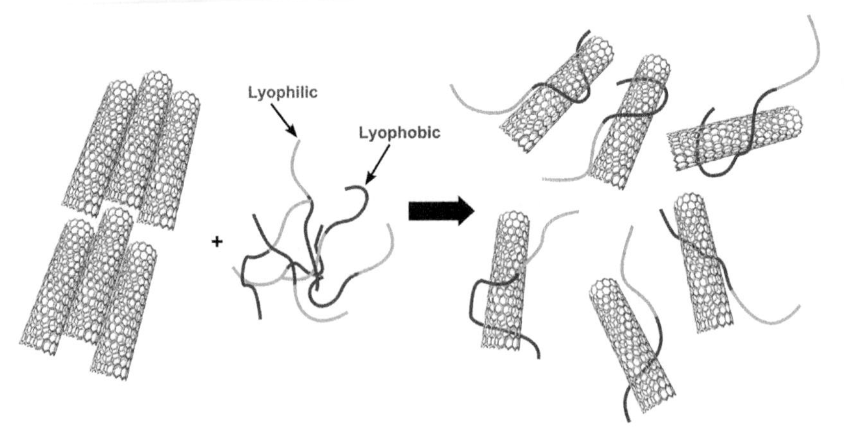

FIGURE 3.16 Schematic illustrating the interaction of nanotubes with block copolymers.

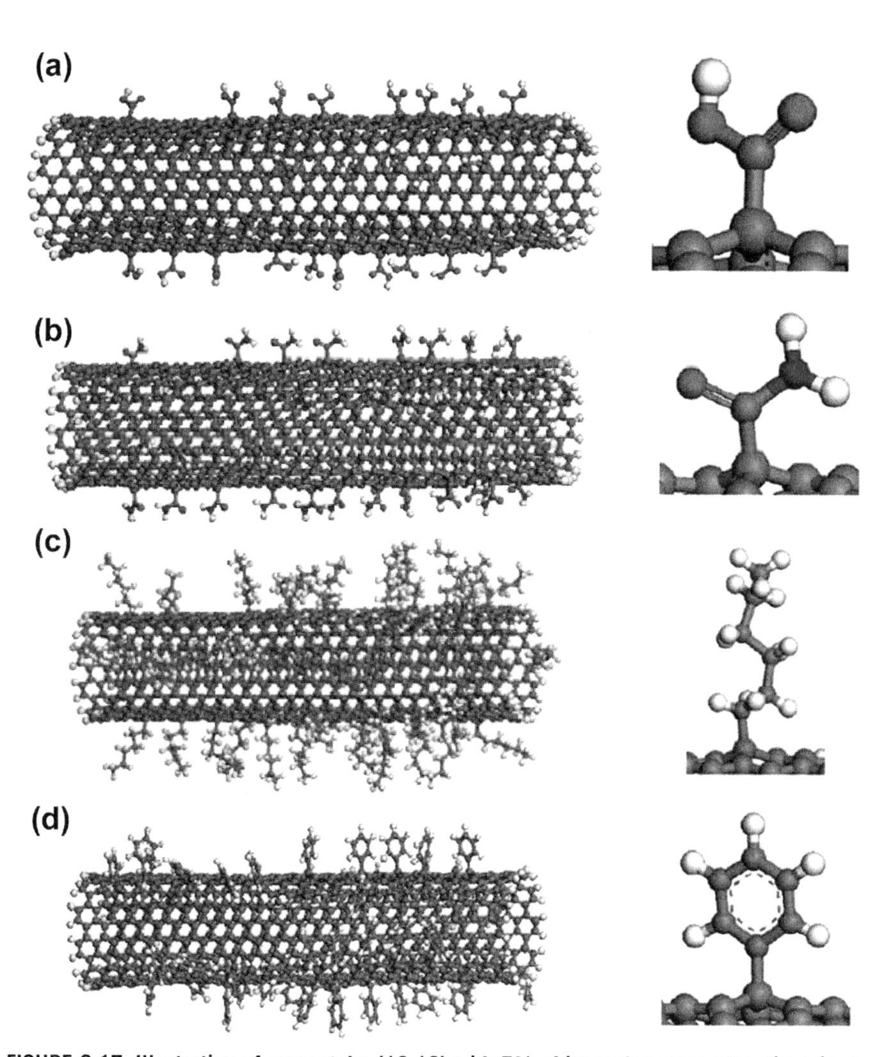

FIGURE 3.17 Illustration of a nanotube (10,10) with 5% of its carbon atoms functionalized with different chemical groups. (a) with carboxylic groups (-COOH); (b) With amide groups (-$CONH_2$); (c) with alkyl groups (-C_6H_{11}); (d) with phenyl groups (-C_6H_5).

Figure sourced from Purification of single-wall carbon nanotubes by microfiltration [64]

To learn more...

We encourage readers to consult the following books:

1. Coelho LAF, Pezzin SH, Loos MR, Pezzin, Amico SC. General issues in carbon nanocomposites technology. "Encyclopedia of polymer composites: properties,

performance and applications". Nova Science Publishers, Inc.; 2009, ISBN-13: 978-1607417170, pp. 389–416.

2. Annick Loiseau, Pascale Launois-Bernede, Pierre Petit, Stephan Roche, Jean-Paul Salvetat, editors. Understanding carbon nanotubes: from basics to applications. 1st ed. Springer; 2010, ISBN-13: 978-3642065989.
3. Meyyappan M. Carbon nanotubes: science and applications. 1st ed. CRC Press; 2004, ISBN-13: 978-0849321115.
4. Hornyak GL, Tibbals HF, Dutta J, Moore JJ. Introduction to nanoscience and nanotechnology. CRC Press; 2008, ISBN-13: 978-1420047790.

References

[1] Kroto HW, Heath JR, O'Brien SC, Curl RF, Smalley RE. C60: Buckminsterfullerene. Nature 1985;318:162–3.

[2] Becker L, Bunch TE. Fullerenes, fulleranes and polycyclic aromatic hydrocarbons in the Allende meteorite. Meteoritics July 1997;32:479–87. http://dx.doi.org/10.1111/j.1945-5100.1997.tb01292.x.

[3] Lahiff E, Leahy R, Coleman JN, Blau WJ. Physical properties of novel free-standing polymer–nanotube thin films. Carbon 2006;44(8):1525–9.

[4] Dresselhaus MS, Lin YM, Rabin O, Jorio A, Souza Filho AG, Pimenta MA, et al. Nanowires and nanotubes. Mater Sci Eng C 2003;23:129–40.

[5] Bethune DS, Kiang CH, Devries MS, Gorman G, Savoy R, Vazquez J. Cobalt-catalysed growth of carbon nanotubes with single-atomic-layer walls. Nature 1993;363:605–7. http://dx.doi.org/10.1038/363605a0.

[6] Iijima S, Ichihashi T. Single-shell carbon nanotubes of 1-nm diameter. Nature 1993; 363:603–5. http://dx.doi.org/10.1038/363603a0.

[7] Ci L, Zhou Z, Tang D, Yan X, Liang Y, Liu D, et al. Double wall carbon nanotubes with an inner diameter of 0.4 nm. Chem Vap Deposition 2003;9:119–21. http://dx.doi.org/10.1002/cvde.200304142.

[8] Iijima S. Helical microtubules of graphitic carbon. Nature 1991;354:56–8. http://dx.doi.org/10.1038/354056a0.

[9] Cumings J, Zettl A. Low-friction nanoscale linear bearing realized from multiwall carbon nanotubes. Science 2000;289:602–4.

[10] Oberlin A, Endo M, Koyama TJ. J Cryst Growth 1976;32:335–49.

[11] Hughes TV, Chambers CR. Manufacture of carbon filaments. US Patent, (405480), 1889.

[12] Schützenberger P, Schuützenberger LCR. Sur quelques faits relatifs à l'histoire du carbone. Acad Sci Paris 1890;111:774–8.

[13] Pélabon C, Pélabon HCR. Sur une variétë de carbone filamenteux. Acad Sci Paris 1903; 137:706–8.

[14] Radushkevich LV, Lukyanovich VM. About the structure of carbon formed by thermal decomposition of carbon monoxide on iron substrate. Zurn Fis Chim 1952;26:88–95.

[15] Monthioux M, Kuznetsov VL. Who should be given the credit for the discovery of carbon nanotubes? Carbon 2006;44:1621–3.

[16] Loos MR, Schulte K. Macromol Theory Simul 2011;20:350–62.

[17] Hornyak GL, Tibbals HF, Dutta J, Moore JJ. Introduction to nanoscience and nanotechnology. Taylor and Francis Group, LLC; 2009, ISBN 9781420047790.

18] http://cobweb.ecn.purdue.edu/mdasilva/structure.html; 2008 [accessed 11.2008].

19] Li Y, Wang K, Wei JQ, Gu Z, Wang ZC, Luo JB, et al. Tensile properties of long aligned double-walled carbon nanotubes strands. Carbon 2005;43:31–5.

20] Qin Z, Qin Q-H, Feng X-Q. Mechanical property of carbon nanotubes with intramolecular junctions: molecular dynamics simulations. Phys Lett A 2008;372:6661–6. http://dx.doi.org/10.1016/j.physleta.2008.09.010.

21] Kelly BT. Physics of graphite. Appl Sci; 1981. http://dx.doi.org/10.1107/S0108767383000422.

22] Robertson DH, Brenner DW, Mintmire JW. Energetics of nanoscale graphitic tubules. Phys Rev B 1992;45:12592–5. http://dx.doi.org/10.1103/PhysRevB.45.12592.

23] Lau KT, Gu C, Hui D. A critical review on nanotube, nanotube/nanoclay related polymer composite materials. Compos. Part B 2006;37:425–36.

24] Gomes D, Loos MR, Wichmann MHG, de la Vega A, Schulte K. Sulfonated polyoxadiazole composites containing carbon nanotubes prepared via in-situ polymerization. Compos. Sci Technol 2009;69:220–7.

25] Collins PG, Arnold MS, Avouris P. Engineering carbon nanotubes and nanotube circuits using electrical breakdown. Science 2001;292:706–9.

26] Wong EW, Sheehan PE, Lieber CM. Nanobeam mechanics: elasticity, strength and toughness of nanorods and nanotubes. Science 1997;277:1971–5.

27] Salvetat J-P, Kulik AJ, Bonard J-M, Briggs D, Stockli T, Metenier K, et al. Elastic modulus of ordered and disordered multiwalled carbon nanotubes. Adv Mater 1999;11:161–5. http://dx.doi.org/10.1002/(SICI)1521-4095(199902)11:2<161::AID-ADMA161>3.0.CO;2-J.

28] Salvetat J-P, Andrew G, Briggs D, Bonard J-M, Basca RR, Kulik A. Elastic and shear moduli of single-walled carbon nanotubes ropes. Phys Rev Lett 1999;82:944–7.

29] Lourie O, Wagner HDJ. Evaluation of Young's modulus of carbon nanotubes by micro-Raman spectroscopy. Mater Res 1998;13:2418–22.

30] Yu MF, Lourie O, Dyer MJ, Moloni K, Kelly TF, Ruoff RS. Strength and breaking mechanism of multiwalled carbon nanotubes under tensile load. Science 2000;287:637–40.

31] Yu MF, Files BS, Arepalli S, Ruoff RS. Tensile loading of ropes of single wall carbon nanotubes and their mechanical properties. Phys Rev Lett 2000;84:5552–5.

32] Lu JP. Elastic properties of carbon nanotubes and nanoropes. Phys Rev Lett 1997;79:1297–300.

33] Yao N, Lordi VJ. Young's modulus of single-walled carbon nanotubes. Appl Phys 1998;84:1939–43.

34] Yakobson BI, Brabec CJ, Bernholc. Nanomechanics of carbon tubes: instabilities beyond linear response. Phys Rev Lett 1996;76:2511–4.

35] Yakobson BI, Smalley RE. Fullerene nanotubes: C-1000000 and beyond. Am Sci 1997;85:324–37.

36] Bernholc J, Brabec CJ, Nardelli MB, Maiti A, Roland C, Yakobson BI. Theory of growth and mechanical properties of nanotubes. Appl Phys A 1998;67:39–46.

37] Despres JF, Daguerre E, Lafdi K. Flexibility of graphene layers in carbon nanotubes. Carbon 1995;33:87–92.

38] Ruoff RS, Lorents DC. Mechanical and thermal properties of carbon nanotubes. Carbon 1995;33:925–30.

39] Iijima S, Brabec C, Maiti A, Bernholc JJ. Structural flexibility of carbon nanotubes. Chem Phys 1996;104:2089–92.

[40] Hirsch A. Functionalization of single-walled carbon nanotubes. Angew Chem Int Ed 2001;41:1853–9. http://dx.doi.org/10.1002/1521-3773(20020603)41:11<1853::AID-ANIE1853>3.0.CO;2-N.
[41] Arepalli S, Nikolaev P, Gorelik O, Hadjiev VG, Holmes W, Files B, et al. Protocol for the characterization of single-wall carbon nanotubes material quality. Carbon 2004;42: 1783–91.
[42] Osawa E. Superaromaticity. Kagaku 1970;25:854–63 [in Japanese].
[43] Tans SJ, Devoret MH, Dai H, Thess A, Smalley RE, Geerligs LJ, et al. Individual single-wall carbon nanotubes as quantum wires. Nature 1997;386:474–7. http://dx.doi.org/10.1038/386474a0.
[44] Saito Y, Nishikubo K, Kawabata K, Matsumoto TJ. Carbon nanocapsules and single-layered nanotubes produced with platinum-group metals (Ru, Rh, Pd, Os, Ir, Pt) by arch discharge. Appl Phys 1996;80:3062–7.
[45] Kim YJ, Shin TS, Choi HD, Kwon JH, Chung YC, Yoon HG. Electrical conductivity of chemically modified multiwalled carbon nanotube/epoxy composites. Carbon 2005;43: 23–30.
[46] Tans SJ, Verschueren ARM, Dekker C. Room-temperature transistor based on a single carbon nanotube. Nature 1998;393:49–52. http://dx.doi.org/10.1038/29954.
[47] Wei BQ. Reliability and current carrying capacity of carbon nanotubes. Appl Phys Lett 2001;79:1172–4.
[48] Fischer JE, Dai H, Thess A, Lee R, Hanjani NM, Dehaas DL, et al. Metallic resistivity in crystalline ropes of single-wall carbon nanotubes. Phys Rev B 1997; 55:4921–4.
[49] Bozhko AD, Sklovsky DE, Nalimova VA, Rinzler AG, Smalley RE, Fischer JE. Resistance vs. pressure of single-wall carbon nanotubes. Appl Phys A 1998;67:75–7.
[50] Ebbesen TW, Lezec HJ, Hiura H, Bennett JW, Ghaemie HF, Thio T. Electrical conductivity of individual carbon nanotubes. Nature 1996;382:54–6. http://dx.doi.org/10.1038/382054a0.
[51] Yang F, Qiu XM, Li Y, Yin Y, Fan Q. Specific heat of super carbon nanotube and its chirality independence. Phys Lett A 2008;372:6960–4.
[52] Hone J. Appl Phys 2001;80:273–86.
[53] Hone J, Whitney M, Zettl A. Thermal conductivity of single-walled carbon nanotubes. Synth Met 1999;103:2498–9.
[54] Hone J, Whitney M, Piskoti C, Zettl A. Thermal conductivity of single-walled carbon nanotubes. Phys Rev B 1999;59:2514–6.
[55] Yi W, Lu L, Zhang DL, Pan ZW, Xie SS. Linear specific heat of carbon nanotubes. Phys Rev B 1999;59:R9015–8.
[56] Kim P, Shi L, Majumdar A, McEuen PL. Thermal transport measurements of individual multiwalled carbon nanotubes. Phys Rev Lett 2001;87(215502):1–4.
[57] Borowiak-Palen E, Pichlera T, Liua X, Knupfera M, Graffa A, Jostd O, et al. Chem Phys Lett 2002;363:567–72.
[58] Ajayan PM, Zhou OZ. Applications of carbon nanotubes. Carbon nanotubes book series: topics in applied physics, vol. 80; 2001.
[59] Damnjanovic M, Milosevic I, Vukovic T, Sredanovic R. Full symmetry, optical activity, and potentials of single-wall and multiwall nanotubes. Phys Rev B 1999;60: 2728–39.
[60] Hou PX, Bai S, Yang QH, Liu C, Cheng HM. Multi-step purification of carbon nanotubes. Carbon 2002;40:81–5.

[61] Hong Chang-Eui, Lee Joong-Hee, Kalappa Prashantha, Advani Suresh G. Effects of oxidative conditions on properties of multi-walled carbon nanotubes in polymer nanocomposites. Compos Sci Technol 2007;67:1027–34.
[62] Banerjee S, Hemraj-Benny T, Wong SS. Covalent surface chemistry of single-walled carbon nanotubes. Adv Mater 2005;17:17–29. http://dx.doi.org/10.1002/adma.200401340.
[63] Balasubramanian K, Burghard M. Chemically functionalized carbon nanotubes. Small 2004;1:180–92. http://dx.doi.org/10.1002/smll.200400118.
[64] Bandow S, Rao AM, Williams KA, Thess A, Smalley RE, Eklund PCJ. Purification of single-wall carbon nanotubes by microfiltration. Phys Chem B 1997;101:8839. http://dx.doi.org/10.1021/jp972026r.
[65] Zheng Qingbin, Xia Dan, Xue Qingzhong, Yan Keyou, Gao Xili, Li Qun. Computational analysis of effect of modification on the interfacial characteristics of a carbon nanotube/polyethylene composite system. Appl Surf Sci 2009;255:3534–43.
[66] Duesberg GS, Burghard M, Muster J, Philipp G, Roth S. Separation of carbon nanotubes by size exclusion chromatography. Chem Commun; 1998:435. http://dx.doi.org/10.1039/A707465D.
[67] Kristic V, Duesberg GS, Muster J, Burghard M, Roth S. Chem Mater 1998;10:2338–40.
[68] Panhuis M, Salvador-Morales C, Franklin E, Chambers G, Fonseca A, Nagy JB, et al. Characterization of an interaction between functionalized carbon nanotubes and enzyme. Nanosci Nanotechnol 2003;3:209–13.
[69] Jayne Wallace E, Sansom Mark SP. Carbon nanotube self-assembly with lipids and detergent: a molecular dynamics study. Nanotechnology 2009;20:045101–7. http://dx.doi.org/10.1088/0957-4484/20/4/045101.

CHAPTER

Production of CNTs and Risks to Health

4

CHAPTER OUTLINE

4.1 Production methods of carbon nanotubes

Carbon nanotubes (CNTs) can be synthesized using various methods. The most common methods are arc discharge [1–5], laser ablation [6], chemical vapor deposition (CVD) [7–11], and conversion of carbon monoxide at high pressure (HiPco) [12].

Various adaptations and improvements of these methods as well as new trends in the production of CNTs are constantly emerging [13–22]. The production of CNTs

Carbon Nanotube Reinforced Composites. http://dx.doi.org/10.1016/B978-1-4557-3195-4.00004-7

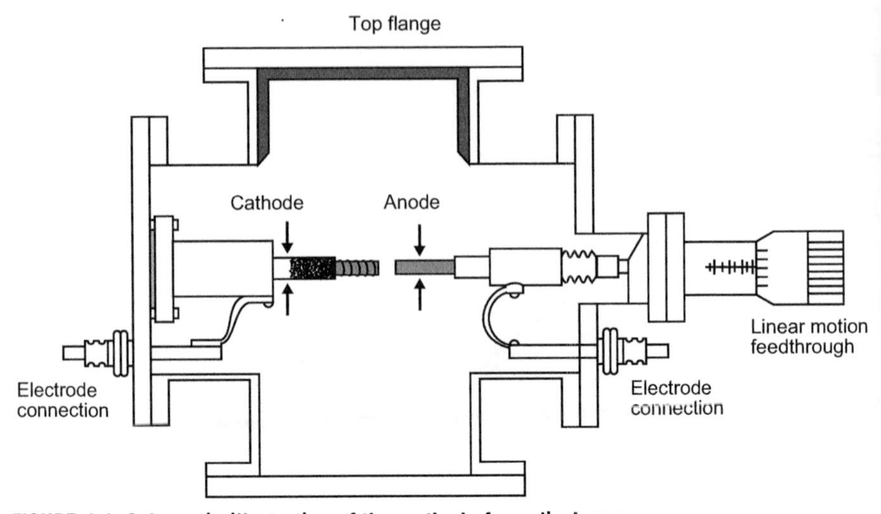

FIGURE 4.1 Schematic illustration of the method of arc discharge.

Figure sourced from High-yield synthesis of single-walled carbon nanotubes with a pulsed arc-discharge technique [15]

with high purity and uniformity, low cost, and large scale remains the biggest concern of the scientific community, and industry has been working on this topic. In this section the most important methods of producing CNTs and their advantages and disadvantages will be reviewed.

4.1.1 Arc discharge

Arc discharge was the method used to prepare multiwalled CNTs (MWCNTs) by Iijima in 1991 [23]. In this method, an AC plasma arc is generated between two electrodes maintained in an inert atmosphere as described in Figure 4.1. The high temperature between the electrodes (3000–4000 °C) causes sublimation of the carbon. The sublimated graphite is deposited at the negative electrode or the walls of the chamber where the process is carried out. These deposits contain CNTs [24,25]. For achieving the single-walled CNT (SWCNT) electrodes are doped with catalyst particles, such as Ni–Co, Co–Y, or Ni–Y [1,25–27]. Nanotubes produced by this method are generally entangled and have varying lengths. However, the tubes are of high quality with low amounts of defects. Figure 4.2 shows an arc discharge system used in the production of SWCNTs in a laboratory.

4.1.2 Laser ablation

Laser ablation was the first method used for synthesis of fullerenes. In this method, a target graphite is vaporized by laser radiation at high temperatures (1200 °C) under constant flow of inert gases [29–31], as shown in Figure 4.3. The laser vaporization produces various carbon species that accumulate in the collector cooled by water

FIGURE 4.2 Lab arc discharge system for single-walled carbon nanotubes (SWCNTs) production.

Figure sourced from Clemson University [28]

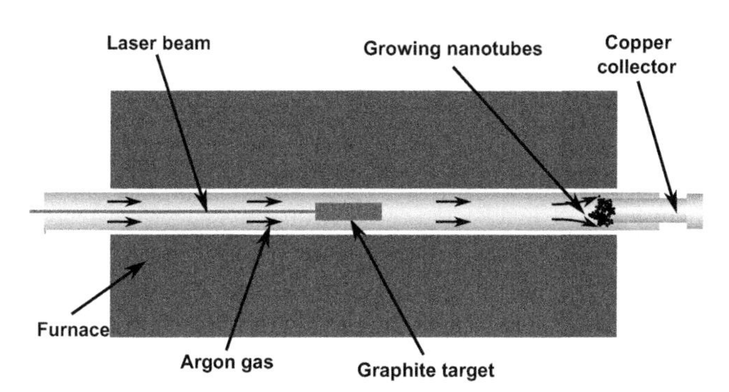

FIGURE 4.3 Schematic of the laser ablation process.

Figure sourced from Advances in the science and technology of carbon nanotubes and their composites [32]

[33]. SWCNTs with a uniform diameter can be produced when the graphite target is doped with small amounts of transition metals such as nickel and cobalt.

4.1.3 Chemical vapor deposition

The CVD has been used for the production of carbon filaments and fibers since the 1960s. In this method, a carbon-rich gas is decomposed into a substrate in the presence of metal catalyst particles at relatively high temperatures (~600 °C or more) [34], as illustrated in Figure 4.4.

During the process, the nanotubes are nucleated and grown by carbon atoms arising from the decomposition of the precursor. The low temperatures used in this process when compared with other methods reduce the cost of production. In addition, deposition of catalyst particles on the substrate allows the formation of

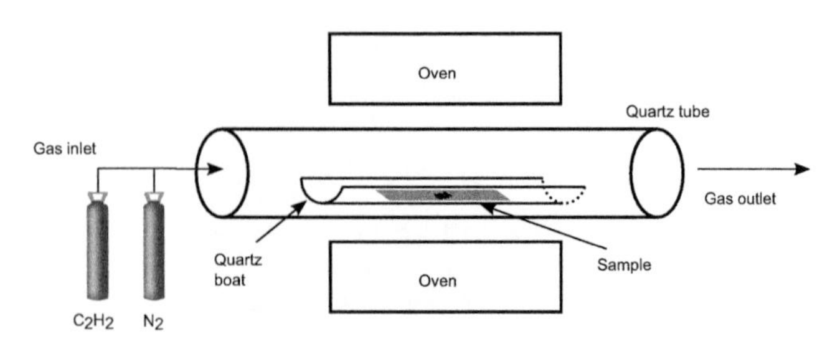

FIGURE 4.4 Schematic of the apparatus used for the synthesis of carbon nanotubes (CNTs) via chemical vapor deposition (CVD).

Figure sourced from Google [35]

new structures. However, CNTs produced via this method usually have a large amount of defects [36]. Figure 4.5 illustrates pictorially the graphene growth via CVD and SWCNTs while Figure 4.6 shows a real example of the synthesis of an MWCNT via CVD. The synthesis can be carried out with different carbon sources (gas), catalysts, substrates, and temperatures.

4.1.4 Conversion of carbon monoxide at high pressure

The method called HiPco was created at Rice University (Texas, USA). This method is used for the production of SWCNTs through the continuous flow of gas (CO) and catalyst precursor $Fe(CO)_5$ in a heated reactor (Figure 4.7) [39]. The average diameter of SWCNT HiPco is about 1.1 nm and the synthetic yield is 70% [12,40]. Generally nanotubes produced by this method have excellent structural integrity, but the production rate is relatively low.

4.2 Cost and production capacity of CNTs

The yield of the synthesis of CNTs, the quantity of amorphous carbon, and the purity of the CNTs vary dramatically with the production method adopted [41]. The methods of arc discharge and laser ablation generally produce nanotubes in grams and so are relatively expensive. Currently, the method of CVD produces the highest amounts of nanotubes and at lower prices, but imperfections in the structure of the CNTs are common.

In 2006, the global production capacity of CNTs was greater than 300 tons per year. Bayer Material Science operated with a production capacity of 60 tons per year in 2007. In 2010 the company opened the largest production facility of CNTs in the world, with a production capacity of 200 tons per year [41]. The production of SWCNTs in 2009 was estimated at around hundreds of kilograms around the world, with a production rate that could reach 50 g/h.

It is expected that in the next 5–10 years, the applications of nanotubes will grow, and as a consequence the demand of quality and production rate of CNTs

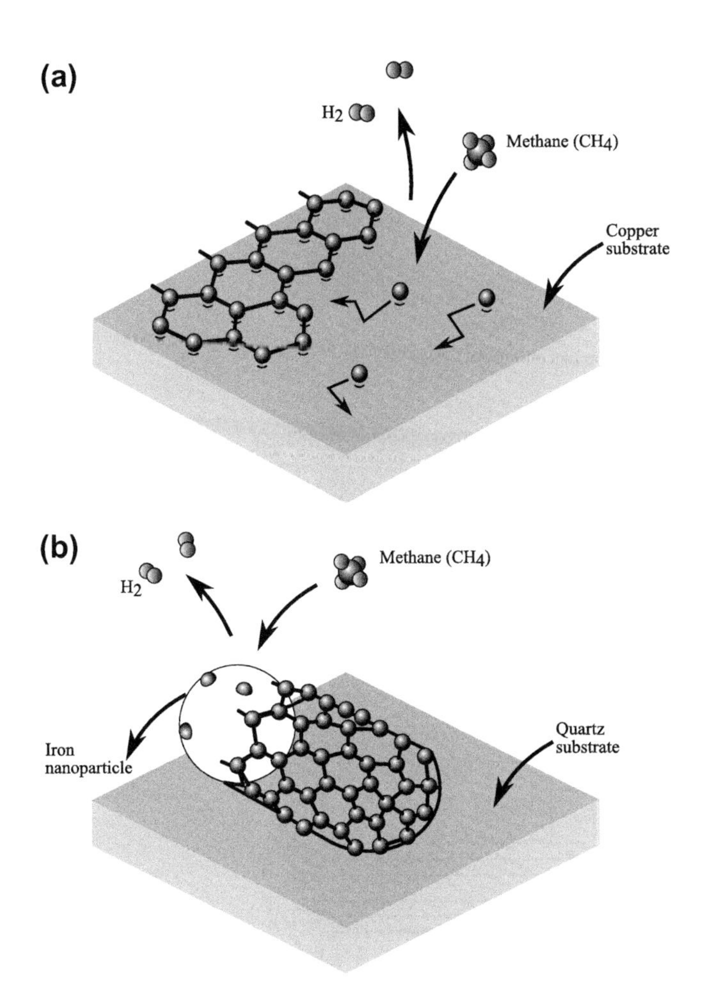

IGURE 4.5 Illustration of the growth of different structures via chemical vapor eposition (CVD).

a) Synthesis of graphene at 1000 °C. (b) Synthesis of carbon nanotubes (CNTs) held at 00 °C in the presence of methane and hydrogen. At the beginning of the synthesis a alf-fullerene is formed on the surface of the catalyst.

Figure sourced from Oregon State University [37]

hould increase, while the price decreases. In fact, the price of CNTs has been educed considerably in recent years, with varying purity and suppliers.

The prices of CNTs are a function of diameter, length, purity, and method of nanufacturing. For SWCNTs, double-walled CNTs (DWCNTs), and MWCNTs, he prices per gram are in the range of \$32–2500, \$21–1600, and \$0.5–136, respectively. For comparison, for carbon fibers (CFs), the price per gram can vary between 0.037 and \$1.8.

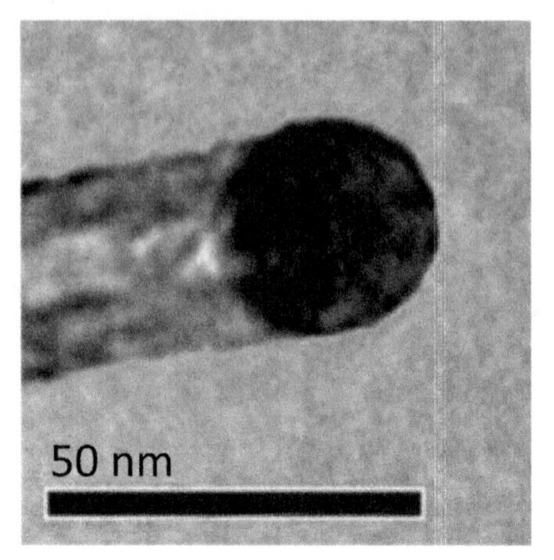

FIGURE 4.6 Nanotube synthesized via chemical vapor deposition (CVD) using nanoparticles of nickel as catalyst. The catalyst can be seen from the end of the nanotube.

Figure sourced from Growth mechanism of vapor phase CVD-grown multi-walled carbon nanotubes [38]

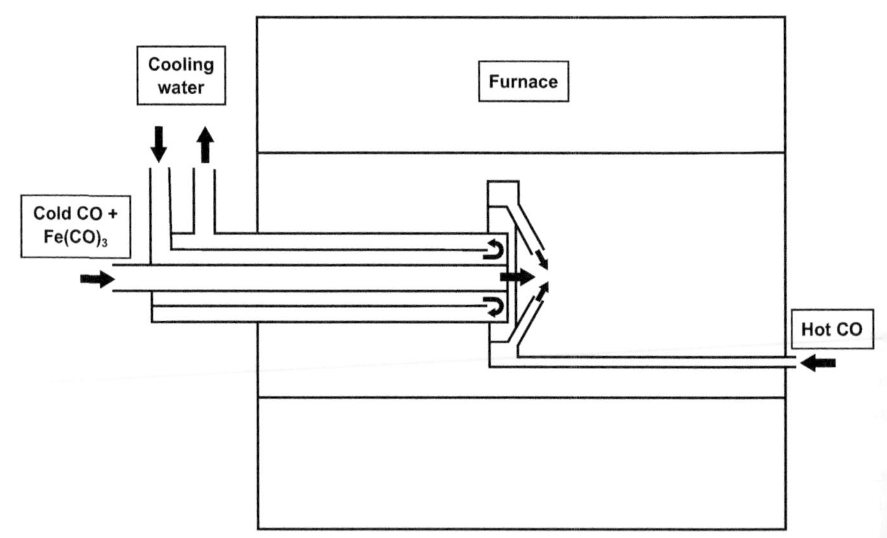

FIGURE 4.7 Schematic of the reactor used for the production of carbon monoxide at high pressure (HiPco) nanotubes.

Figure sourced from Growth mechanism of vapor phase CVD-grown multi-walled carbon nanotubes [12]

4.3 CNTs: risks to health, safe disposal, and environmental concerns

All materials may present a risk to health. Even water, one of the basic elements for life, can be toxic [42]. When large amounts of water are consumed, electrolyte imbalance can occur, which leads to hypotony, a condition in which the muscle tone is abnormally low and may lead to death. Thus, one of the key parameters to understand the toxicity of materials is knowing how much of the substance, i.e., the dose, may be dangerous. Certainly you should not stop drinking water because it can be harmful. Instead, you will consume water as you always have: in adequate amounts and not excessively. This concept, introduced by Paracelsus in the fourteenth century, is the basis for modern toxicology: "The dose makes the poison."

There are several ways to study the effect of substances in the body. Typically, different substances are tested at different doses in laboratory animals to determine whether side effects occur and at what dosage they occur. Because they are mammals, mice may indicate effects that may occur in humans. In addition, some people are different and are more sensitive than others. A person may be unaffected when exposed to certain substances while another suffers an adverse effect such as allergy, for example.

The widespread use of nanoscale particles raised the debate on environmental and toxicological aspects resulting from direct or indirect exposure to these materials. The concern over the use of CNTs is based on their similarity to asbestos. Inhalation of asbestos fibers is recognized as a cause of asbestosis (a disease in the lungs), lung cancer, and mesothelioma [43–45]. This mineral silicate has a fibrous form, such as nanofibers and consequently size, aspect ratio, and the surface can influence the toxicity [43,46].

Since asbestos have a diameter on the micrometer scale and not nanometers, the comparison between nanotubes and asbestos should be exercised with caution. Harmful effects of nanotubes may arise from the combination of high surface area and surface intrinsic toxicity [47]. Unlike conventional materials, nanoparticles may be more toxic to the lung and even modify the structure of proteins. Therefore, nanoparticles can activate immune and inflammatory responses that could affect the body and tissue function [47]. Indeed, investigations of the toxicity of SWCNTs in the lungs of mice concluded that severe exposure to inhalation of SWCNTs is a serious risk of occupational health [48].

In a recent study, Poland [49] showed that exposure of the mesothelium of the cavity of mice to long MWCNTs results in a pathogenic behavior dependent on length, such as asbestos. This includes inflammation and the formation of lesions known as granulomas. The results suggest the need for further research and precaution to avoid problems in cases of prolonged exposure.

The effects of the ingestion of SWCNTs were analyzed in mice as a function of dose, length, and surface [50]. High doses of CNTs, up to 1000 mg/kg of body weight, were used to assess the toxicity of these materials. The results showed that at the doses tested, CNTs can coalesce within the body of rats and form fibrous structures. The use of CNTs with a length (l) greater than 10 μm resulted in the formation of granulomas. Figure 4.8 shows the interior of a rat 14 days after ingestion of an overdose of CNTs [50].

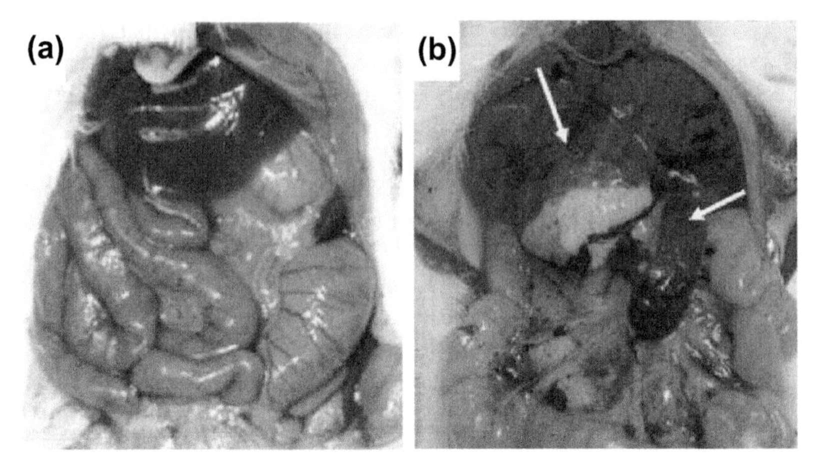

FIGURE 4.8 The left image is the reference and the right was obtained 14 days after injecting carbon nanotubes (CNTs) in the mouse. The dose used was extremely high: 1000 mg/kg of mouse. The arrows indicate the adhesion of CNTs to organs and a gray film of cells infiltrated by CNTs.

Figure sourced from In vivo behavior of large doses of ultrashort and full-length single-walled carbon nanotubes after oral and intraperitoneal administration to Swiss mice [50]

The results obtained by Ryman-Rasmussen have demonstrated that MWCNTs reach the sub-pleura of mice after exposure to a single inhalation of 30 mg/m^3 for 6 h [51]. The study suggests that inhalation of CNTs should be minimized during handling until exposure studies are conducted.

Currently, it is known that nanoparticles can enter the human body through the intestine and lungs (if ingested). Although skin penetration is less likely, the chance that nanoparticles reach the dermis cannot be ruled out [43].

The results above make it clear that the toxicological effects of nanoparticles are not completely understood. Thus, until consistent results are available, nanotubes should be considered a hazardous material! Direct contact with nanoparticles and air pollution by them should be avoided. The handling of nanoparticles also requires special care. Personal protective equipment such as lab coat, goggles, gloves, and masks should be used. Nanoparticles should not be stored or moved in open pots or tubes. The weighing and transferring of nanoparticles should preferably take place in a glove box, as shown in Figure 4.9. The risk associated with the use of CNTs is greatly reduced when they are encapsulated by polymers (in a composite) or a solution. It is also suggested that CNTs should be stored in containers with screw caps.

As far as environmental precaution concerns, it is suggested that spilled material or wash water containing CNTs should not be allowed to enter drains, sewers, surface water, or groundwater. According to the literature

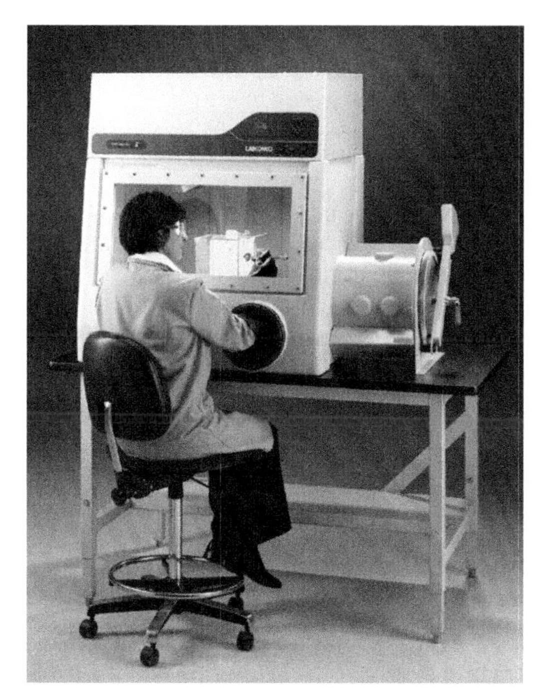

IGURE 4.9 Typical glove box used for weighing and transfer of nanotubes.

Figure sourced from Labconco.com [52]

53–58], it is suggested that before disposal of CNTs, one should consider the ollowing:

- Disposal of CNTs is not allowed by federal, state, and local government regulations.
- Offer surplus and nonrecyclable solutions to a licensed disposal company.
- Contact a licensed professional waste disposal service to dispose of CNTs.
- CNTs must be destroyed by a suitable method, including, but not limited to, incineration.
- Volatile dust must be collected during incineration.
- Dissolve or mix CNTs with a combustible solvent and burn in a chemical incinerator equipped with an afterburner and scrubber.
- Liquids containing significant amount of CNTs must be filtered before being released to the sewer.
- Contaminated packaging should be disposed of as unused product.

Different companies selling CNTs provide different Material Safety Data Sheets MSDS), some more complete than others as expected. After analyzing many MSDS from several companies, a "new" MSDS, as complete as possible, has been prepared by this author. The aim was to compile useful information for people dealing with CNTs. This MSDS is presented below.

CARBON NANOTUBES

Material Safety Data Sheet

This document has been prepared based on the MSDS from several companies and, as the information on the potential health effects of CNTs is continually developing, this MSDS shall be used as a guide only.

SECTION 1: PRODUCT AND COMPANY IDENTIFICATION

PRODUCT NAME: CNT, single-walled, double-walled, multi-walled; CNT friable solids.

SECTION 2: COMPOSITION/INFORMATION ON INGREDIENTS

INGREDIENT:

Substance	CAS No.	% WT
Carbon nanotubes (CNTs)	308068-56-6 [0]	40–100
Carbon	7440-44-0	<10%
Metallic impurities (including molybdenum, iron, magnesium)		0–10 wt%

SECTION 3: HAZARDS IDENTIFICATION

CLASSIFICATION OF THE SUBSTANCE OR MIXTURE

Classification according to Regulation (EC) No 1272/2008 [EU-GHS/CLP].

Eye irritation (Category 2)

Specific target organ toxicity - single exposure (Category 3)

Classification according to EU Directives 67/548/EEC or 1999/45/EC.

Irritating to eyes and respiratory system.

LABEL ELEMENTS

Labeling according Regulation (EC) No 1272/2008 [CLP]

Signal word — Warning

Hazard statement(s)

H319 — Causes serious eye irritation.

H335 — May cause respiratory irritation.

Precautionary statement(s)

P261 — Avoid breathing dust/fume/gas/mist/vapors/spray.

P305 + P351 + P338 — IF IN EYES: Rinse cautiously with water for several minutes. Remove contact lenses, if present and easy to do. Continue rinsing.

Supplemental hazard — None

Statements according to European Directive 67/548/EEC as amended.

Risk-phrase(s)

R36/37 — Irritating to eyes and respiratory system.

Safety-phrase(s)

S22 — Do not breathe dust

S25 — Avoid contact with eyes

S26 — In case of contact with eyes, rinse immediately with plenty of water and seek medical advice

S29 — Do not empty into drains

S35 — This material and its container must be disposed of in a safe way

S36/37/39 — Wear suitable protective clothing, gloves, eye/face protection.

Clean contaminated surfaces using water and detergent.

Caution—substance not yet tested completely.

OTHER HAZARDS

none

CARBON NANOTUBES—cont'd

SECTION 4: FIRST AID MEASURES

DESCRIPTION OF FIRST AID MEASURES

EYES: Rinse thoroughly with plenty of water for at least 15 min. Ensure complete irrigation of the eye. Keeping eyelids apart and away from eye during irrigation. Removal of contact lenses should only be undertaken by skilled personnel. Get medical attention if irritation develops or persists.

SKIN: Wash off with soap and plenty of water for at least 15 min. Get medical attention if irritation develops or persists.

INGESTION: If swallowed, rinse mouth with water provided person is conscious. Never give anything by mouth to an unconscious person. Get medical attention if irritation develops or persists.

INHALATION: If breathed in, move person into fresh air. Encourage patient to blow nose and clear passage for breathing. If not breathing, give artificial respiration. Get medical attention if irritation develops or persists.

MOST IMPORTANT SYMPTOMS AND EFFECTS, BOTH ACUTE AND DELAYED

To the best of our knowledge, the chemical, physical, and toxicological properties have not been thoroughly investigated.

INDICATION OF ANY IMMEDIATE MEDICAL ATTENTION AND SPECIAL TREATMENT NEEDED

Treat symptomatically.

GENERAL ADVICE

Consult a physician. Show this safety data sheet to the doctor in attendance.

SECTION 5: FIREFIGHTING MEASURES

EXTINGUISHING MEDIA

Suitable extinguishing media:

- Water spray or fog—Large fires only.
- Alcohol-resistant foam.
- Dry chemical powder.
- Carbon dioxide.
- BCF (where regulations permit).

SPECIAL HAZARDS ARISING FROM THE SUBSTANCE OR MIXTURE

Carbon monoxide, Carbon dioxide, Carbon oxides.

ADVICE FOR FIREFIGHTERS

As in any fire, wear NIOSH/MSHA-approved positive-pressure self-contained breathing apparatus and full protective clothing. Airborne dust from the dried dispersion in an enclosed space and in the presence of an ignition source may constitute an explosion hazard.

FIRE/EXPLOSION HAZARD

- Solid that exhibits difficult combustion or is difficult to ignite.
- Avoid generating dust as dusts may form an explosive mixture with air, and any source of ignition, i.e., flame or spark, may cause fire or explosion.
- Dust clouds generated by the fine grinding of the solid are a particular hazard; accumulations of fine dust (420 μm or less) may burn rapidly and fiercely if ignited; once initiated, larger particles up to 1400 μm diameter will contribute to the propagation of an explosion.
- A dust explosion may release large quantities of gaseous products, resulting in a subsequent pressure rise of explosive force capable of damaging plant and buildings and injuring people.

Continued

CARBON NANOTUBES—cont'd

- Dry dust can be charged electrostatically by turbulence, pneumatic transport, pouring, in exhaust ducts, and during transport.
- Build-up of electrostatic charge may be prevented by bonding and grounding.
- Powder handling equipment may require additional protection measures such as explosion venting.

FIREFIGHTING

- Alert Fire Brigade and tell them location and nature of hazard.
- Wear breathing apparatus, protective gloves, and eye protection.
- Determine whether to evacuate or isolate the area according to your local emergency plan.
- Prevent spillage from entering drains or water courses.
- Cool fire-exposed containers with water spray.
- If safe to do so, remove containers from path of fire.
- Equipment should be thoroughly decontaminated after using wet cleanup procedure.

FIRE INCOMPATIBILITY

Avoid contamination with oxidizing agents i.e., nitrates, oxidizing acids, chlorine bleaches, pool chlorine, etc., as ignition may result.

SECTION 6: ACCIDENTAL RELEASE MEASURES

PERSONAL PRECAUTIONS, PROTECTIVE EQUIPMENT, AND EMERGENCY PROCEDURES

- Use appropriate personal protection during clean-up.
- Avoid dust formation.
- Avoid breathing vapors, mist, dust, or gas.
- Avoid contact with skin and eyes.
- Ensure adequate ventilation.
- Evacuate personnel to safe areas.

ENVIRONMENTAL PRECAUTIONS

Do not allow spilled material or wash water to enter drains, sewers, surface water, or groundwater.

METHODS AND MATERIALS FOR CONTAINMENT AND CLEANING UP

- Remove mechanically by a method that minimizes the generation of airborne dust (HEPA-equipped vacuum, wet mopping, etc.).
- Use wet clean-up procedures and avoid generating dust: Mist area with water containing mild detergent.
- Wipe with light-colored disposable wipes until clear, then double bag wipes for disposal.
- Absorb material in labeled plastic bags, double bagged, or other sealed container for waste disposal.
- Ventilate area and wash spill site, exposed skin, and clothing after material pickup is complete.

SECTION 7: HANDLING AND STORAGE

PRECAUTIONS FOR SAFE HANDLING

- Avoid contact with skin and eyes.
- If possible, use in a closed, well-ventilated area (e.g., fume hood).
- Avoid formation of dust and aerosols.
- Provide appropriate exhaust ventilation at places where dust is formed.
- Normal measures for preventive fire protection.

CONDITIONS FOR SAFE STORAGE, INCLUDING ANY INCOMPATIBILITIES

- To maintain product quality, store between 40 °F (4 °C) and 100 °F (38 °C) in a dry and well-ventilated place.

CARBON NANOTUBES—cont'd

- Store in original container.
- Keep in clearly labeled, sealed non-breakable container.
- Keep container tightly closed.
- Packing as recommended by manufacturer.

SUITABLE CONTAINER

- Double bag and place into a dust tight container.
- Suitable container materials include but not limited to metal, polyethylene, or polypropylene.
- If CNTs are dispersed in liquid or mixed with other chemical, ensure container compatibility.
- Check all containers are clearly labeled and free from leaks.

NOTES:

Detailed information on handling CNTs may be found at the ASTM Standard E 2535 - 07, "Standard Guide for Handling Unbound Engineered Nanoscale Particles in Occupational Settings". West Conshohocken, PA; ASTM International, www.astm.org.

SECTION 8: EXPOSURE CONTROLS/PERSONAL PROTECTION

ENGINEERING CONTROLS

- General room ventilation is adequate for storage and ordinary handling.
- Use local exhaust at points of use to maintain exposure below the PEL/TLV exposure limits.
- Current knowledge indicates that a well-designed exhaust ventilation system using a high efficiency particulate air (HEPA) filter should efficiently capture CNTs.
- Safety shower and eye bath must be available.
- Whenever practicable all manipulations of free CNTs should be carried out in an exhaust fume hood or glove box or other such device equipped with HEPA filtration.
- If CNTs have the potential to contaminate the air exhausted from a containment device, the exhaust air must pass through a HEPA filter.
- Handle in accordance with good industrial hygiene and safety practice.
- Wash hands before breaks and at the end of workday.

PERSONAL PROTECTIVE EQUIPMENT:
EYE/FACE PROTECTION

- Safety glasses with side-shields conforming to EN166. Use equipment for eye protection tested and approved under appropriate government standards such as NIOSH (US) or EN 166(EU) or ANSI Z87.1.
- Contact lenses may pose a special hazard. A written policy document, describing the wearing of lens or restrictions on use, should be created for each workplace or task, taking into account the nature of the work and chemicals being used. In the event of chemical exposure, begin eye irrigation immediately and remove contact lens as soon as practicable if safe to do so without causing further damage.

SKIN PROTECTION

- Handle with chemical-resistant gloves (suitable types include nitrile, latex, vinyl).
- Gloves must be inspected prior to use.
- Use proper glove removal technique (without touching glove's outer surface) to avoid skin contact with this product.
- Dispose of contaminated gloves after use in accordance with applicable laws and good laboratory practices.
- Wash and dry hands.
- The selected protective gloves have to satisfy the specifications of EU Directive 89/686/EEC and the standard EN 374 derived from it.

Continued

CARBON NANOTUBES—cont'd

BODY PROTECTION

- Impervious clothing.
- The type of protective equipment must be selected according to the concentration and amount of the dangerous substance at the specific workplace.

RESPIRATORY PROTECTION

- For nuisance exposures, use type P95 (US) or type P1 (EU EN 143) particle respirator. For higher level protection, use type OV/AG/P99 (US) or type ABEK-P2 (EU EN 143) respirator cartridges or N100cartridges or better.
- Use respirators and components tested and approved under appropriate government standards such as NIOSH (US) or CEN (EU). The respirator must be used in accordance with the OSHA respiratory protection standard (29 CFR 1910.134).

MINIMUM PROTECTION

- Minimum protection P2 respirator, dust coat (disposable preferred), eye protection, shoes.

SECTION 9: PHYSICAL AND CHEMICAL PROPERTIES

INFORMATION ON BASIC PHYSICAL AND CHEMICAL PROPERTIES

APPEARANCE FORM:	Black powder
PHYSICAL STATE:	Solid
DENSITY:	1.4–1.9 g/cm^3
DENSITY (BULK):	0.1 g/cm^3
MOLECULAR WEIGHT:	not applicable
ODOR:	no data available
ODOR THRESHOLD:	no data available
pH:	pH: 4–10
DECOMPOSITION TEMPERATURE:	>600 °C
WATER SOLUBILITY:	insoluble
INITIAL BOILING POINT AND RANGE:	no data available
FLASH POINT:	not applicable
FLAMMABILITY (SOLID, GAS):	no data available
VAPOR PRESSURE:	no data available
AUTO-IGNITION TEMPERATURE:	no data available
VISCOSITY:	not applicable
EXPLOSIVE PROPERTIES:	no data available
OXIDIZING PROPERTIES:	no data available

SECTION 10: STABILITY AND REACTIVITY

REACTIVITY

no data available.

CHEMICAL STABILITY

Thermal decomposition or combustion may produce smoke of carbon oxide and metal oxide.

POSSIBILITY OF HAZARDOUS REACTIONS

No hazardous reaction when used as directed.

CARBON NANOTUBES—cont'd

CONDITIONS TO AVOID

Avoid excessive heating.

INCOMPATIBLE MATERIALS

- Reducing agents.
- Acids.
- Halogens.
- Strong oxidizing agents.

HAZARDOUS DECOMPOSITION PRODUCTS

Thermal decomposition products may include carbon monoxide, carbon dioxide, and oxides of metallic impurities (including molybdenum and cobalt).

SECTION 11: TOXICOLOGICAL INFORMATION

IMMEDIATE (ACUTE) TOXICITY

no data available.

SKIN CORROSION/IRRITATION

No data available.

SERIOUS EYE DAMAGE/EYE IRRITATION

no data available.

RESPIRATORY OR SKIN SENSITIZATION

no data available.

GERM CELL MUTAGENICITY

no data available.

CARCINOGENICITY

IARC: No component of this product present at levels greater than or equal to 0.1% is identified as probable, possible, or confirmed human carcinogen by IARC.

REPRODUCTIVE TOXICITY

no data available.

SPECIFIC TARGET ORGAN TOXICITY—SINGLE EXPOSURE

Inhalation—May cause respiratory irritation.

SPECIFIC TARGET ORGAN TOXICITY—REPEATED EXPOSURE

no data available.

ASPIRATION HAZARD

no data available.

POTENTIAL HEALTH EFFECTS

Inhalation: May be harmful if inhaled. Causes respiratory tract irritation.
Ingestion: May be harmful if swallowed.
Skin: May be harmful if absorbed through skin. May cause skin irritation.
Eyes: Causes serious eye irritation.

SIGNS AND SYMPTOMS OF EXPOSURE

To the best of our knowledge, the chemical, physical, and toxicological properties have not been thoroughly investigated.

SECTION 12: ECOLOGICAL INFORMATION

TOXICITY

no data available.

PERSISTENCE AND DEGRADABILITY

no data available.

Continued

CARBON NANOTUBES—cont'd

BIOACCUMULATIVE POTENTIAL
no data available.

MOBILITY IN SOIL
no data available.

RESULTS OF PBT AND VPVB ASSESSMENT
no data available.

OTHER ADVERSE EFFECTS
no data available.

SECTION 13: DISPOSAL CONSIDERATIONS

WASTE DISPOSAL METHOD

Product

- Disposal of this product is not allowed by federal, state, and local government regulations.
- Offer surplus and non-recyclable solutions to a licensed disposal company.
- Contact a licensed professional waste disposal service to dispose of this material.
- It must be destroyed by a suitable method including, but not limited to, incineration.
- Volatile dust must be collected during incineration.
- Liquids containing significant amount of CNTs must be filtered before their release to the sewer.
- Dissolve or mix the material with a combustible solvent and burn in a chemical incinerator equipped with an afterburner and scrubber.

Contaminated packaging

- Dispose of as unused product.

SECTION 14: TRANSPORT INFORMATION

UN NUMBER
ADR/RID: IMDG: IATA:

UN PROPER SHIPPING NAME
ADR/RID: Not dangerous goods.
IMDG: Not dangerous goods.
IATA: Not dangerous goods.

TRANSPORT HAZARD CLASS(ES)
ADR/RID: IMDG: IATA:

PACKAGING GROUP
ADR/RID: IMDG: IATA:

ENVIRONMENTAL HAZARDS
ADR/RID: no IMDG Marine Pollutant: no IATA: no.

SPECIAL PRECAUTIONS FOR USER

- Use precaution during transport in order to prevent accidental spill.
- Not dangerous cargo.
- Keep separated from foodstuffs.

SECTION 15: REGULATORY INFORMATION

U.S. FEDERAL REGULATIONS

TOXIC SUBSTANCE CONTROL ACT:

To the best of our knowledge, this material is not included in the Toxic Substances Control Act (TSCA). Inventory, and is defined as a new chemical substance that cannot be imported or manufactured for commercial purposes without complying with the Pre-Manufacturing notice (PMN) requirements codified at 40CFR Part 720.

CARBON NANOTUBES—cont'd

SARA TITLE III (SUPERFUND AMENDMENTS AND REAUTHORIZATION ACT):
"Reportable Quantities" (RQs) and/or "Threshold Planning Quantities" (TPQs) exist for the following ingredients.

SECTION 16: OTHER INFORMATION

none.

DISCLAIMER

The information contained in this document is believed to be correct but does not purport to be all-inclusive. The information has been prepared using the current knowledge of CNTs presented in the MSDS of different companies. It does not represent a guarantee that the information is complete or correct in all respects. Information on the potential health effects of CNTs is continually developing. This information shall be used as a guide only! A precautionary approach should be considered in all aspects of the use of CNTs, including their manufacture, storage, transportation, handling, and disposal.

4.4 Commercially available CNTs

As the applications of CNTs grow, the number of companies producing CNTs also grows. However, many companies do not stay in business for long time. With the objective of guiding researchers interested in buying CNTs, we present here an extensive, up-to-date list of manufacturers of CNTs.

Advanced nanopower Inc. www.anp.com.tw/English/main02.htm
Alfa Aesar www.alfa.com/
American Dye Source, Inc. www.adsdyes.com/index.html
ApNano Materials www.apnano.com/
Arkema www.arkema.com/en/index.html
Arry www.arry-nano.com/
Bayer AG www.bayer.com/en/homepage.aspx
buckyusa http://buckyusa.com/
Canatu Oy www.canatu.com
Carbon Design Innovations www.carbondesigninnovations.com/
Carbon Nano-Material Technology: www.carbonnano.co.kr/english/
Carbon NT&F 21 www.carbon-nanofiber.com
Carbon Solutions, Inc. www.carbonsolution.com/
Catalytic Materials www.catalyticmaterials.com/
Cheap Tubes Inc. www.cheaptubesinc.com/
Cheap Tubes www.cheaptubes.com/default.htm
CNano Technology Limited http://cnanotechnology.com/english.htm
Haydale Limited www.haydale.com/
HeJi Inc www.nanotubeseu.com/index.html
Helix Material Solutions, Inc. www.helixmaterial.com/
Hyperion Catalysis International www.hyperioncatalysis.com

Iljin www.iljin.co.kr
Materials and Electrochemical Reserch Corporation www.mercorp.com
Materials Technologies Research www.mtr-ltd.com/
MicroTechNano www.microtechnano.com/
Nano CS www.nanocs.com
Nano Integris www.nanointegris.com/en/home/
Nano Lab www.nano-lab.com/home.html
Nano Powder R&D Center www.nanotubes.cn/
Nano Thinx www.nanothinx.com/home.html
Nano-C http://nano-c.com/
Nanocyl www.nanocyl.com/
NanoGap
Nanoshel www.nanoshel.com
Nanostructured & Amorphous Materials www.nanoamor.com
Shanghai Feibo Chemical Technology www.arknano.com/
Sigma Aldrich www.sigmaaldrich.com/
Time nano www.timesnano.com/en/
Unidym www.unidym.com
XinNano Material www.xinnanomaterials.com/

To learn more...

We encourage readers to consult the following books:

1. Coelho LAF, Pezzin SH, Loos MR, Pezzin, Amico, SC. General issues in carbon nanocomposites technology. Book chapter "Encyclopedia of Polymer Composites: Properties, Performance and Applications". Nova Science Publishers, Inc; 2009, ISBN-13: 978-1,607,417,170:p. 389–416.

References

[1] Saito Y, Nishikubo K, Kawabata K, Matsumoto TJ. Carbon nanocapsules and single-layered nanotubes produced with platinum-group metals (Ry, Rh, Pd, Os, Ir, Pt) by arc discharge. Appl Phys 1996;80:3062–7.
[2] Qian D, Wagner GJ, Liu WK, Yu M-F, Ruoff RS. Mechanics of carbon nanotubes. Appl Mech Rev 2002;55:495–533. http://dx.doi.org/10.1115/1.1490129.
[3] Zijiong Li, Wei L, Zhang Y. Appl Surf Sci 2008;254:5247–51.
[4] Jiang B, Wei J, Ci L, Wu D. Chin Sci Bull 2004;49:107–10.
[5] Roch A, Jost O, Schultrich B, Beyer E. High-yield synthesis of single-walled carbon nanotubes with a pulsed arc-discharge technique. Phys Stat Sol 2007;244:3907–10.
[6] Rinzler AG, Liu J, Dai H, Nikolaev P, Huffman CB, Rodriguez-Macias FJ. Large-scale purification of single-wall carbon nanotubes: process, product, and characterization. Appl Phys a 1998;67:29–37.

[7] Ren ZF, Huang ZP, Xu JW, Wang DZ, Wen JG, Wang JH. Growth of a single free-standing multiwall carbon nanotube on each nanonickel dot. Appl Phys Lett 1999;75: 1086–8.
[8] Ren ZF, Huang ZP, Xu JW, Wang JH, Bush P, Siegal MP. Synthesis of large arrays of well-aligned carbon nanotubes on glass. Science 1998;282:1105–7.
[9] Huang ZP, Xu JW, Ren ZF, Wang JH, Siegal MP, Provencio PN. Growth of highly oriented carbon nanotubes by plasma-enhanced hot filament chemical vapor deposition. Appl Phys Lett 1998;73:3845–7.
[10] Queipo P, Nasibulin AG, Shandakov SD, Jiang H, Gonzalez D, Kauppinen EI. CVD synthesis and radial deformations of large diameter single-walled CNTs. Curr Appl Phys 2009;9:301–5.
[11] Mondal KC, Strydom AM, Erasmus RM, Keartland JM, Coville NJ. Physical properties of CVD boron-doped multiwalled carbon nanotubes. Mat Chem Phys 2008;111: 386–90.
[12] Nikolaev P, Bronikowski MJ, Bradley RK, Rohmund F, Colbert DT, Smith KA, et al. Gas-phased catalytic growth of single-walled carbon nanotubes from carbon monoxide. Chem Phys Lett 1999;313:91–7.
[13] Edgar K, Spencer JL. Aerosol-based synthesis of carbon nanotubes. Curr Appl Phys 2004;4:121–4.
[14] Shao M, Wang D, Yu G, Hu B, Yu W, Qian Y. The synthesis of carbon nanotubes at low temperature via carbon suboxide disproportionation. Carbon 2004;42:183–5.
[15] Paradise M, Goswami T. Carbon nanotubes-production and industrial applications. Mater Des 2007;28:1477–89.
[16] Yokomichiand H, Ichihara M, Nimori S, Kishimoto N. Morphology of carbon nanotubes synthesized by thermal CVD under high magnetic field up to 10 T. Vacuum 2008;83:625–8.
[17] Rao CNR, Govindaraj A, Gundiah G, Vivekchand SRC. Nanotubes and nanowires. Chem Eng Sci 2004;59:4665–71.
[18] Li M-W, Hu Z, Wang X-Z, Wu Q, Chen Y, Tian Y-L. Low-temperature synthesis of carbon nanotubes using corona discharge plasma at atmospheric pressure. Diam Relat Mater 2004;13:111–5.
[19] Li Y-L, Kinloch IA, Shaffer MSP, Geng J, Johnson B, Windle AH. Synthesis of single-walled carbon nanotubes by a fluidized-bed method. Chem Phys Lett 2004;384: 98–102.
[20] Qiu J, Li Y, Wang Y, Li W. Production of carbon nanotubes from coal. Fuel Process Technol 2004;85:1663–70.
[21] Livage J. Sol-gel chemistry and electrochemical properties of vanadium oxide gels. Solid State Ionics 1996;86:935–42.
[22] Montoro LA, Lofrano RCZ, Rosolen JM. Synthesis of single-walled and multi-walled carbon nanotubes by arc-water method. Carbon 2005;43:200–3.
[23] Iijima S. Helical microtubules of graphitic carbon. Nature 1991;354:56–8.
[24] Journet C, Bernier P. Production of carbon nanotubes. Appl Phys A 1998;67:1–9.
[25] Journet C, Maser WK, Bernier P, Loiseau A, delaChapelle ML, Lefrant S, et al. Large-scale production of single-walled carbon nanotubes by the electric-arc technique. Nature 1997;388:756–8.
[26] Bethune DS, Kiang CH, Devries MS, Gorman G, Savoy R, Vazquez J. Cobalt-catalysed growth of carbon nanotubes with single-atomic-layer walls. Nature 1993; 363:605–7.

[27] Shi Z, Lian Y, Liao FH, Zhou X, Gu Z, Zhang Y, et al. Large scale synthesis of single wall carbon nanotubes by arc discharge method. J Phys Chem Solids 2000;61:1031–6.
[28] www.clemson.edu/ces/lemt/Arc-Discharge.htm.
[29] Microstructural design of fiber composites. Cambridge University Press; 1992. ISBN-10: 052135482X; ISBN-13: 978-0521354820.
[30] Thess A, Lee R, Nikolaev P, Dai HJ, Petit P, Robert J. Crystalline ropes of metallic carbon nanotubes. Science 1996;273:483–7.
[31] Zhang Y, Iijima S. Formation of single-wall carbon nanotubes by laser ablation of fullerenes at low temperatures. Appl Phys Lett 1999;75:3087–9.
[32] Thostenson ET, Renb ZF, Choua TW. Advances in the science and technology of carbon nanotubes and their composites: a review Compos Sci Technol 2001;61:1899–912.
[33] Guo T, Nikolaev P, Thess A, Colbert DT, Smalley RE. Catalytic growth of single-walled nanotubes by laser vaporization. Chem Phys Lett 1995;243:49–54.
[34] Xie S, Li W, Pan Z, Chang B, Sun L. Carbon nanotube arrays. Mater Sci Eng A 2000; 286:11–5.
[35] https://Sites.google.com/site/cntcomposites/production-methods.
[36] Lahiff E, Leahy R, Coleman NJ, Blau WJ. Small but strong: a review of the mechanical properties of carbon nanotube-polymer composites. Carbon 2006;44: 1624–52.
[37] www.science.oregonstate.edu/minote/wiki/doku.php?id=start.
[38] Deck CP, Vecchio K. Growth mechanism of vapor phase CVD-grown multi-walled carbon nanotubes. Carbon 2005;43:2608–17.
[39] Daenen M, de Fouw RD, Hamers B, Janssen PGA, Schouteden K, Veld MAJ. The wondrous world of carbon nanotubes 'a review of current carbon nanotube technologies'. Technische Universiteit Eindhoven; 2003. p. 1–96.
[40] Smalley RE, Yakobson BI. The Future of the fullerenes. Solid State Commun 1998;107: 597–606.
[41] Bierdel M, Buchholz S, Michele V, Mleczko L, Rudolf R, Voetz M, et al. Industrial production of multiwalled carbon nanotubes. Phys Stat Sol 2007;244:3939–43.
[42] Hornyak GL, Tibbals HF, Dutta J, Moore JJ. Introduction to nanoscience and nanotechnology. Taylor and Francis Group, LLC; 2009. ISBN-10: 1420047795; ISBN-13: 978-1420047790.
[43] Brayner R. The toxicological impact of nanoparticles. Nanotoday 2008;3:48–55.
[44] LaDou J. Environ Health Perspect 2004;112:285–90.
[45] Huczko A, Lange H, Calko E, Grubek-Jaworska H, Droszcz P. Physiological testing of carbon nanotubes: are they asbestos-like? Fuller Nanotu Car 2001;9:251–4.
[46] Rogers RA, Antonini JM, Brismar H, Lai J, Hesterberg TW, Oldmixon EH, et al. In situ microscopic analyis of asbestos and synthetic vitreous fibers retained in hamster lungs following inhalation. Environ Health Perspect 1999;107:367–75.
[47] Donaldson K, Stone V, Tran CL, Kreyling W, Borm PJ. Nanotoxicology. Occup Environ Med 2004;61:727–8.
[48] Lam CW, James JT, McCluskey R, Hunter RL. Pulmonary toxicity of single-wall carbon nanotubes in mice 7 and 90 days after intratracheal instillation. Toxicol Sci 2004; 77:126–34.
[49] Poland CA, Dun R, Kinloch I, Maynard A, Wallace WAH, Seaton A, et al. Carbon nanotubes introduced into the abdominal cavity of mice show asbestos-like pathogenicity in a pilot study. Nat Nanotechnol 2008;3:423–8.

50] Kolosnjaj-Tabi J, Hartman KB, Boudjemaa S, Ananta JS, Morgant G, Szwarc H, et al. In vivo behavior of large doses of ultrashort and full-length single-walled carbon nanotubes after oral and intraperitoneal administration to Swiss mice. ACS Nano 2010;4: 1481−92.

51] Ryman-Rasmussen JP, Cesta MF, Brody AR, Shipley-Phillips JK, Everitt JI, Tewksbury EW, et al. Inhaled carbon nanotubes reach the subpleural tissue in mice. Nat Nanotechnoly 2009;4:747−51.

52] www.labconco.com/scripts/editc20.asp?catid=9.

53] www.swentnano.com/tech/docs/SWeNT_MSDS_SMW.pdf [accessed on 08/08.13].

54] www.nanothinx.com; [accessed on 08.08.13].

55] www.sigmaaldrich.com; [accessed on 08.08.13].

56] www.cheaptubesinc.com/cntmaterialsafetydatasheet.htm; [accessed on 08.08.13].

57] www.tedpella.com/msds_html/Nanotubes-MSDS.pdf; [accessed on 08.08.13].

58] www.csiro.au/Organisation-Structure/Divisions/CMSE/Fibre-Science/Multi-walled-Carbon-Nanotubes-safety.aspx; [accessed on 08.08.13].

[50] [illegible] Morgan C, Sywan [illegible] and full length single walled carbon nano- [illegible] function for Swiss mice. ACS Nano 2011 [illegible]

[51] Ryman [illegible] Shipley [illegible] Evans [illegible] carbon nanotubes reach the subpleural tissue [illegible] Nat Nano [illegible]

[52] [illegible]

[53] [illegible] SNW [illegible]

[54] [illegible]

[55] [illegible] accessed on [illegible]

[56] [illegible]

[57] [illegible]

CHAPTER

Fundamentals of Polymer Matrix Composites Containing CNTs

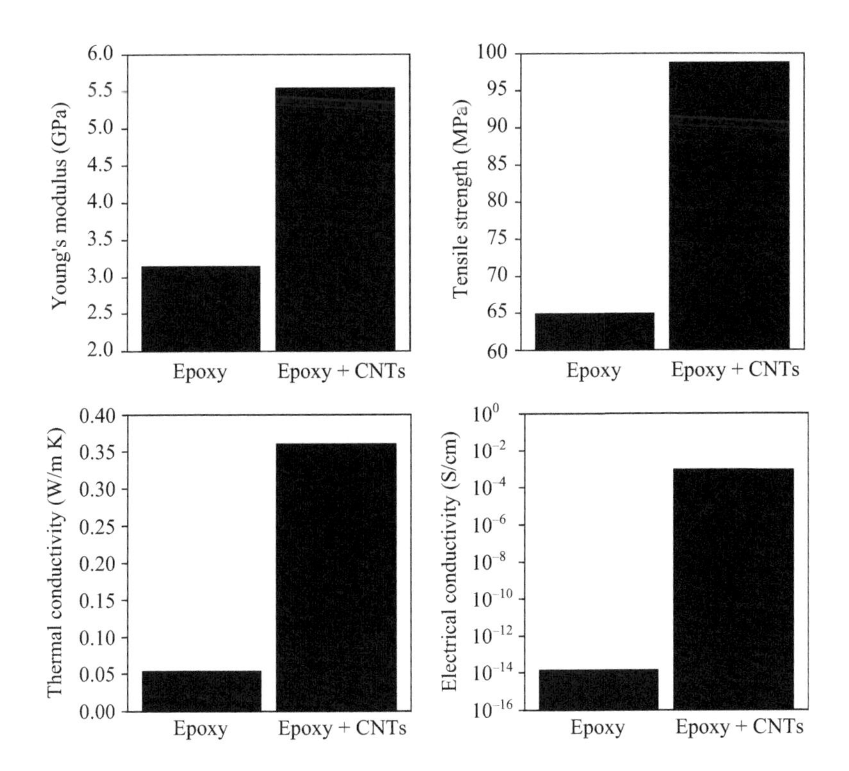

CHAPTER OUTLINE

Carbon Nanotube Reinforced Composites. http://dx.doi.org/10.1016/B978-1-4557-3195-4.00005-9

5.1 Use of CNTs for improvement of polymer properties

The exceptional properties of carbon nanotubes (CNTs) make them a promising reinforcement to improve the mechanical, electrical, and thermal properties of polymers. The main challenge is to transfer these properties of nanotubes to a polymer matrix during composite manufacturing. Due to long-range van der Waals interactions, nanotubes tend to agglomerate and form bundles or ropes, with a highly entangled structure. Thus, although the high specific surface area of the nanotubes can act as a desirable interface for the transfer of stress, it may undesirably raise attractive forces between the nanotubes themselves, leading to excessive agglomeration behavior. In this respect, two main issues are of extreme importance for acquisition of composites with improved properties due to the addition of nanotubes. These issues are [1,2]:

1. The interfacial interaction between polymer and reinforcement.
2. The proper dispersion and distribution of the reinforcement in the polymer matrix.

Before we go any further with our discussion, it is necessary to define the exact aggregation state of reinforcements in nanocomposites. The distribution of the filler describes the homogeneity throughout the sample, and the dispersion describes the level of agglomeration. Figure 5.1 illustrates schematically possible states of the aggregation of nano-fillers.

Homogeneous distribution and dispersion of fillers within a host polymer matrix are common prerequisites for composite materials. It is also vital to stabilize the dispersion to prevent reagglomeration of the filler.

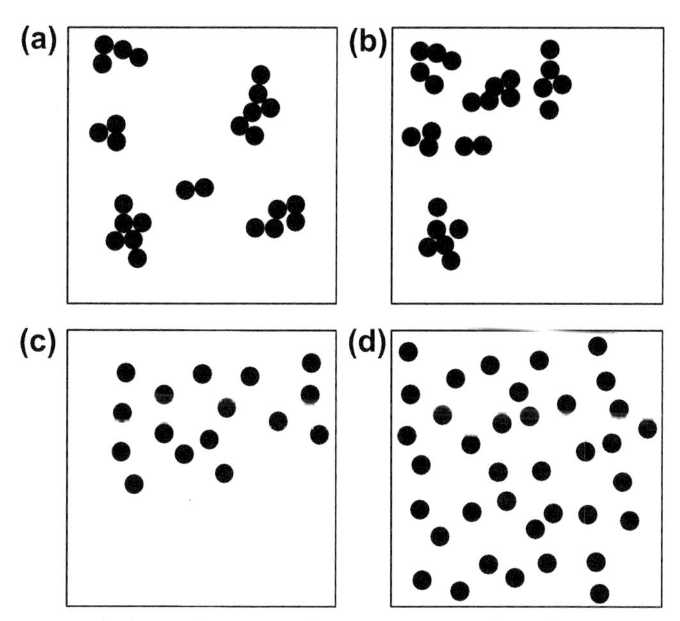

FIGURE 5.1 Schematic illustration of the difference between dispersion and distribution.

Good distribution and poor dispersion (a); poor distribution and poor dispersion (b); poor distribution and good dispersion (c); and good distribution and good dispersion (d).

Figure courtesy of Prof. Ica Manas-Zloczower, Case Western Reserve University

Mixing by shear is the primary method used to disperse CNTs whilst chemical modification can help stabilize the dispersion and allows CNTs to engage with the polymeric matrix. Thus, chemical modification seems attractive since this may improve adhesion (interaction) between the reinforcement phase (CNTs) and the matrix, so that external stresses imposed can be efficiently transferred to the nanotubes.

Molecular simulation results (Figure 5.2) show that the shear strength of the interface between nanotube and polymer with weak links or interactions can be raised by an order of magnitude by introducing a relatively low density ($<1\%$) of chemical bonds between nanotubes and the matrix [3]. This target is currently a major challenge for chemists and materials scientists in the field.

There are numerous methods and approaches for the functionalization of CNTs and additional processing of composites containing CNTs. Details about treatment of nanotubes (purification, oxidation, and functionalization) were discussed in Section 3.3, while the processing of nanocomposites is the subject of Chapter 6. As we shall see, even before the preparation of nanocomposites it is possible to estimate the effect of the addition of CNTs in mechanical properties as well as electrical and thermal conductivities. This will be the topic of the following sections.

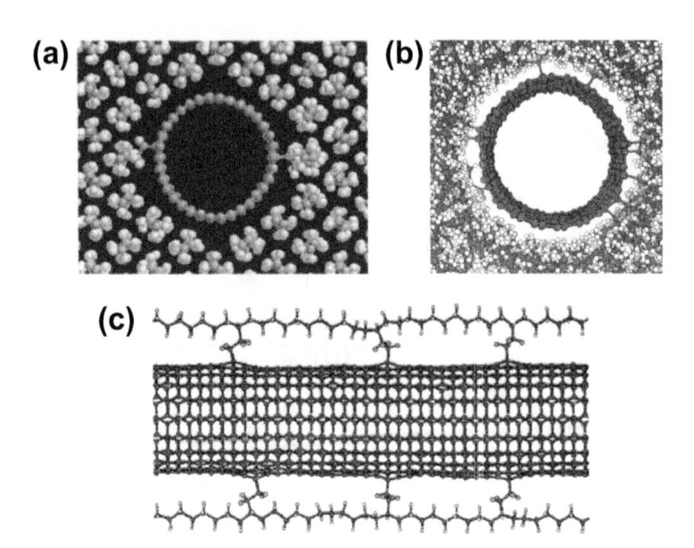

FIGURE 5.2 Simulation results illustrating molecular chemical bonds between nanotubes and matrix: crystalline matrix (a, c) and amorphous matrix (b).

Figure sourced from Molecular simulation of the influence of chemical cross-links on the shear strength of carbon nanotube-polymer interfaces [3]

5.2 Mechanical properties of composites containing CNTs

CNTs are regarded as the most promising reinforcement for improving the mechanical properties of polymeric matrices. In Chapter 2, the properties of various fibers used to reinforce composite materials have been presented. It was clear that carbon fibers (CFs) are a great option for making materials stronger and lighter. However, the mechanical properties of CNTs introduced in Chapter 3 indicate that nanotubes are a very attractive alternative for use in composites. Table 5.1 shows the mechanical properties of various polymers.

The potential of CNT to improve the mechanical properties of composites can be investigated using different theoretical models. Before presenting these models, it is important to introduce the concept of critical aspect ratio and briefly discuss the relationship between density and mass and volumetric fractions of nanotubes.

5.2.1 Critical length and aspect ratio

In Section 2.5 we learned that the length of conventional fibers used in composites has great influence on the ultimate strength and stiffness of the material. In fact, in order for an effective load transfer from the matrix phase to the reinforcement phase to take place, it is necessary that the fiber length exceeds the critical length l_c. The same criterion applies to CNT-reinforced composites where the critical length, l_c is, in principle, represented by

Table 5.1 Mechanical Properties of Polymers and Density

Polymer	ρ (g/cm^3)	E (GPa)	σ_m (MPa)	ε (%)
Acrylonitrile butadiene styrene (ABS)	1.03–1.06	2–2.8	30–50	15–30
Styrene-butadiene rubber (SBR)	0.93–1	1–2	1.4–2.8	450–600
Styrene-acrylonitrile copolymer (SAN)	1.07–1.09	3.4–3.7	55–75	2–5
Epoxy	1.2–1.3	2.8–4.2	55–130	
Poly(*p*-phenylene sulfide) (PPS)	1.35	3.6	65–75	1–2
Polycaprolactone (PCL) (Nylon 6)	1.13	3	80	50–120
Poly(*cis*-isoprene)	0.93	1–2	17–30	650–900
Polyether ether ketone (PEEK)	1.32	3.6	90–200	50
Poly(hexamethylene adipamide) (Nylon 6,6)	1.14	3.4	75–90	20
Polybutylene terephthalate (PBT)	1.31	2.3–2.5	50–60	120–200
Polycarbonate (PC)	1.2	2.1–2.4	70–90	100–120
Polyvinyl chloride (PVC)	1.32–1.58	1–3.5	40–75	30–80
Polyester (thermoset)	1.04–1.46	2–4.5	30–40	1.5–2.5
Polystyrene (PS)	1.04–1.05	2.4–3.2	30–60	1–4
Polyethersulfone (PES)	1.37	2.5	80–90	40–80
High density polyethylene (HDPE)	0.94–0.97	0.7–1.4	20–40	100–1000
Low density polyethylene (LDPE)	0.915–0.93	0.14–0.3	7–17	200–900
Ultra high molecular weight Polyethylene (UHMWPE)	0.93–0.94	0.1–0.7	20–40	200–500
Poly(methyl methacrylate) (PMMA)	1.17–1.20	2.5–3.3	55–75	3–5
Polypropylene (PP)	0.90–0.91	1.1–2	30–40	100–600
Polyethylene terephthalate (PET)	1.29–1.40	3	50	50–300
Polytetrafluoroethylene (PTFE)	2.15–2.20	0.41	7–30	200–400

ρ: Density; E: Young's modulus; σ_m: Tensile strength; ε: Elongation at break.
sourced from Physical Properties of Polymers Handbook *[4]*

$$l_c = \frac{\sigma_{\mathrm{CNT}} d}{2\tau_{\mathrm{CNT}}} \tag{5.1}$$

where

σ_{CNT} is the ultimate strength of the CNTs;
d is the outer diameter of the CNTs;
τ_{CNT} is the interfacial shear strength between nanotubes and matrix.

Equation (5.1) considers the nanotube as a rigid fiber and not tubular (hollow), making their applicability questionable. To take into account the effect of the tubular structure of CNTs [5], a model based on the of Kelly–Tyson approximation [6] was proposed:

$$l_c = \frac{\sigma_{\mathrm{CNT}} d}{2\tau_{\mathrm{CNT}}}\left(1 - \frac{d_i^2}{d^2}\right) \tag{5.2}$$

where

d_i is the internal diameter of the CNT.

According to the model in Eqn (5.2), CNTs can reach their ultimate strength under tension when their aspect ratio (α) is worth at least

$$\alpha_c = \frac{l_c}{d} \tag{5.3}$$

where

α_c is the critical aspect ratio of the CNTs.

Experimental results show that for multiwalled CNTs (MWCNTs), l_c ranges from 50 to 500 nm [7] and the maximum resistance ranges between 20 and 150 GPa [8–12].

Assuming that for single-walled CNTs (SWCNTs) $\sigma_{\mathrm{CNT}} = 50$ GPa and its outer and inner diameter satisfy the relation $d = d_{\mathrm{i}} + 0.68$ nm, we can analyze the effect of l_c and d in the τ_{CNT}. We can also compare Eqn (5.1) and the model proposed in Eqn (5.2). The findings are plotted in Figure 5.3.

Although the model proposed in Eqn (5.2) requires experimental demonstration, its importance is clear. While Eqn (5.1) overestimates the shear strength (τ_{CNT} increases linearly with the diameter of the nanotube), Eqn (5.2) shows that τ_{CNT} does not vary with the diameter of CNTs for diameters greater than 2–3 nm.

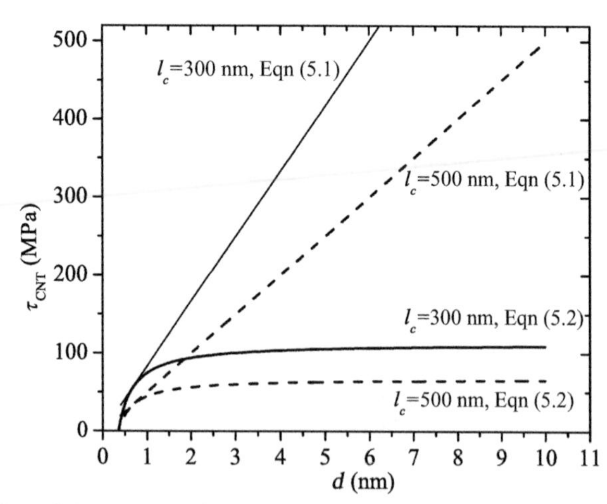

FIGURE 5.3 Effect of the external diameter of single-walled carbon nanotubes (SWCNTs) in the interfacial shear strength resistance, τ_{CNT}, according to Eqns (5.1) and (5.2). Simulations performed considering (a) $\sigma_{\mathrm{CNT}} = 50$ GPa and (b) the wall thickness of SWCNTs = $(d - d_i)/2 = 0.34$ nm.

Considering an $l_c = 100$ nm and a maximum value of $\sigma_{CNT} = 150$ GPa, we ascertain that $\tau_{CNT} = 980$ MPa. This value is far superior to the shear strength measured for epoxy composites reinforced with CF, where τ ranges from 50 to 100 MPa [5].

5.2.2 Density, volumetric concentration, and fraction of CNTs

One of the most important factors for modeling composites is the volume fraction (V_f) of the reinforcement phase. However, during the preparation of composites reinforced with CNTs in laboratories, it is much easier to deal with the weight concentration ($wt_{CNT}\%$ or wt%). Thus it is often necessary to convert mass fraction to volume fraction. To do this one must know the density of the polymer matrix and CNTs selected. Table 5.2 summarizes the different notation used to represent the amount of CNTs in the composite.

For CNT-reinforced composites, the volume fraction can be calculated based on the density of the constituents according to

$$V_{CNT} = \frac{\rho_C}{\rho_{CNT}} wt_{CNT} \tag{5.4}$$

$$\rho_C = \rho_{CNT} V_{CNT} + \rho_m V_m \tag{5.5}$$

where ρ_{CNT}, V_{CNT}, and wt_{CNT} represent the density, volume, and weight fraction of CNTs, respectively;

ρ_c is the density of the composite; and
ρ_m and V_m are the density and volume fraction of the matrix. Note that $V_m = 1 - V_{CNT}$.

Substituting Eqn (5.5) in Eqn (5.4), we find an equation for converting weight fraction to volume fraction that depends on the densities of the constituents

$$V_{CNT} = \left(\frac{\rho_r}{wt_{CNT}} - \rho_r + 1 \right)^{-1} \tag{5.6}$$

where

$\rho_r = \frac{\rho_{CNT}}{\rho_m}$ is the ratio between the densities of the CNTs and the matrix.

Table 5.2 Notation Used to Represent Fractions and Concentrations of CNTs

Quantity	Representation	Unit	Domain
Volume fraction	V_{CNT}	–	(0.1)
Volume concentration	$V_{CNT}\%$ or vol%	%	(0.100)
Weight fraction	wt_{CNT}	–	(0.1)
Weight concentration	$wt_{CNT}\%$ or wt%	%	(0.100)

EXAMPLE 5.1

An epoxy matrix composite is manufactured by addition of 0.01, 0.1, and 0.5 wt% of MWCNTs. Knowing that the density of the epoxy and CNTs are 1.25 and 1.50 g/cm^3, respectively, calculate the volume fraction of CNTs in the composites.

Solution

The weight fraction of CNTs ($wt_{CNT} = wt_{CNT}\%/100$) can be converted to volume fraction (V_{CNT}) applying Eqn (5.6). The values of the variables needed were provided in the problem, so for $wt_{CNT}\% = 0.01$ wt% we have

$$V_f = \left(\frac{(1.50\text{g/cm}^3/1.25\text{g/cm}^3)}{(0.01/100)} - (1.50\text{g/cm}^3/1.25\text{g/cm}^3) + 1 \right)^{-1}$$

$$= 0.000083$$

Repeating the same procedure for $wt_{CNT}\% = 0.1$ wt% and $wt_{CNT}\% = 0.5$ wt% we obtain $V_f = 0.00083$ and $V_f = 0.0042$, respectively.

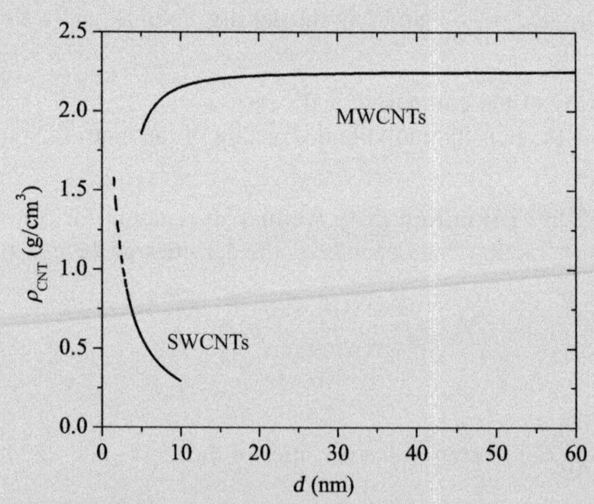

FIGURE 5.4 Effect of density in the outer diameter of single-walled carbon nanotubes (SWCNTs) and multiwalled carbon nanotubes (MWCNTs) according to Eqn (5.7). Simulations considering SWCNTs with d_i of at least 0.82 nm up to 10 nm, the wall thickness of SWCNTs = $(d - d_i)/2$ = 0.34 nm, and MWCNTs with d_i minimum of 2.1 nm and d maximum of 60 nm.

The density of CNTs can be calculated assuming that the walls of the CNTs have the density of graphite ($\rho_g = 2.25$ g/cm^3). Thus, taking into account the tubular structure of the nanotubes, we have [13]

$$\rho_{CNT} = \rho_g \left(\frac{d^2 - d_i^2}{d^2} \right) \tag{5.7}$$

As shown in Figure 5.4, the density of SWCNTs decreases with the increase in diameter whereas the density of MWCNTs increases with the number of walls (diameter) reaching an approximately constant value after $d \sim 20$ nm.

EXAMPLE 5.2

A sample of CNTs comprises nanotubes with length $l = 10$ μm and outer and inner diameter of 4 nm and 16 nm, respectively. Calculate the density of CNTs.

Solution

The density of CNTs can be obtained using Eqn (5.7). Hence, assuming the density of graphite as $\rho_g = 2.25$ g/cm^3, we have

$$\rho_{CNT} = 2.25\text{g/cm}^3 \left(\frac{(16\text{nm})^2 - (4\text{nm})^2}{(16\text{nm})^2} \right) = 2.11\text{g/cm}^3$$

Due to their structure and presence of various walls, MWCNTs' density is generally higher than that of SWCNTs and may approach the density of graphite.

5.2.3 Elastic modulus of nanocomposites

There are several theoretical models to estimate the effect of CNT addition on the properties of composites. Some of these models are based only on the properties of the constituent materials and their volume fraction while others consider the geometry of the reinforcement and even its packaging in the polymer matrix [14]. Various models proposed for the calculation of the elastic modulus of composites containing nanotubes are presented as follows. Some of these models do not apply to fiber composites or have several limitations that prevent their use; however, some of these models are presented, for didactic purposes, in order to clarify the need for more realistic or appropriate models.

5.2.4 The Einstein equation

One of the first theories for composite materials has been developed for reinforced elastomers and is based on Einstein's equation [15]. According to the Einstein's equation, the viscosity of a suspension of rigid spherical inclusions, at low concentrations, is given by

$$\eta_c = \eta_m(1 + K_E V_p) \tag{5.8}$$

where

η_c and η_m is the viscosity of the suspension (composite) and the matrix, respectively;
K_E is the Einstein coefficient; and
V_p is the volume fraction of particles.

For spherical particles, $K_E = 2.5$. It is assumed that Eqn (5.8) represents the changes in the composite module ($\eta_c/\eta_m = E_c/E_m$) [16,17]; so for the Young's modulus, this equation can be rewritten as

$$E_c = E_m(1 + K_E V_p) \tag{5.9}$$

where

E_c and E_m are the Young's modulus of the composite and matrix phase, respectively.

As can be seen in Eqn (5.9) the reinforcement effect of the filler is independent of its size and its own rigidity. Only their volume fraction is considered. Also this equation is usable only for low concentrations of fillers because with high concentrations, the interaction between fillers takes place and this effect is not considered.

Several modifications of Eqn (5.9) have been proposed to try to define and consider the interaction between fillers in a composite [18,19]. Guth [20] generalized Eqn (5.9) by introducing an interaction term inter-particles:

$$E_c = E_m\left(1 + K_E V_p + 14.1 V_p^2\right) \tag{5.10}$$

This equation is applicable only to elastomers reinforced with spherical inclusions. Figure 5.5 compares the Einstein model (Eqn (5.9)) and the Guth model (Eqn (5.10)).

For nonspherical particles, Eqn (5.10) was modified by Guth, taking into account the aspect ratio of the filler ($\alpha = l/d$)

$$E_c = E_m\left(1 + 0.67\alpha V_f + 1.62\alpha^2 V_f^2\right)\ p \gg 1 \tag{5.11}$$

where

V_f is the volume fraction of fillers.

Figure 5.6 shows the effect of varying the aspect ratio of the filler, i.e., size, on the stiffness of a composite. Note that fixing the volume fraction of particles, the modulus increases dramatically with increasing aspect ratio.

Several other empirical models have been proposed for the calculation of E_c, many of which have variables used to fit experimental curves. The models presented to this point considered that the modulus of the reinforcement was far greater than that of the matrix (elastomers). The following models are applicable to rigid matrices.

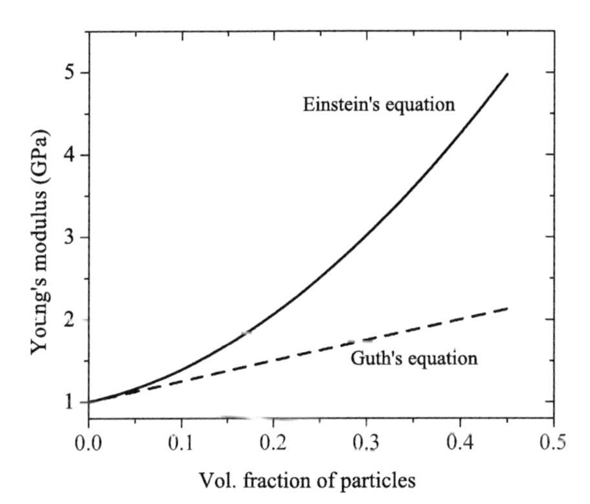

FIGURE 5.5 Comparison between the Einstein and Guth models. During simulations it was considered a polymeric matrix with modulus of 1 GPa.

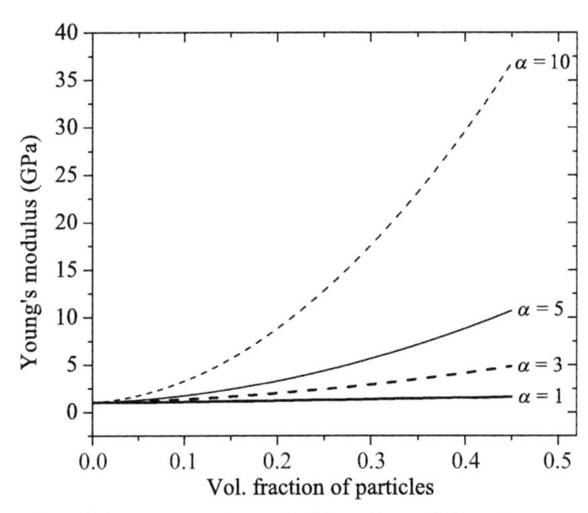

FIGURE 5.6 Effect of particle aspect ratio on the Young's modulus of a composite according to the modified Guth model (Eqn (5.11)). During simulations a polymeric matrix with modulus equal to 1 GPa was considered.

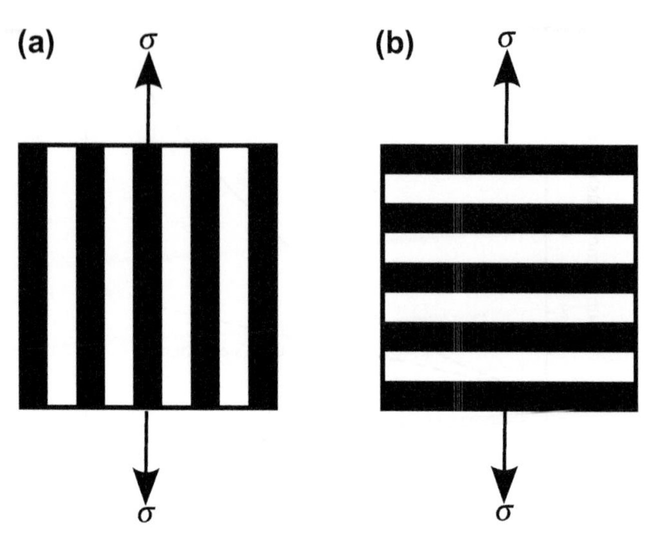

FIGURE 5.7 Possible arrangements of the two phases in a composite: (a) parallel and (b) series.

5.2.5 The series and parallel models

For a two-phase material, the simplest way to predict the Young's modulus is through the series and parallel models based on the rule of mixtures. The composite is then described by a two-phase model. The possible arrangements of these phases (series or parallel) are shown in Figure 5.7.

In the series model the tension is assumed to be uniform for both phases, whereas in the parallel model deformation is considered constant. According to these models, the Young's modulus should lie between an upper limit given by the model in parallel (known as the Voigt model)

$$E_c = E_m V_m + E_f V_f \tag{5.12}$$

and a lower limit given by the series model (known as Reuss model)

$$E_c = \frac{E_f E_m}{E_f V_m + E_m V_f} \tag{5.13}$$

where E_f and V_f are the Young's modulus and the volume fraction of filler, respectively.

Equation (5.12) emphasizes the importance of using fibers with high stiffness aligned along the direction of the load applied to the composite.

In general, the above models can be applied to particulate composites, whereas similar equations have been derived for calculating the longitudinal and transversal modulus of composites reinforced with long fibers. The series and parallel models assume that the individual phases of the composite are under uniform tension and

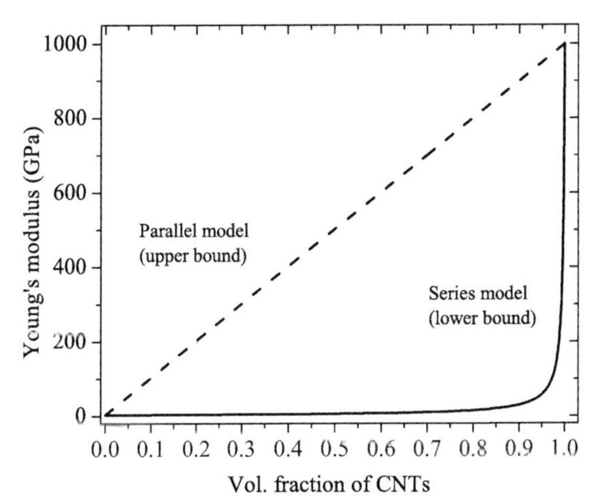

FIGURE 5.8 Effect of addition of carbon nanotubes (CNTs) on the Young's modulus of an epoxy matrix composite in accordance with the series and parallel models. The simulations were conducted considering epoxy and nanotubes with a Young's modulus of 3 and 1000 GPa, respectively.

deformation. In practice, at the microscopic level, the filler cannot be completely separated from each other and consideration of uniform stress/strain is a simplification. Nevertheless, this model has become popular because of its simplicity.

The series and parallel models have been used to estimate the Young's modulus of composites containing CNTs. The results obtained with these models give an idea of the lower (Eqn (5.13)) and upper limits (Eqn (5.12)) for the modulus of the composite. In this case the results obtained using the parallel model can be interpreted as the maximum possible value for the composite material, and this implies nanotubes aligned in a single direction. At the other extreme, due to difficulties during processing, the modulus obtained for composites can be even lower than the calculated using the series model. In fact, in cases where CNTs are extremely agglomerated, the module may be lower than that of the polymer matrix itself. This is because agglomerates of CNTs act as stress concentration points and consequently the composite may fail prematurely.

Figure 5.8 shows an example of application of the series and parallel models to an epoxy matrix composite reinforced with CNTs.

5.2.6 Cox model

The model proposed by Cox [21,22], a modification of the rule of mixtures, emphasizes the importance of fillers with high aspect ratio through a length efficiency factor (η_l). For composites with preferably aligned fibers, the modulus of elasticity is given by

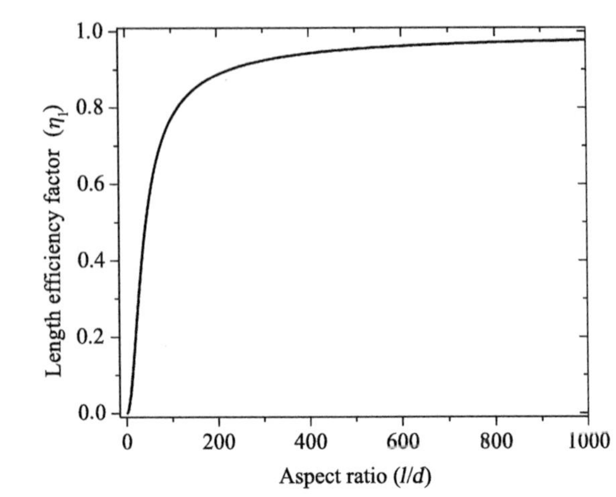

FIGURE 5.9 Variation of the length efficiency factor η_l as a function of the aspect ratio of carbon nanotubes (CNTs). The simulations were conducted assuming an epoxy matrix composite ($E_m = 3$ GPa) reinforced with 10 vol% of CNTs ($E_f = 1000$ GPa).

$$E_c = (\eta_l E_f - E_m) V_f + E_m \tag{5.14}$$

$$\eta_l = 1 - \frac{\tan h(al/d)}{al/d} \tag{5.15}$$

$$a = \sqrt{\frac{-3E_m}{2E_f \ln(V_f)}} \tag{5.16}$$

The length efficiency factor η_l increases with the aspect ratio of the filler and approaches 1 for large aspect ratios (Figure 5.9). This fact highlights the importance of using fillers with high aspect ratio.

In the case of composites with fibers randomly aligned, Eqn (5.14) can be written in the general form

$$E_c = (\eta_0 \eta_l E_f - E_m) V_f + E_m \tag{5.17}$$

where

η_0 is the orientation efficiency factor: $\eta_0 = 1$ for aligned fibers; $\eta_0 = 3/8$ for fibers aligned in the plane, and $\eta_0 = 1/5$ for randomly oriented fibers [23]. The Cox model is often used to calculate the modulus of CNT-reinforced composites. In Figure 5.10 we have an example where the Cox model is applied to predict the modulus of a composite with CNTs. Note how the stiffness of the composite is increased by the alignment of the CNTs.

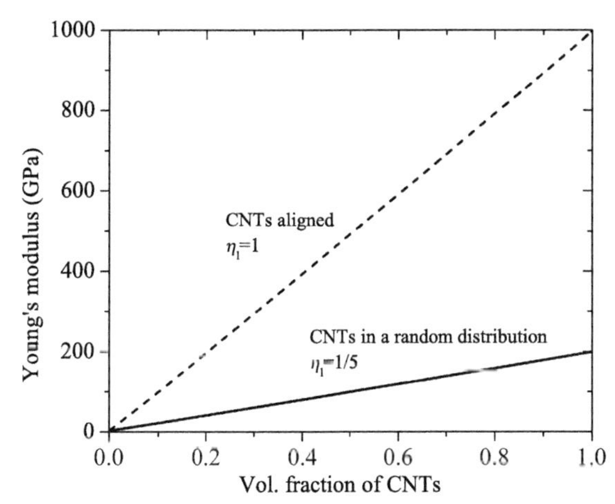

IGURE 5.10 Effect of the addition of aligned carbon nanotubes (CNTs) on the Young's odulus of an epoxy matrix composite according to the Cox model. Curves are shown for a omposite with aligned CNTs and CNTs in a random distribution. The simulations were onducted considering $E_m = 3$ GPa, $E_f = 1000$ GPa, $l = 10$ μm, and $d = 1$ nm.

.2.7 Halpin–Tsai model

'he Halpin–Tsai model equations [24,25] have been used for a long time to predict he properties of composites reinforced with short fibers. These equations were riginally developed for long-fiber composites [26].

For composites with aligned short fibers, the Halpin–Tsai model describes the odulus of elasticity as

$$E_c = E_m\left(\frac{1+\zeta\eta V_f}{1-\eta V_f}\right) \tag{5.18}$$

$$\eta = \frac{E_f/E_m - 1}{E_f/E_m + \zeta} \tag{5.19}$$

is a parameter that depends on the geometry of the filler as follows:

$\zeta = 2l/d$ for calculation of the longitudinal modulus;
$\zeta = 2$ for calculation of the transversal modulus.

When the value of ζ becomes very small ($\zeta \rightarrow 0$), the Halpin–Tsai model reduces ɔ the series model.

$$E_c = \frac{E_f E_m}{E_f V_m + E_m V_f} \tag{5.20}$$

For composites with very large ζ values ($\zeta \rightarrow \infty$), the Halpin–Tsai model reduces to the parallel model

$$E_c = E_m V_m + E_f V_f \tag{5.21}$$

For composites reinforced with fibers randomly oriented, the Halpin–Tsai model equations are more elaborate and take the form

$$E_c = E_m \left[\frac{3}{8} \left(\frac{1 + \zeta \eta_L V_f}{1 - \eta_L V_f} \right) + \frac{5}{8} \left(\frac{1 + 2\eta_T V_f}{1 - \eta_T V_f} \right) \right] \tag{5.22}$$

where

$$\eta_L = \frac{E_f/E_m - 1}{E_f/E_m + 2l/d} \tag{5.23}$$

$$\eta_T = \frac{E_f/E_m - 1}{E_f/E_m + 2} \tag{5.24}$$

$$\zeta = 2l/d \tag{5.25}$$

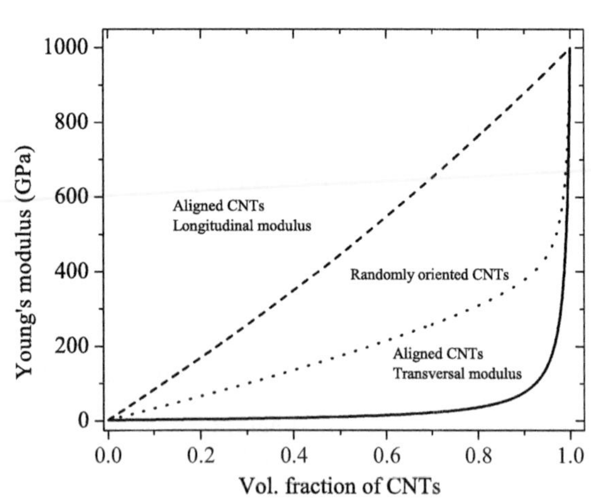

FIGURE 5.11 Young's modulus of an epoxy/carbon nanotube (CNT) composite according to the Halpin–Tsai model. Curves are shown for the modulus of a composite with aligned CNTs and CNTs with random distribution. The simulations were conducted considering $E_m = 3$ GPa, $E_f = 1000$ GPa, $l = 10$ μm, and $d = 15$ nm.

The Halpin–Tsai model is known to fit some experimental results at low-volume fractions and underestimate the values for high-volume fractions of filler [27]. This fact motivated the creation of Lewis and Nielsen model, which will be discussed later.

Equations (5.18) and (5.22) can be used to estimate the modulus of reinforced composites. Figure 5.11 shows the modulus of a CNT-reinforced composite according to the Halpin–Tsai model. The curves show the modulus for the case of aligned and randomly oriented CNTs.

EXAMPLE 5.3

A composite with a vinyl ester (VE) matrix contains 2 vol% CNTs with the following characteristics: $l = 7$ μm, $d = 50$ nm, $d_i = 4$ nm, and $E_{CNT} = 900$ GPa. The modulus of the VE matrix is 1.9 GPa.

1. Estimate the maximum modulus of elasticity for this composite.
2. What is the magnitude of the composite's modulus according to the Halpin–Tsai model? Assume that the CNTs are in a random distribution.

Solution

1. The maximum modulus of elasticity is achieved through the parallel model (Voigt)

$$E_c = E_m V_m + E_f V_f$$

Note that when applying the above equation, the quantity of CNTs should be considered as the volume fraction (0–1) and not volume concentration (0–100%). In this case 2 vol% is equivalent to a volume fraction of $2/100 = 0.02$. Thus, the maximum modulus of elasticity is

$$E_c = (1.9\text{GPa})(1 - 0.02) + (900\text{GPa})(0.02) = 36.62\text{GPa}$$

2. Unlike the series model, the Halpin–Tsai model considers the aspect ratio of the filler during the calculation of the elastic modulus. Applying the values given in the problem we get:

$$\eta_L = \frac{E_f/E_m - 1}{E_f/E_m + 2l/d} = \frac{(900\text{GPa}/1.9\text{GPa}) - 1}{(900\text{GPa}/1.9\text{GPa}) + 2(7 \times 10^{-6}/50 \times 10^{-9}\text{m})} = 0.63$$

similarly

$$\eta_T = \frac{E_f/E_m - 1}{E_f/E_m + 2} = \frac{(900\text{GPa}/1.9\text{GPa}) - 1}{(900\text{GPa}/1.9\text{GPa}) + 2} = 0.99$$

Therefore the module is

$$E_c = E_m\left[\frac{3}{8}\left(\frac{1 + \zeta\eta_L V_f}{1 - \eta_L V_f}\right) + \frac{5}{8}\left(\frac{1 + 2\eta_T V_f}{1 - \eta_T V_f}\right)\right]$$

$$E_c = 1.9\text{GPa}\left[\frac{3}{8}\left(\frac{1+2(7\times10^{-6}/50\times10^{-9}\text{m})(0.63)(0.02)}{1-(0.63)(0.02)}\right)\right.$$

$$\left.+\frac{5}{8}\left(\frac{1+2(0.99)(0.02)}{1-(0.99)(0.02)}\right)\right]$$

$$E_c = 4.53\text{GPa}$$

Note that the value obtained above is eight times lower than that calculated in part (a). This difference is mainly due to the different orientation of the CNTs. The parallel model provided the longitudinal modulus; (CNTs aligned), while in the calculation with the Halpin–Tsai model a random distribution was assumed.

According to the Halpin—Tsai model, one way to increase the modulus of composites is by increasing the aspect ratio of the filler used. Figure 5.12 shows how the aspect ratio of the nanotubes influences the modulus of the composite. The curves depict composite materials with three different volume fractions of CNTs.

5.2.8 Nielsen model

Nielsen derived a macroscopic model that is considered the most versatile for composites reinforced with short fibers and particles. The model considers the properties

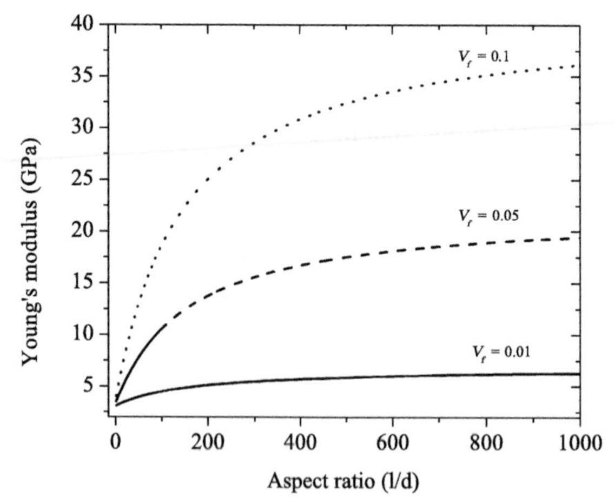

FIGURE 5.12 Young's modulus of an epoxy/carbon nanotube (CNT) composite, according to the Halpin—Tsai model. A random distribution of CNTs is assumed. The simulations were conducted considering E_m = 3 GPa and E_f = 1000 GPa.

and concentrations of the constituents. As in the Halpin–Tsai model, the aspect ratio of the filler is included and the difference is that the Nielsen model considers the packing of the filler. The Nielsen model, a modification of the Halpin–Tsai model, is given by

$$E_c = E_m\left(\frac{1 + A\eta_N V_f}{1 - \psi\eta_N V_f}\right) \tag{5.26}$$

where

$$A = k_E - 1 \tag{5.27}$$

$$\eta_N = \frac{E_f/E_m - 1}{E_f/E_m + A} \tag{5.28}$$

$$\psi = 1 + \left(\frac{1 - \phi_{max}}{\phi_{max}^2}\right)V_f \tag{5.29}$$

where ϕ_{max} is the maximum packing fraction of filler.

The parameter A in Eqn (5.27) is related to the Einstein coefficient, which is a function of the aspect ratio and orientation of the filler, while ϕ_{max} is based on the shape of the filler (sphere, tube, etc.) and packing. In Eqn (5.29), factor ψ is related to the maximum filler packing. Nielsen proposed an alternative to Eqn (5.29) for the calculation of ψ:

$$\psi = \frac{1}{V_f}\left[1 - e^{\left(\frac{-V_f}{1 - \frac{V_f}{\phi_{max}}}\right)}\right] \tag{5.30}$$

To apply the Nielsen model it is necessary to know the values of A and ϕ_{max}. Tables 5.3 and 5.4 present these values for different systems.

Table 5.3 Values for *A* for Nielsen's Model

Filler Type	Distribution	Aspect Ratio	A
Cubes	–	1	2
Spheres	–	1	1.5
Fibers	Random	2	1.58
Fibers	Random	4	2.08
Fibers	Random	6	2.80
Fibers	Random	10	4.93
Fibers	Random	15	8.38
Fibers	Uniaxially oriented	–	2 (*l*/*d*)

sourced from Tensile Modulus Modeling *[28]*

Table 5.4 Values for Maximum Packing Fraction (ϕ_{max})

Filler Shape	Type of Packing	ϕ_{max}
Spheres	Hexagonal close	0.7405
Spheres	Face centered cubic	0.7405
Spheres	Body centered cubic	0.60
Spheres	Simple cubic	0.524
Spheres	Random loose	0.601
Spheres	Random close	0.637
Irregular particles	Random close	~0.637
Fibers	Three-dimensional random	0.52
Fibers	Uniaxial hexagonal close	0.907
Fibers	Uniaxial simple cubic	0.785
Fibers	Uniaxial random	0.82

Figure 5.13 shows the variation of the Young's modulus of the composites as a function of the concentration of nanotubes according to the Nielsen model. Simulated are composites with CNTs aligned and CNTs randomly oriented, and calculations are performed using the two equations put forward by Nielsen. Note in Figure 5.13 that the maximum concentration of CNTs does not exceed the value

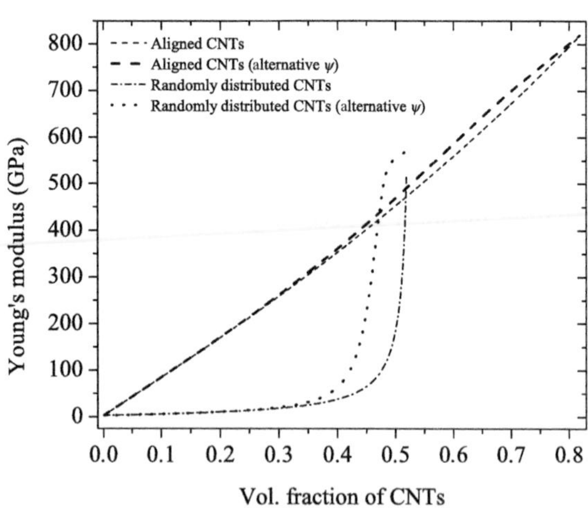

FIGURE 5.13 Young's modulus of an epoxy/carbon nanotube (CNT) composite according to the Nielsen model. Shown are curves for the modulus of a composite with aligned and randomly distributed CNTs. The value of ψ was calculated using Eqn (5.29) and Eqn (5.30) (alternative ψ). The simulations were conducted considering $E_m = 3$ GPa, $E_f = 1000$ GPa, $l = 10$ μm, and $d = 15$ nm. For aligned CNTs, $A = 2\ l/d$ and $\phi_{max} = 0.82$. For CNTs in random distribution, $A = 8.38$ and $\phi_{max} = 0.52$.

of the maximum packing fraction (ϕ_{max}). In addition, note that the results obtained using Eqns (5.29) and (5.30) are similar.

5.2.9 Tensile strength of nanocomposites

Similar to the Young's modulus, the tensile strength of composites can also be estimated using appropriate equations. The model used varies according to the length of the fiber, i.e., CNTs.

For composites with long fibers, where $l > \sim 10l_c$ the tensile strength is described by

$$\sigma_c = \sigma_m + (\sigma_f - \sigma_m)V_f \tag{5.31}$$

where σ_c, σ_m, and σ_f are the tensile strength of the composite, matrix, and fiber, respectively.

Equation (5.31) is similar to the rule of mixtures. The tensile strength decreases with fiber length. Thus, for short fibers with medium length ($l > l_c$), the tensile strength is given by

$$\sigma_c = \sigma_m + (\eta_s\sigma_f - \sigma_m)V_f \tag{5.32}$$

$$\eta_s = 1 - l_c/2l \tag{5.33}$$

where η_s is the strength efficiency factor.

In the case of $l > l_c$, fiber fractures when high load is applied to the composite. However, when $l < l_c$, it is not possible to transfer enough load from the matrix phase to the fibers and so the composite fails with the fibers being pulled out of the matrix. In this situation, the tensile strength can be described by

$$\sigma_c = \sigma_m + (\tau_{CNT}l/d - \sigma_m)V_f \tag{5.34}$$

where

τ_{CNT} is the interfacial shear strength between fiber (nanotube) and matrix and
d is the outer diameter of the fiber.

Figure 5.14 shows the tensile strength of CNT-reinforced composites, where $l > \sim 10l_c$, $l > l_c$, $l < l_c$. Notice how the increase in length of CNTs reflects an increase in tensile strength of the composites.

EXAMPLE 5.4

Scientists in a laboratory plan to increase the tensile strength of a polymer whose current strength is $\sigma_m = 60$ MPa. Three grades of CNT with different lengths are available: CNT-1, $l_1 = 25$ μm; CNT-2, $l_2 = 4$ μm; and CNT-3, $l_3 = 1$ μm. CNTs will be used in a volume fraction equal to 0.05. Knowing that $\sigma_{NTC} = 50$ GPa, $d = 16$ nm, $d_i = 4$ nm, and $\tau_{CNT} = 150$ MPa, find out which grade of CNT is the most promising for the experiment.

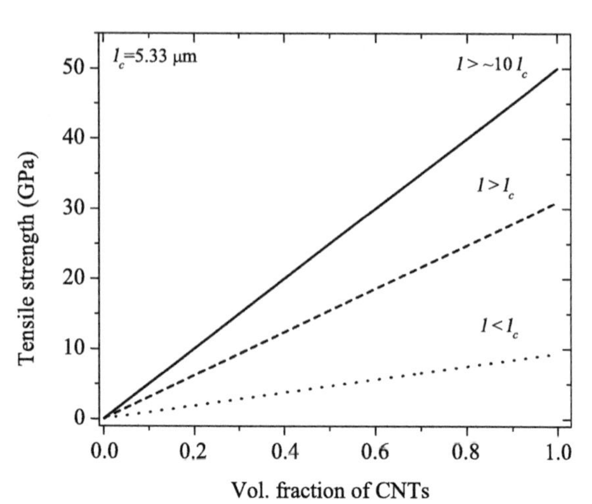

FIGURE 5.14 Tensile strength of composites with carbon nanotubes (CNTs) of different lengths. Curves are shown for composites where $l > \sim 10 l_c$, $l > l_c$, and $l < l_c$. The simulations were conducted assuming $\sigma_m = 60$ MPa, $\sigma_{CNT} = 50$ GPa, $l_c = 5.33$ μm, $d = 16$ nm, and $\tau_{CNT} = 150$ MPa.

Solution

The tensile strength of the composite will vary with the length of the CNTs. Three equations can be used to calculate the tensile strength of the composite and the appropriate choice is based on the critical length of the CNTs (l_c). The l_c value can be obtained through the equation

$$l_c = \frac{\sigma_{CNT} d}{2\tau_{CNT}}\left(1 - \frac{d_i^2}{d^2}\right) = \frac{(50 \times 10^9 \text{Pa})(16 \times 10^{-9}\text{m})}{2(150 \times 10^6 \text{Pa})}\left(1 - \frac{(4 \times 10^{-9}\text{m})^2}{(16 \times 10^{-9}\text{m})^2}\right)$$
$$= 2.51 \mu\text{m}$$

Based on the value obtained earlier we can select the appropriate equation for each type of CNT:

$$l_1 > \sim 10 l_c \rightarrow \sigma_{c1} = \sigma_m + (\sigma_f - \sigma_m) V_f$$
$$l_2 > l_c \rightarrow \sigma_{c2} = \sigma_m + ((1 - l_c/2l)\sigma_f - \sigma_m) V_f$$
$$l_3 < l_c \rightarrow \sigma_{c3} = \sigma_m + (\tau_{CNT} l/d - \sigma_m) V_f$$

The resistance of the composites is then:

$$\sigma_{c1} = 60 \times 10^6 \text{Pa} + (50 \times 10^9 \text{Pa} - 60 \times 10^6 \text{Pa})(0.05) = 2.56 \text{GPa}$$

$$\sigma_{c2} = 60 \times 10^6\text{Pa} + \left[\left(1 - \left(\frac{2.51\,\mu\text{m}}{2.4\mu\text{m}}\right)\right)(50 \times 10^9\text{Pa}) - (60 \times 10^6\text{Pa})\right](0.05)$$
$$= 1.77\text{GPa}$$

$$\sigma_{c3} = 60 \times 10^6\text{Pa} + \left[\left(\frac{(150 \times 10^6\text{Pa})(1 \times 10^{-6}\text{m})}{16 \times 10^{-9}\text{m}}\right) - (60 \times 10^6\text{Pa})\right](0.05)$$
$$= 0.53\text{GPa}$$

As we could ascertain, the grade of nanotubes CNT-1 is the most suitable to increase the tensile strength of the polymer matrix and the enhancement could be of up to 400%. It should be noted that the equation used to calculate σ_{c1} considers that the CNTs are well dispersed and interacting with the polymer matrix so that any load applied to the composite is effectively transferred to the reinforcement phase. In practice, CNTs tend to be agglomerated and the CNT–polymer interaction will not be perfect, and hence the value of 2.56 GPa should be seen as an upper limit.

5.2.10 Effective properties of CNTs and the modified Halpin–Tsai model

ome of the major differences between micro- and nanocomposites (reinforced with NTs) are the structure of the fibers themselves. In the case of composites reinforced vith MWCNTs, for example, it is necessary to consider how the structure of the nultilayer nanotube interacts with the polymer matrix. In fibrous composites, the oad transfer from the matrix to the reinforcing phase occurs through shear stress t the fiber polymer interface. As discussed earlier, the interaction between the arious concentric layers of MWCNTs arises from van der Waals interactions, nd is weak. As a result, the transfer of the load applied to the outer layer to the inner ayers of MWCNTs is not effective. Thus, the outer layer of the nanotube will withtand practically all the load transferred in the nanotube–matrix interface. For this eason, it is necessary to define the effective properties of CNTs, which take into account these characteristics of nanotubes. Next, we define the effective Young's nodulus and tensile strength of nanotubes dispersed in a matrix.

The calculation of the elastic properties of CNTs requires an accurate definition f the cross-sectional area to eliminate arbitrariness and allow comparison of ifferent types of nanotubes. Usually, the hollow part of the CNTs is ignored to alculate the cross-sectional area, since this area does not support any load. However, when CNTs are dispersed in a polymer matrix to form composites, nicromechanics-based models, used to predict their properties, consider the entire olume occupied by CNTs. Hence the need to determine the effective elastic

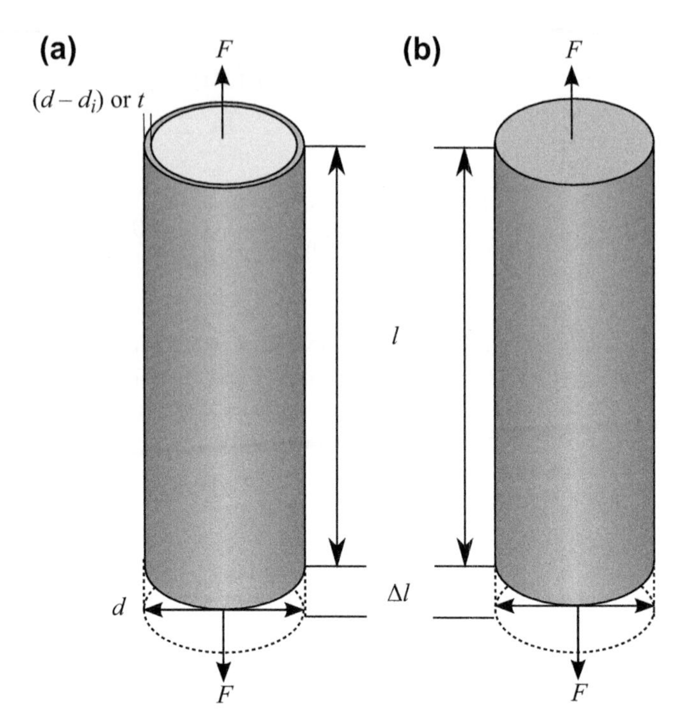

FIGURE 5.15 Schematic of a nanotube (a) and effective fiber (b) used for modeling the elastic properties of a nanotube embedded in a composite. *F* is the external force applied, *d* is the diameter of the fiber, $(d - d_i) = t$ is the wall thickness, and *l* is the length of the fiber.

Figure sourced from On the elastic properties of carbon nanotube-based composites [13]

properties of CNTs embedded in a composite, applying the ability of transfer from the outer layer of CNTs for all its cross-sectional area [13].

Thus, the Young's modulus of a nanotube with outer diameter d, an internal diameter d_i, and length l (Figure 5.15(a)) is modeled considering that the wall of the nanotube acts as an effective solid fiber (Figure 5.15(b)) with even deformation behavior, and even diameter (d) and length.

An external force $\boldsymbol{F}$ applied to the nanotube and effective fiber (f_{ef}) will result in a condition of isostrain

$$\varepsilon_{\text{CNT}} = \varepsilon_{f_{ef}} \tag{5.35}$$

Using Eqn (5.35), the elastic properties of the nanotube may be related to an effective fiber through

$$E_{f_{ef}} = \frac{\sigma_{f_{ef}}}{\sigma_{\text{CNT}}} E_{\text{CNT}} \tag{5.36}$$

As the external force applied is the same, the effective modulus can be expressed in terms of the ratio of the cross-sectional areas

$$E_{f_{ef}} = \frac{A_{\mathrm{CNT}}}{A_{f_{ef}}} = E_{\mathrm{CNT}} \tag{5.37}$$

where

$$A_{f_{ef}} = \frac{\pi d^2}{4} \tag{5.38}$$

and

$$A_{\mathrm{CNT}} = \pi \frac{\left(d^2 - d_i^2\right)}{d^2} \tag{5.39}$$

Substituting Eqns (5.38) and (5.39) in Eqn (5.37), we can express the modulus of the effective fiber as

$$E_{f_{ef}} = \frac{\left(d^2 - d_i^2\right)}{d^2} E_{\mathrm{CNT}} \tag{5.40}$$

Similar to the module, the tensile strength of the effective fiber $\sigma_{f_{ef}}$ can be related to the cross-sectional area of the nanotube by

$$\sigma_{f_{ef}} = \frac{4F}{\pi d^2} \tag{5.41}$$

$$\sigma_{\mathrm{CNT}} = \frac{4F}{\pi\left(d^2 - d_i^2\right)} \tag{5.42}$$

$$\sigma_{f_{ef}} = \frac{\left(d^2 - d_i^2\right)}{d^2} \sigma_{\mathrm{CNT}} \tag{5.43}$$

The effective Young's modulus and tensile strength for various types of nanotubes can be calculated using Eqns (5.40) and (5.43). Two cases can be considered for double-walled carbon nanotubes (DWCNTs) and MWCNTs: (1) only the outer wall supports the load and (2) all of the walls bear the applied load. For case (1), $d - d_i = t$ where t is the thickness of the outer layer of the nanotube and has value $t = 0.34$ nm. In this case the equations can be simplified and

$$A_{\mathrm{CNT}} = \pi(d - d_i)d \tag{5.44}$$

$$E_{f_{ef}} = 4\frac{(d - d_i)}{d} E_{\mathrm{CNT}} = \frac{4t}{d} E_{\mathrm{CNT}} \tag{5.45}$$

$$\sigma_{f_{ef}} = 4\frac{(d - d_i)}{d} \sigma_{\mathrm{CNT}} = \frac{4t}{d} \sigma_{\mathrm{CNT}} \tag{5.46}$$

Table 5.5 Effective Properties of Different CNTs: Only the External Walls Bear the Load

CNT Type	d (nm)	d_i (nm)	$E_{f_{ef}}$ (GPa)	$\sigma_{f_{ef}}$ (GPa)
SWCNTs	1.5	0.81	907	118
DWCNTs	2.8	2.1	486	63
MWCNTs	15	4	91	12

Table 5.6 Effective Properties of Different CNTs: All the Walls Bear the Load

CNT Type	d (nm)	d_i (nm)	$E_{f_{ef}}$ (GPa)	$\sigma_{f_{ef}}$ (GPa)
SWCNTs	1.5	0.81	907	118
DWCNTs	2.8	2.1	486	57
MWCNTs	15	4	929	120

For case (2) where all the nanotubes support the applied load, the effective properties must be calculated using Eqns (5.40) and (5.43).

Tables 5.5 and 5.6 provide mechanical properties of various CNTs. The values were obtained considering $E_{CNT} = 1$ TPa and $\sigma_{CNT} = 130$ GPa.

As can be seen from these results, the mechanical properties of CNTs are highly dependent on their structure and the interaction between external and internal layers. Since the Young's modulus of the effective fiber is calculated in the elastic region (where Hooke's law is obeyed), it is probable that all walls of the nanotube deform and then the amount calculated using Eqn (5.40) must be considered. In the case of the tensile strength, slipping is likely to occur between the inner walls before the final fracture of the nanotube, and thus the inner walls of DWCNTs and MWCNTs do not contribute fully for the resistance of the nanotube. For this reason, the value of tensile strength calculated using Eqn (5.46), or at least between this value and the values obtained using Eqn (5.43), must be considered.

Considering the effective modulus of CNTs due to their tubular structure, the parameters η_L and η_T (Eqns (5.23) and (5.24)) may be rewritten as

$$\eta'_L = \frac{(E_f/E_m) - (d/4t)}{(E_f/E_m) + (l/2t)} \tag{5.47}$$

$$\eta'_T = \frac{(E_f/E_m) - (d/4t)}{(E_f/E_m) + (d/2t)} \tag{5.48}$$

where $t = 0.34$ nm is the thickness of a graphite layer.

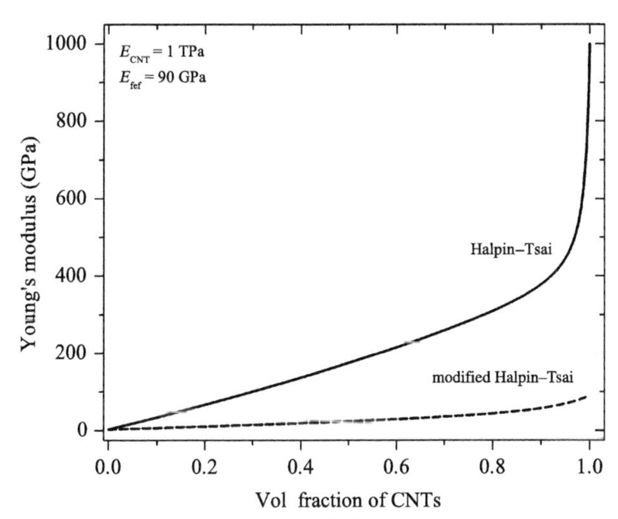

IGURE 5.16 Young's modulus of a composite reinforced with carbon nanotubes (CNTs) andomly distributed according to the Halpin–Tsai model and modified Halpin–Tsai model. imulations were carried out considering E_m = 3 GPa, E_f = 1000 GPa, l = 10 μm, and = 15 nm.

Thus the Halpin–Tsai model (Eqn (5.22)) can be adapted with Eqns (5.47) and 5.48) for the calculation of Young's modulus of composites containing nanotubes. 'he Halpin–Tsai model is compared with the modified model including the concept f effective fiber in Figure 5.16.

.2.11 Final considerations about the mechanical properties

ome of the basic considerations of the models used to predict the elastic properties f composites containing CNTs are as follows:

The fillers and the matrix are linearly elastic, the matrix is isotropic, and fillers are isotropic or transversely isotropic;
The fillers are identical in size and shape and are characterized by an aspect ratio α;
The fillers and the matrix are connected through an interface and remain connected during deformation. Thus, interfacial slip, rupture of the connection filler/matrix, and matrix microfracture are not considered.

However, when applying the models discussed, it should be borne in mind that:

. CNTs exhibit curvature and treating them as fully aligned fibers is an approximation;
. Hardly all CNTs are homogeneously dispersed in the matrix;
. Nanotube–nanotube interactions may occur;

4. The presence of CNTs can vary matrix properties such as crystallinity;
5. CNTs are characterized by a diameter and length distribution.

For all these reasons mentioned, the results obtained during the simulation of the effects of CNTs on the mechanical properties of composites should be interpreted with caution.

5.3 Thermal conductivity of composites containing CNTs

As shown in Table 5.7, most polymers have low thermal conductivity. Compared to the conductivity of metals such as aluminum, the thermal conductivity of polymers can be up to 1000 times lower. This is a consequence of its structure and the way the atoms are connected on these materials. Glass and organic fibers, such as Kevlar and polyethylene, and low-modulus CFs, have a reduced thermal conductivity. If the thermal conductivity of polymers could be improved without sacrificing other properties, many applications would be available for new polymeric materials.

Due to their exceptional thermal properties, CNTs are considered very promising fillers for improving the thermal conductivity of polymers. The calculated conductivity at room temperature for SWCNTs is 6600 W/m K [29] while experimental results have shown a conductivity of up to 3500 W/m K for SWCNTs [30] and 3000 W/m K for MWCNTs [31]. Based on these values, theoretical models predict that even the addition of a small-volume fraction of SWCNTs could result in a significant increase in thermal conductivity of composites. However, experimental results for polymer composites containing SWCNTs are generally smaller than expected [32–35].

There are several theoretical models to predict the thermal conductivity of two-phase composites. In the following sections we present the main models used to predict the thermal conductivity of polymer matrix composites containing CNTs.

5.3.1 Series and parallel models

For a two-phase composite, the series and parallel models are the simplest ones. The parallel model (rule of mixtures) is a weighted average of the thermal conductivity of the filler and matrix and is given by

$$\kappa_c = (1 - V_f)\kappa_m + V_f\kappa_f \tag{5.49}$$

where κ_c, κ_m, and κ_f are the thermal conductivity of the composite, matrix, and filler, respectively.

The parallel model maximizes the contribution of the conductive phase and implicitly assumes perfect contact between particles in a percolating network of CNTs. This model provides good results for composites with unidirectional continuous fiber but usually underestimates the conductivity for other composites.

Table 5.7 Thermal Conductivity (κ) of Polymers

Polymer	κ (W/m K)
Polyvinyl acetate (PVAc)	0.16
Acrylonitrile butadiene styrene (ABS)	0.33
Polyvinyl alcohol (PVA)	0.20
Polydimethylsiloxane (PDMS)	0.20
Epoxy	0.19
Polyneoprene	0.19
Poly(*p*-phenylene sulfide) (PPS)	0.30
Polycaprolactone (PCL) (Nylon 6)	0.24
Poly(*cis*-isoprene)	0.13
Polyether ether ketone (PEEK)	0.25
Poly(hexamethylene adipamide) (Nylon 6,6)	0.24
Polybutylene terephthalate (PBT)	0.29
Polyacetal (POM)	0.30
Polyacrylonitrile (PAN)	0.26
Polycarbonate (PC)	0.20
Polyvinyl chloride (PVC)	0.21
Polystyrene (PS)	0.11
Polyethersulfone (PES)	0.18
High-density polyethylene (HDPE)	0.52
Low-density polyethylene (LDPE)	0.33
Polyimide—thermoset resin	0.23–0.50
Polyimide—thermoplastic resin	0.11
Poly(methyl methacrylate) (PMMA)	0.21
Polypropylene (PP)	0.12
Polyethylene terephthalate (PET)	0.15
Polytetrafluoroethylene (PTFE)	0.25
Polyurethane thermoset (PU)	0.21

sourced from Physical Properties of Polymers Handbook *[4]*

The series model (inverse rule of mixtures) is given by

$$\frac{1}{\kappa_c} = \frac{(1 - V_f)}{\kappa_m} + \frac{V_f}{\kappa_f} \tag{5.50}$$

The series model does not consider the contact between particles and thus the contribution of the particle is confined to the region of the matrix surrounding the particle.

Alternatively, these models can be interpreted as the two phases of the composite were arranged in series or in parallel with respect to the heat flow. These two models provide upper and lower limit of the effective thermal conductivity, respectively. The

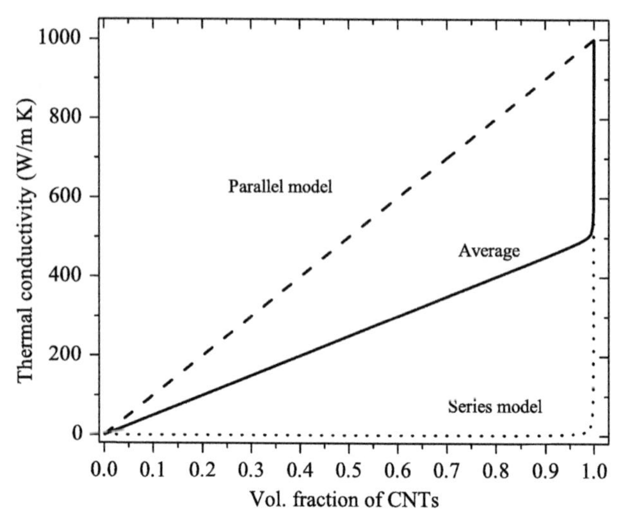

FIGURE 5.17 Thermal conductivity of a polymer composite containing carbon nanotubes (CNTs) according to the series, parallel, and an average of both models. The simulations were performed considering $\kappa_m = 0.18$ W/m K/ and κ_f 1000 W/m K.

parallel model generally overestimates the thermal conductivity while the series model tends to underestimate the thermal conductivity of nanocomposites. This is because these models are based on the concentration of fillers only and do not consider its geometry and size. However, experimental results are closer to the values obtained with the model in series. More realistic figures are provided by the average of these extremes, which describes a parallel arrangement of Eqns (5.49) and (5.50) [36]:

$$\kappa_c = \frac{\kappa_{\text{series}} + \kappa_{\text{parallel}}}{2} \tag{5.51}$$

Figure 5.17 shows the variation of the thermal conductivity of a composite containing CNTs in accordance with the series and parallel models and the average of both models.

5.3.2 Geometric model

The geometric model, as its name suggests, uses the geometric mean conductivity of both components of the composite [37]

$$\kappa_c = \kappa_m^{1-V_f} \kappa_f^{V_f} \tag{5.52}$$

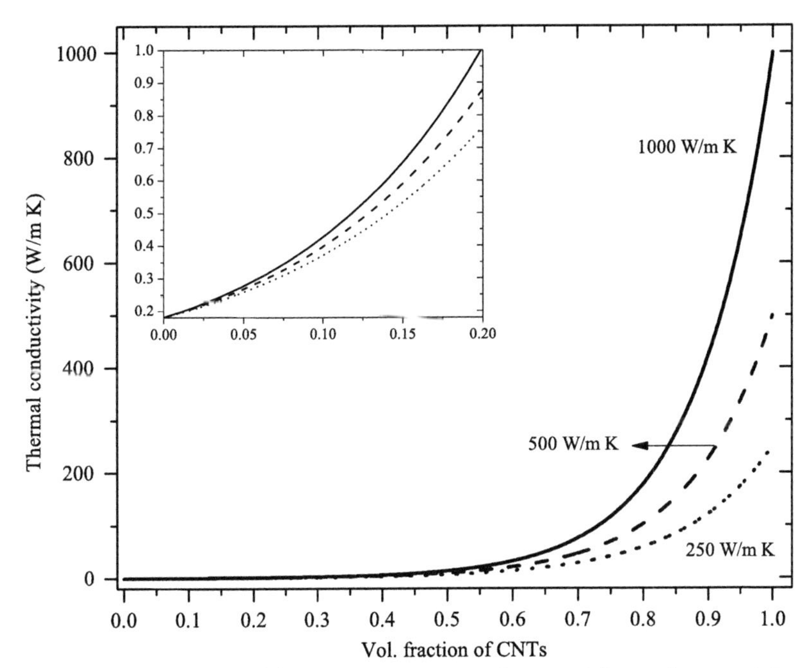

FIGURE 5.18 Thermal conductivity of a polymer composite containing carbon nanotubes (CNTs) according to the geometric model. The thermal conductivity of CNTs was varied between 250 W/m K and 100 W/m K. It was assumed $\kappa_m = 0.18$ W/m K.

Although there is no physical basis for this model, satisfactory results can be obtained. Values obtained with the geometric model generally lie between those obtained with the series and parallel models. Figure 5.18 shows the results obtained with the geometric model.

EXAMPLE 5.5

Physicists in a laboratory used CNTs to increase the thermal conductivity of the polymer polyamide (PA). The results obtained are shown in the following table.

MWCNTs vol%	κ (W/m K)	Standard Deviation
0	0.233	0.18
5	0.278	0.22
10	0.314	0.24
19	0.366	0.24
29	0.453	0.26
39	0.479	0.3

1. Plot a graphic of the thermal conductivity as a function of volume fraction of CNTs.
2. Using the geometric model, vary the value of the thermal conductivity of the CNTs to fit the experimental results.
3. Given that the thermal conductivity of CNTs experimentally measured ranges between ~200 and 3000 W/m K, discuss the results obtained in (2).

Solution

- The figure shows a graph of the experimental results with the standard deviation of the measurements. The graph was plotted using the software OriginPro®.

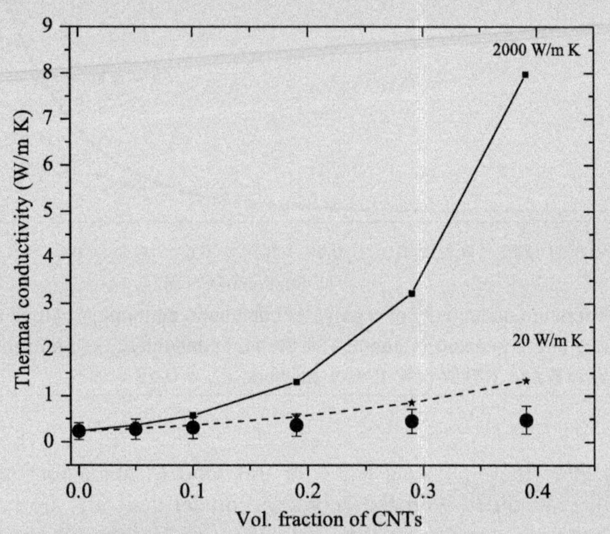

Figure: Graph showing the experimental results (•). The curves plotted represent simulated results and the method used to obtain these results is discussed in the following.

1. To fit the data, the software Microsoft Excel® was adopted. As shown in the following, the equation representing the geometric model was "converted" to the Excel format

$$\kappa_c = \kappa_m^{1-V_f} \kappa_f^{V_f} \Rightarrow \kappa_c = ((\kappa_m)\wedge(1 - V_f)) * ((\kappa_f)\wedge(V_f))$$

The calculations were made considering $\kappa_{CNT} = 2000$ W/m K and $\kappa_{CNT} = 20$ W/m K. The results are presented in the updated table below, and the data are already included in the graph. Note that the best fitting is obtained when $\kappa_{CNT} = 20$ W/m K.

MWCNTs vol%	κ (W/m K)	Standard Deviation	Simulated $\kappa_{CNT} = 2000$ W/m K	Simulated $\kappa_{CNT} = 20$ W/m K
0	0.233	0.18	0.233	0.233
5	0.278	0.22	0.366	0.291
10	0.314	0.24	0.576	0.364
19	0.366	0.24	1.302	0.543
29	0.453	0.26	3.222	0.847
39	0.479	0.30	7.971	1.323

- The value $\kappa_{CNT} = 20$ W/m K used in the calculations is much smaller than that measured for individual CNTs. Furthermore, experimental results shown in the table are less than the calculated, considering $\kappa_{CNT} = 2000$ W/m K. Although CNTs have extremely high thermal conductivity, their use in composites generally is not reflected in increases of the same magnitude. One of the reasons for this behavior observed is their high surface area, which results in a large interfacial resistance to heat flow between the surface of nanotubes and the polymer matrix. By having nanometer size, the number of CNTs present in a composite is extremely high, implying a huge number of interfaces. The quality of the dispersion of CNTs and their interaction with the polymer are other reasons for the increase in thermal conductivity of composites to be lower than expected. Finally, other reasons for the discrepancy between theory and experiment include purity and quality of CNTs, non-uniformity in size, and resistance to motion of phonons from one nanotube to another.

5.3.3 Nan model

Assuming a perfect thermal contact between nanotube and matrix, Nan and colleagues proposed a new model for the thermal conductivity of composites [38]:

$$\frac{\kappa_c}{\kappa_m} = 1 + \frac{V_f \kappa_f}{3\kappa_m} \tag{5.53}$$

This equation is valid for volume fractions less than 0.02 ($V_f < 0.02$). If there is a continuous network of CNTs in the composite (i.e., above the percolation threshold), Eqn (5.53) will underestimate the thermal conductivity of the composite. In Figure 5.19, we see how different volume fractions of CNTs affect the thermal conductivity of a polymer matrix according to the Nan model.

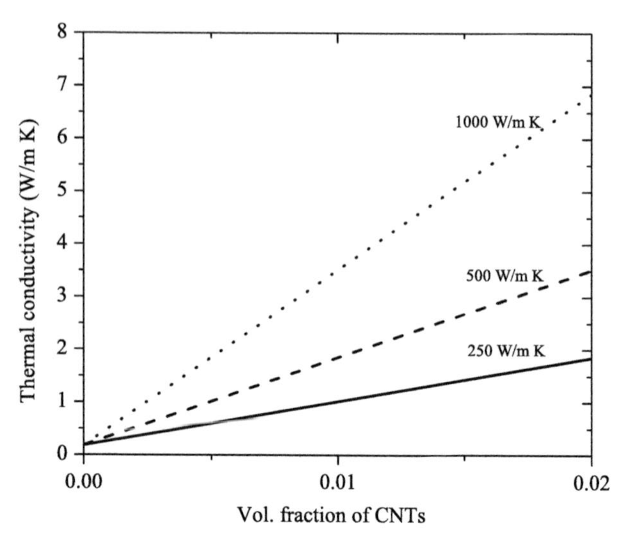

FIGURE 5.19 Thermal conductivity according to the Nan model. The thermal conductivity of carbon nanotubes (CNTs) was varied between 250 W/m K and 1000 W/m K. It was assumed $\kappa_m = 0.18$ W/m K.

5.3.4 Hatta model

Hatta and colleagues proposed a more sophisticated method for calculation of the effective thermal conductivity of composites with short fibers randomly dispersed. The method of equivalent inclusion adopted takes into account the interaction between fibers in different directions and is given by [39]

$$\frac{\kappa_c}{\kappa_m} = 1 + V_f\left[\frac{\left(\kappa_f - \kappa_m\right)\left(2S_{33} + S_{11}\right) + 3\kappa_m}{J}\right] \tag{5.54}$$

$$J = 3\left(1 - V_f\right)\left(\kappa_f - \kappa_m\right)S_{11}S_{33} + \kappa_m\left[3\left(S_{11} + S_{33}\right) - V_f\left(2S_{11} + S_{33}\right)\right] + \frac{3\kappa_m^2}{\left(\kappa_f - \kappa_m\right)} \tag{5.55}$$

where the tensors for thermal conduction, S_{ij}, as a function of the aspect ratio α of the fiber are given by

$$S_{11} = \frac{\alpha}{2\left(\alpha^2 - 1\right)^{3/2}}\left\{\alpha\left(\alpha^2 - 1\right)^{1/2} - \cos h^{-1}\alpha\right\} \tag{5.56}$$

$$S_{33} = 1 - 2S_{11} \tag{5.57}$$

Note that the aspect ratio of the filler is taken into account in this model. Simulations obtained using the Hatta model are presented in Figure 5.20.

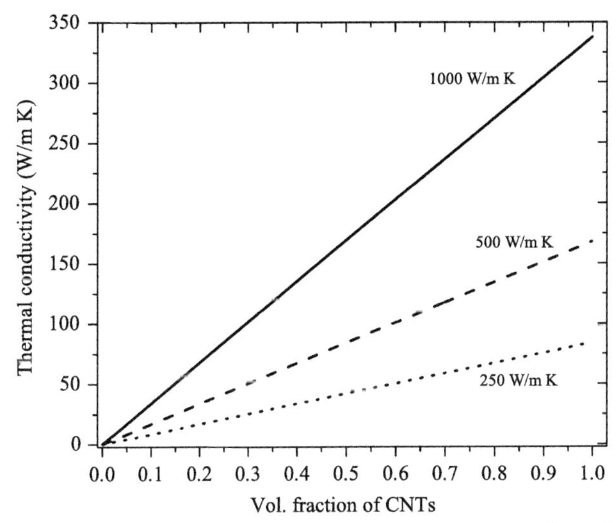

FIGURE 5.20 Thermal conductivity according to the Hatta model. The simulations were conducted considering $\kappa_m = 0.18$ W/m K, $\kappa_f = 250-1000$ W/m K, $l = 10$ μm, and $= 15$ nm.

5.3.5 Nielsen model

The Nielsen model is one of the most versatile for systems containing conductive particles or short fibers. Several important factors such as conductivity of the constituents, their volume fraction, aspect ratio, orientation, and packing of the filler are taken into consideration. The model is based on the analogy of the equations describing the elastic modulus of the composite, and uses the Halpin–Tsai equations as basis. For a two-phase system, the equations are given by [40,41]

$$\frac{\kappa_c}{\kappa_m} = \frac{1 + ABV_f}{1 - B\psi V_f} \tag{5.58}$$

$$A = K_E - 1 \tag{5.59}$$

$$B = \frac{\kappa_f/\kappa_m - 1}{\kappa_f/\kappa_m + A} \tag{5.60}$$

$$\psi = 1 + \left(\frac{1 - \phi_{\max}}{\phi_{\max}^2}\right) V_f \tag{5.61}$$

where the constant A is related to the generalized Einstein coefficient (K_E) and depends on the shape of the filler dispersed and its orientation with respect to heat flow. Factor B considers the relative thermal conductivity of the filler and matrix, whereas ψ determines the shape of the curve thermal conductivity versus volume fraction of filler. The maximum packing fraction, $\phi_{\max}$, is defined as the true volume of the particle divided by the apparent volume occupied by them when the packing is maximum. The parameters A and ϕ_{max} can be obtained in Tables 5.3 and 5.5. This model explains the effect of the shape and type of packaging of the particles in a two-phase system [42].

The conductivity of a two-phase system increases dramatically at concentrations close to the maximum packing fraction due to the high number of particle–particle contacts. Furthermore, it should be noted that the value of $\phi_{max} = 0.52$ for random particle distribution was obtained for fillers with low aspect ratio (<10) and therefore the values obtained during the simulation of composites containing CNTs can be inaccurate.

Equations (5.58)–(5.61) include a vast number of rules of mixtures. When $A \to \infty$ and $\phi_{max} = 1$, the equations become the rule of mixtures (parallel model)

$$\kappa_c = \left(1 - V_f\right)\kappa_m + V_f \kappa_f \tag{5.62}$$

When $A \to 0$ equations become the inverse rule of mixtures (series model)

$$\frac{1}{\kappa_c} = \frac{\left(1 - V_f\right)}{\kappa_m} + \frac{V_f}{\kappa_f} \tag{5.63}$$

In Figure 5.21 we see how the thermal conductivity of composites varies with different volume fractions of aligned CNTs according to the Nielsen model.

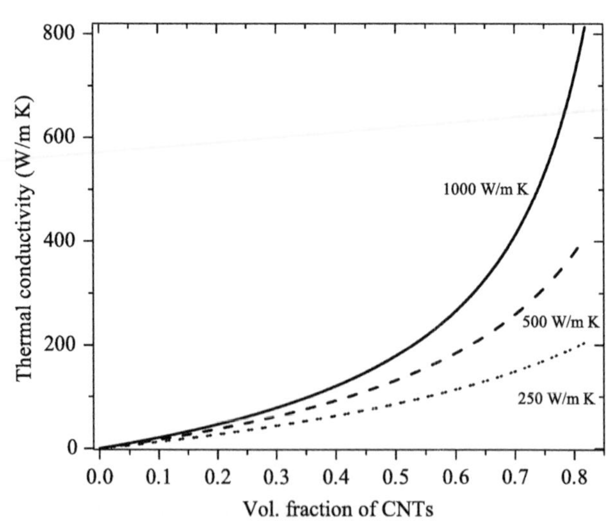

FIGURE 5.21 Thermal conductivity of a composite with aligned carbon nanotubes (CNTs) according to the Nielsen model. The simulations were conducted considering $\kappa_m = 0.18$ W/m, $\kappa_f = 250$–1000 W/m K, $A = 2\ l/d$, $\phi_{max} = 0.82$, $l = 10$ μm, and $d = 15$ nm.

In summary, the discrepancy between theoretical and experimental results for the thermal conductivity of composites containing CNTs is mainly due to:

1. The fact that analytical models are based on conventional reinforcements (large particles, fibers, etc.), and may thus not be suitable for nanoscale reinforcements;
2. Nonuniformity of the length and diameter of the CNTs;
3. Defects and purity of CNTs;
4. High thermal resistance between the nanotube–matrix interface or nanotube–nanotube;
5. Nonhomogeneity of the dispersion/distribution of nanotubes;
6. Low efficiency of heat transfer between the particles and the matrix (interfacial resistance).

5.4 Electrical conductivity of composites containing CNTs

The use of conductive particles in an insulating polymer matrix can induce electrical conductivity at low volume fraction, exceeding the antistatic limit of $\sim 10^{-6}$ S/m. This occurs because conductive fillers alter the electrical properties of the host matrix. In contrast to the mechanical reinforcement of composites using CNTs, where a homogeneous dispersion and adhesion between filler and matrix are desired, the electrical conductivity is based on the formation of percolation paths of conductive particles. Thus, to some degree reagglomeration can be a positive factor.

The electrical conductivity of polymeric composites can be well explained by the percolation theory [43]. A critical volume of fillers, V_c, should be used to render an insulating composite conductor. In this critical volume, which is known as the percolation threshold, the electrical conductivity of the composite is suddenly elevated by several orders of magnitude. Generally, in concentrations equal to the percolation threshold, the fillers form a continuous network within the polymer matrix and additional increase of the volume of fillers has little effect on the conductivity of the composite.

The critical volume for conductive spherical particles, randomly distributed in a polymer matrix, is estimated to be 16 vol%. Note that spherical particles have an aspect ratio $\alpha \sim 1$. CNTs on the other hand have an aspect ratio $\alpha >> 1$ and are therefore of great advantage; once using this filler much lower critical volumes as low as 0.021 vol% could be observed. This occurs because the percolation threshold decreases with increasing aspect ratio of the filler [44]. Also, experimental results show that the electrical conductivity of CNTs can reach 10^7 S/m.

The percolation threshold (V_c) can be estimated by [44]

$$V_c = \frac{0.5}{\alpha} \tag{5.64}$$

Figure 5.22 shows how the percolation threshold varies according to the aspect ratio of the nanotubes.

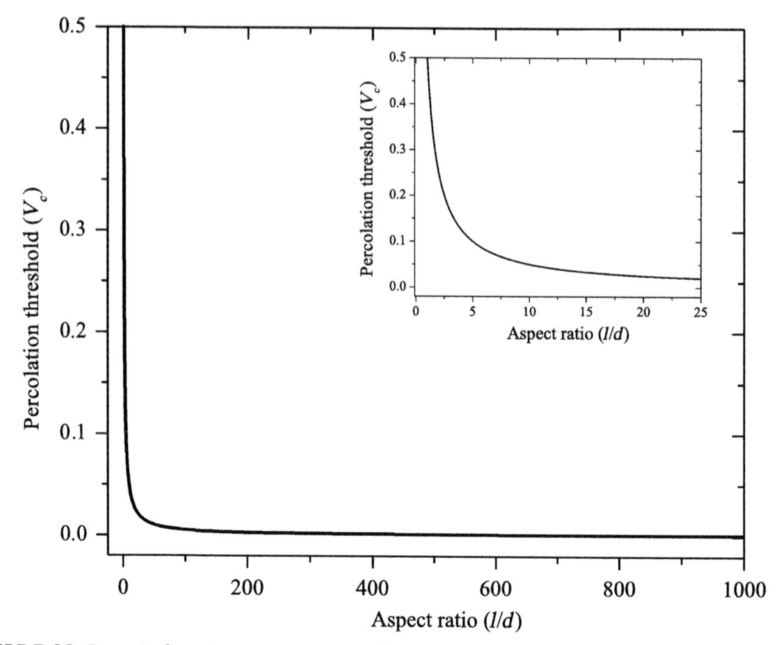

FIGURE 5.22 Percolation threshold as a function of aspect ratio for carbon nanotubes (CNTs).

Figure 5.23 shows schematically the formation of a percolation path (threshold) for composites with CNTs and carbon black (CB). According to Figure 5.23(a), the particles remain isolated (1) until the concentration exceeds a certain value, which is enough to form a small conductive network. The occurrence of conduction paths leads to a sudden increase of the conductivity by several orders of magnitude (2). The further increase in concentration of fillers results in a small increase in conductivity approaching a maximum value (3). The contact resistance between fillers limits the maximum achievable conductivity of the composite. Following the same logic, Figure 5.23(b) shows the effect of CB on the resistivity of a composite. Note how the amount of filler needed to reach the percolation threshold varies for different shapes of fillers.

According to the percolation theory, the relationship between the conductivity and the volume fraction of the composite filler is given by [45]

$$\sigma = \sigma_0 (V_f - V_c)^t \tag{5.65}$$

Where σ is the composite conductivity (S/m);

V_f is the volume fraction of filler;
V_c is the percolation threshold (critical volume);
σ_0 is a constant (S/m);
t is the critical exponent.

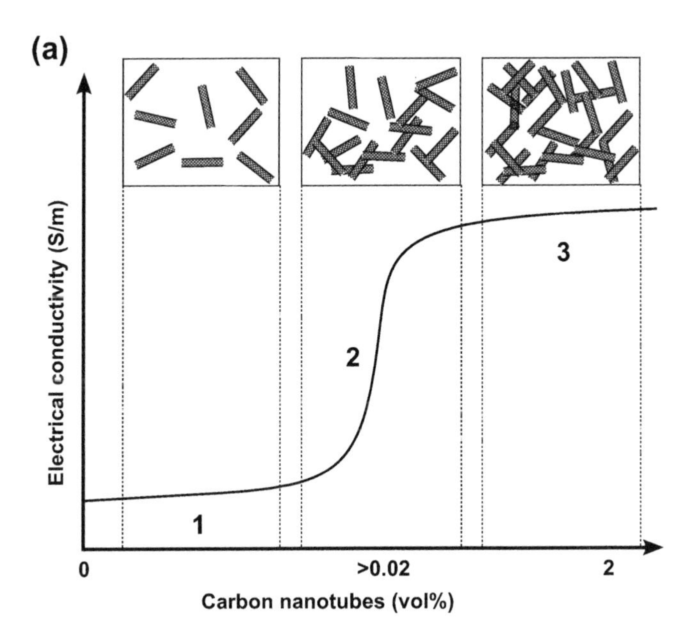

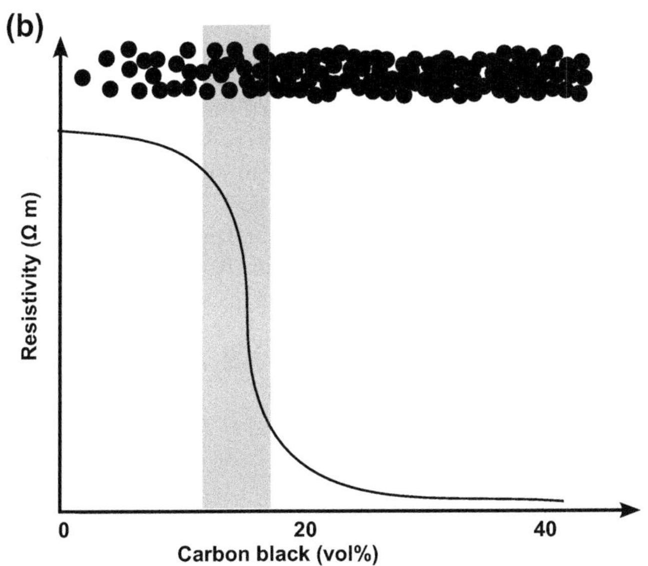

FIGURE 5.23 (a) Dependence of the electrical conductivity of a composite with the concentration of carbon nanotubes (CNTs) and (b) resistivity of a carbon black reinforced composite (spheres). Recall that the conductivity σ is defined as the inverse of the resistivity ρ.

This equation is valid for concentrations higher than the percolation threshold, i.e., $V_f > V_c$. The value of the critical exponent depends on the aspect ratio of the filler. Theoretical values of t range between 1.3 and 2.0 [46–48] while experimental values range from 0.7 to 3.1 [47,48]. It should be emphasized that the calculated value for t depends on the simulation method used by different authors. Usually a graph $\log \sigma \times \log(V_f - V_c)$ is made and the value of V_c is varied until the best linear fit is obtained. Thus one can estimate the value of V_c and t.

The contact resistance between CNTs and their clusters in the system decreases the specific conductivity of CNTs. Also, in CNT/polymer composites, conductive nanotubes are separated by the insulating polymer as a potential barrier, so that electrical conductivity is limited by tunneling of charge carriers between CNTs. This means that the electrons "jump" from one electronic state to another. Still, the percolation threshold of composites containing CNTs increases with the curvature of the CNTs [44]. Also, CNTs are dispersed generally in the form of bundles. The effect of the size of bundles in the percolation threshold was calculated and results show that, as expected, the percolation threshold increases with the size of the bundles.

It is extremely important to reduce the percolation threshold of CNT-reinforced composites, since the price of CNT is much higher than the polymer matrix itself. For a conductive network to be formed by CNTs in a polymer matrix, there is a need for homogeneous distribution of these fillers. Remember that the words distribution and dispersion do not have the same meaning. Figure 5.24 outlines the conditions necessary for the formation of a conduction path of fibers (1D) in a plane (2D) through different possibilities of distribution and dispersion of the filler. Figures 5.24(a) and 5.24(b) show that a conducting network is not formed in case of nonuniform dispersion. In Figure 5.24(d) it can be noted that a uniform distribution of filler homogeneously dispersed increases the distance between fillers. The formation of conduction paths is favored in the case of preferential distribution of fillers dispersed homogeneously, as shown in Figure 5.24(c).

In summary, the electrical conductivity of polymeric composites containing CNTs depends on several factors such as:

1. Aspect ratio of CNTs;
2. Dispersion;
3. Distribution;
4. Conductivity of CNTs;
5. Crystallinity of the matrix;
6. Surface tension of the polymer;
7. Integrity of CNTs;
8. Cure conditions in the case of thermosets;
9. Curvature of CNTs;
10. Functionalization of CNTs.

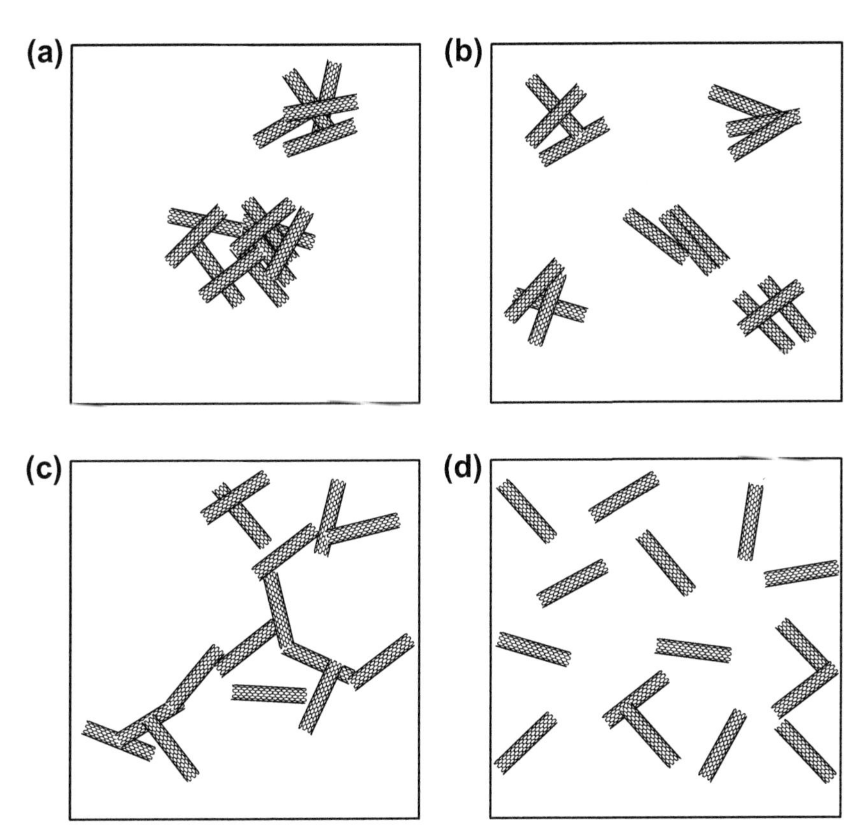

FIGURE 5.24 Schematic showing the effect of 1D fillers on the conductivity of polymeric composites: (a) null conductivity: poor dispersion and distribution; (b) null conductivity: homogeneous distribution, poor dispersion; (c) high conductivity: a conductive network of carbon nanotubes (CNTs) is formed; (d) null conductivity: homogeneous distribution, poor dispersion.

EXAMPLE 5.6

The electrical conductivity of polyamide (PA) is 6.3×10^{-18} S/cm. To apply PA as antistatic film, its conductivity should be higher than 1×10^{-8} S/cm. Based on this information, a company interested in conducting polymer prepared composites of PA with CNTs and measured the electrical conductivity. The results obtained are shown in the table below.

Sample	CNTs (vol%)	σ (S/cm)
0	0	6.3×10^{-18}
1	0.02	3.0×10^{-17}
2	0.1	1.6×10^{-8}
3	0.2	7.0×10^{-8}
4	0.5	3.0×10^{-7}
5	1	1.0×10^{-6}

1. Plot a graph of electrical conductivity of the samples as a function of the volume concentration of CNTs. Discuss the graph obtained.
2. Fit the experimental data and determine σ_0 and t.

Solution

1. Figure below shows the graph plotted based on the experimental data presented in the table.

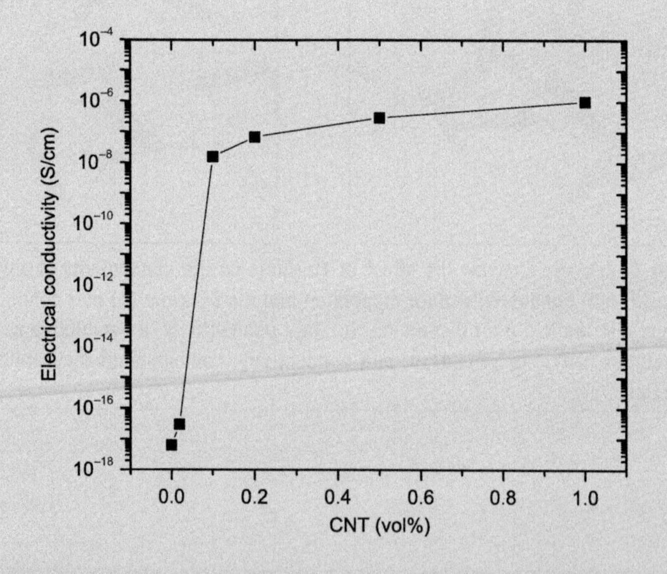

A sharp increase in the conductivity is observed between 0.02 and 0.1 vol% where the conductivity changed from 3.0×10^{-17} to 1.6×10^{-8} S/cm. For CNTs concentrations in excess of 0.1 vol%, conductivity increased only moderately. The composite with 0.5 vol% of CNTs have a conductivity of 3.0×10^{-7} S/cm, which is 10 orders of magnitude higher than the value for composites with 0.02 vol% of CNTs. This behavior is indicative of a percolation transition.

According to percolation theory, there is a certain concentration of fillers called the critical concentration or percolation threshold at which a conduction path is formed in the composite material turning the insulator material to a conductor. According to the graph, the percolation threshold lies between 0.02 and 0.1 vol%. Note that at a concentration of 0.1 vol%, the conductivity of the composite is greater than the criterion for antistatic applications.

- The first step to fit the experimental results is to assume an appropriate value for V_c. Based on the figure above we assume $V_c = 0.05$ vol%. Then we calculate the values of $V_{CNT} - V_c$. As the equation based on percolation theory is valid only for $V > V_c$, we consider in the graph only volumetric concentrations of CNTs above the percolation threshold. Thus, we have

$$V_2 - V_c = 0.1 - 0.05 = 0.05$$

$$V_3 - V_c = 0.2 - 0.05 = 0.15$$

$$V_4 - V_c = 0.5 - 0.05 = 0.45$$

$$V_5 - V_c = 1.0 - 0.05 = 0.95$$

The above data are then used to plot a graph of σ versus $(V_{CNT} - V_c)$. For the data to appear in a linear fashion, a logarithmic scale was used on both axes, as shown in figure below.

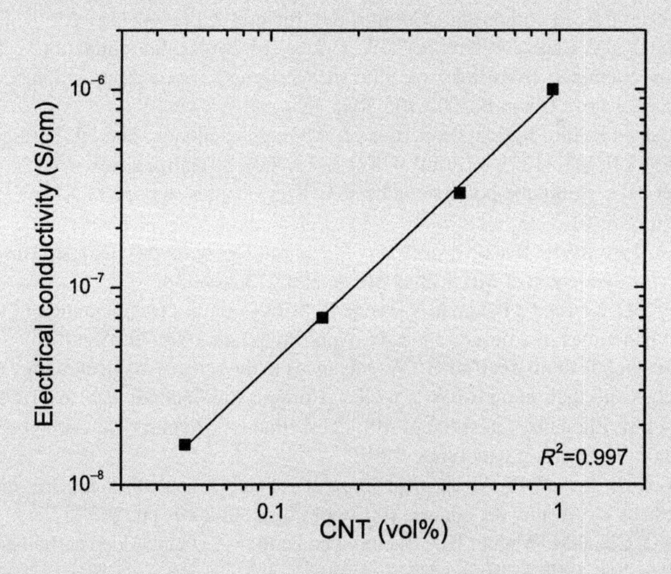

The graph shows the best fit obtained, and such fitting resulted in values of $\sigma_0 = 1.0 \times 10^{-6}$ S/cm and $t = 1.39$. The accuracy of fitting is confirmed by the value of $R^2 = 0.997$. Remember that a perfect fit occurs when $R^2 = 1$.

To learn more...

We encourage readers to consult the following references:

1. Jinsong Leng, Alan Kin-tak Lau. Multifunctional polymer nanocomposites. 1st ed. CRC Press; 2010, ISBN-13: 978-1,439,816,820.
2. Coleman JN, Umar Khan, Blau WJ, Gun'ko YK. Small but strong: A review of the mechanical properties of carbon nanotube-polymer composites. Carbon 2006;44:1624–1652.
3. Wolfgang Bauhofer, Kovacs JZ. A review and analysis of electrical percolation in carbon nanotube polymer composites. Compos Sci Technol 2009;69: 1486–1498.
4. Sushant Agarwal, M. Masud K. Khan, Gupta RK. Thermal conductivity of polymer nanocomposites made with carbon nanofibers. Polym Eng Sci 2008; 48:2474–2481.

References

[1] Gojny FH, Wichmann MHG, Fiedler B, Schulte K. Influence of different carbon nanotubes on the mechanical properties of epoxy matrix composites – a comparative study. Compos Sci Technol 2005;65:2300–13.

[2] Fiedler B, Gojny FH, Wichmann MHG, Nolte MCM, Schulte K. Fundamental aspects of nano-reinforced composites. Compos Sci Technol 2006;66:3115–25.

[3] Frankland SJV, Caglar A, Brenner DW, Griebel M. Molecular simulation of the influence of chemical cross-links on the shear strength of carbon nanotube-polymer interfaces. J Phys Chem B 2002;106:3046–8.

[4] Mark James E. In: Physical properties of polymers handbook. 2nd ed. Springer; 2002. ISBN: 978-0-387-31235-4 (Print) 978-0-387-69002-5 (Online).

[5] Wagner HD. Nanotube-polymer adhesion: a mechanics approach. Chem Phys Lett 2002;361:57–61.

[6] Kelly A, Tyson WR. Tensile properties of fiber-reinforced metals: Copper/Tungsten and Copper/Molybdenum. J Mech Phys Solids 1965;13:329–38.

[7] Wagner HD, Lourie O, Feldman Y, Tenne R. Stress-Induced fragmentation of multiwall carbon nanotubes in a polymer matrix. Appl Phys Lett 1998;72:188–90.

[8] Thostenson ET, Renb ZF, Choua TW. Advances in the science and technology of carbon nanotubes and their composites: a review. Compos Sci Technol 2001;61:1899–912.

[9] Ajayan PM, Zhou OZ, Dresselhaus MS, Dresselhaus G, Avouris Ph. Carbon nanotubes: applications of carbon nanotubes; 2001.

[10] Yakobson BI, Avouris Ph. Mechanical properties of carbon nanotubes. In: Dresselhaus MS, Dresselhaus G, Avouris Ph, editors. Top Appl Phys, 80; 2001. pp. 287–327.

[11] Lourie O, Cox DM, Wagner HD. Buckling and collapse of embedded carbon nanotubes. Phys Rev Lett 1998;81:1638–41.

[12] Bacon R. Growth, structure, and properties of graphite whiskers. J Appl Phys 1960;31: 283–90.

[13] Thostenson ET, Chou TW. On the elastic properties of carbon nanotube-based composites: modeling and Characterization. J Phys D Appl Phys 2003;36:573–82.

[14] Ahmed S, Jones FR. A review of particulate reinforcement theories for polymer composites. J Mater Sci 1990;25:4933–42.
[15] Einstein A. Investigation on theory of Brownian motion (English translation). New York; 1956.
[16] Smallwood HM. Limiting law of the reinforcement of rubber. J Appl Phys 1944;15: 758–66.
[17] Hansen TC. Influence of aggregate and voids on modulus of elasticity of concrete, cement mortar, and cement paste. J Amer Conc Inst 1965;62:193–216.
[18] Mooney M. The viscosity of a concentrated suspension of spherical particles. J Colloid Sci 1951;6:162–70.
[19] Brodnyan JG. The concentration dependence of the Newtonian viscosity of prolate ellipsoids. Trans Soc Rheol 1959;3:61–8.
[20] Guth E. Theory of filler reinforcement. J Appl Phys 1945;16:20. http://dx.doi.org/10.1063/1.1707495.
[21] Cox HL. The elasticity and strength of paper and other fibrous materials. J Appl Phys; 1952;3:72–9.
[22] Carman GP, Reifsnider KL. Micromechanics of short-fiber composites. Compos Sci Technol 1992;43:137–46.
[23] Coleman JN, Khan U, Blau WJ, Gun'ko YK. Small but strong: a review of the mechanical properties of carbon nanotube-polymer composites. Carbon 2006;44:1624–52.
[24] Ashton JE, Halpin JC, Petit PH. Primer on composite materials: analysis. Technomic Stamford Conn; 1969.
[25] Halpin JC. Stiffness and expansion estimates for oriented short fiber composites. J Compos Mater 1969;3:732–4.
[26] Tucker III CL, Liang E. Stiffness predictions for unidirectional short-fiber composites. Compos Sci Technol 1999;59:655–71.
[27] Nielsen LE. Generalized equation for the elastic moduli of composite materials. J Appl Phys 1970;41:4626–7.
[28] Keith JM, King JA, Miskioglu I, Roache SC. Tensile modulus modeling of carbon-filled liquid crystal polymer composites. Polym Compos 2009;30(8):1166–74.
[29] Berber S, Kwon Y-K, Tomnek D. Unusually high thermal conductivity of carbon nanotubes. Phys Rev Lett 2000;84:4613–6.
[30] Pop E, Mann D, Wang Q, Goodson K, Dai H. Thermal conductance of an individual single-wall carbon nanotube above room temperature. Nano Lett 2006;6:96–100.
[31] Kim P, Shi L, Majumdar A, McEuen PL. Thermal transport measurements of individual multiwalled carbon nanotubes. Phys Rev Lett 2001;87(215502):1–4.
[32] Choi SUS, Zhang ZG, Yu W, Lockwood FE, Grulke EA. Anomalous thermal conductivity enhancement in nanotube suspensions. Appl Phys Lett 2001;79:2252–4.
[33] Biercuk MJ, Laguno MC, Radosavljevic M, Hyun JK, Johnsond AT. Carbon nanotube composites for thermal management. Appl Phys Lett 2002;80:2767–9.
[34] Choi ES, Brooks JS, Eaton DL, Al-Haik MS, Hussaini MY, Garmestan H, et al. Enhancement of thermal and electrical properties of carbon nanotube polymer composites by magnetic field processing. J Appl Phys 2003;94:6034–9.
[35] Xie HQ, Leand H, Youn W, Choi M. Nanofluids containing multiwalled carbon nanotubes and their enhanced thermal conductivities. J Appl Phys 2003;94:4967–71.
[36] Horai K. Thermal conductivity of Hawaiian basalt: a new interpretation of Robertson and Peck's data. J Geophy Res 1991;96:4125–32.

[37] Agarwal S, Khan MMK, Gupta RK. Thermal conductivity of polymer nanocomposites made with carbon nanofibers. J Appl Polym Sci 2008;48:2474–81.
[38] Nan C-W, Shi Z, Lin Y. A simple model for thermal conductivity of carbon nanotube-based composites. Chem Phys Lett 2003;375:666–9.
[39] Hatta H, Taya M, Kulacki FA, Harder JF. NÃO ACHEI. J Compos Mater 1992;26:612–25.
[40] Guthy C, Du F, Brand S, Winey KI, Fischer JE. Thermal conductivity of single-walled carbon Nanotube/PMMA nanocomposites. J Heat Trans 2007;129:1096–9.
[41] Nielsen LE. The thermal and electrical conductivity of two-phase systems. Ind Eng Chem Fundam 1974;13:17–20.
[42] Song YS, Youn JR. Evaluation of effective thermal conductivity for carbon nanotube/polymer composites using control volume finite element method. Carbon 2006;44:710–7.
[43] Kirkpatrick S. Percolation and conduction. Rev Mod Phys 1973;45:574–88.
[44] Bauhofer W, Kovacs JZ. A review and analysis of electrical percolation in carbon nanotube polymer composites. Compos Sci Technol 2009;69:1486–98.
[45] Jiang X, Bin Y, Matsuo M. Electrical and mechanical properties of polyimide-carbon nanotubes composites fabricated by in situ polymerization. Polymer 2005;46:7418–24.
[46] Kim YJ, Shin TS, Choi HD, Kwon JH, Chung YC, Yoon HG. Electrical conductivity of chemically modified multiwalled carbon nanotube/epoxy composites. Carbon 2005;43:23–30.
[47] Stauffer D. Introduction to the percolation theory. London; 1991. ISBN-10: 0748402535 | ISBN-13: 978–0748402533.
[48] Hu G, Zhao C, Zhang S, Yang M, Wang Z. Low percolation threshold of electrical conductivity and rheology in poly(ethylene terephthalate) through the networks of multi-walled carbon nanotubes. Polymer 2006;47:480–8.

CHAPTER

Processing of Polymer Matrix Composites Containing CNTs

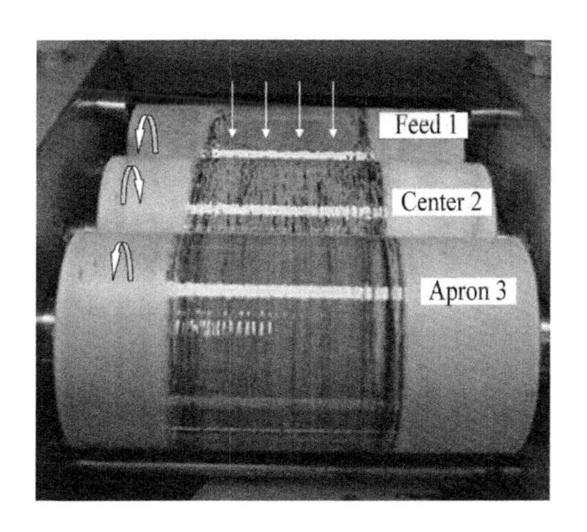

CHAPTER OUTLINE

arbon Nanotube Reinforced Composites. http://dx.doi.org/10.1016/B978-1-4557-3195-4.00006-0

6.1 Processing of polymer matrix composites containing CNTs

Although great advances have been made in recent years to overcome the difficulties with the manufacturing of polymer composites, the processing remains key to fully using the properties of nanosized reinforcements [1]. The first difficulty is to achieve an appropriate distribution and dispersion of nanotubes in a composite. Without adequate dispersion, the fillers tend to agglomerate and act as defects that limit the mechanical performance; such agglomerates also influence the physical properties of composites. Figure 6.1 shows a possible route for the production of a polymer matrix composite reinforced with carbon nanotubes (CNTs). The dispersion technology is of the utmost importance for obtaining composites containing adequately dispersed CNTs.

When particles are dispersed in a medium of low viscosity, diffusion processes and particle–particle interactions and particle matrix are of great importance [3]. For example, a load of 15 vol% particles of 10 nm diameter in a medium results in an interparticle distance of only 5 nm [4]. This distance is comparable to the radius of gyration of a typical polymer molecule. This basically shows that, in a composite, the entire polymeric matrix can behave as if it were part of an interface. This effect should influence both the processing and final properties of the polymer phase.

Generally, the most used methods for preparation of polymer matrix composites containing CNTs are:

- Solution processing;
- Melt processing;
- In situ processing.

In *solution processing*, CNTs are usually dispersed in a solvent and then mixed with a polymer solution using techniques such as shear mixing, magnetic or mechanical stirring, or sonication [5]. Subsequent to the dispersion of the nanotubes, the composite can be obtained by vaporization of the solvent. Figure 6.2 shows the flowchart for the process in solution.

This method is considered effective for preparing composites with CNTs homogeneously distributed and is often used to prepare films [6]. However, it should be emphasized that this method is based on the efficiency of dispersion of the CNTs in the solvent used. Solvent choice is made based on the solubility of the polymer. Furthermore, unmodified nanotubes cannot be homogeneously dispersed in most solvents. Again the importance of functionalization to assist in dispersion and prevent reagglomeration is established.

Melt processing is a versatile alternative to insoluble polymers and is especially useful in the case of thermoplastics. This technique relies on the fact that thermoplastic polymers normally melt when heated. Usually, the processing involves melting of polymer pellets to form a viscous liquid. Any additive such as CNTs

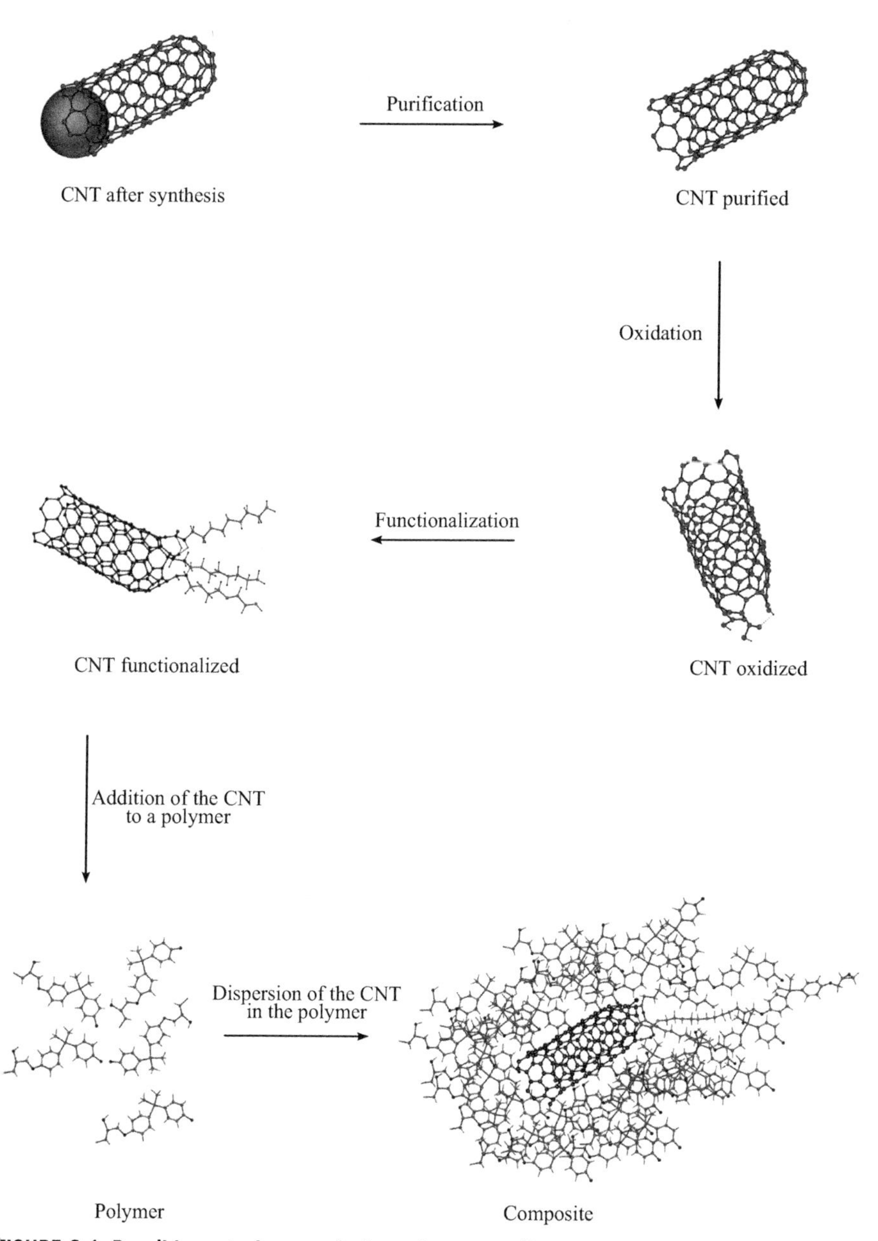

FIGURE 6.1 Possible route for manufacture of a composite.

After synthesis, the carbon nanotubes (CNTs) may contain catalyst particles that are eliminated through the purification process. In the next stage, the CNTs are oxidized and carboxylic groups can be created at the ends of the nanotubes. These groups may serve as an "anchor" for the growth of specific macromolecules (functionalization) that can interact with a particular polymer. CNTs are then dispersed in a polymer matrix to obtain a composite. This latter step involves the dispersion technology.

Figure sourced from Fundamental aspects of nano-reinforced composites [2]

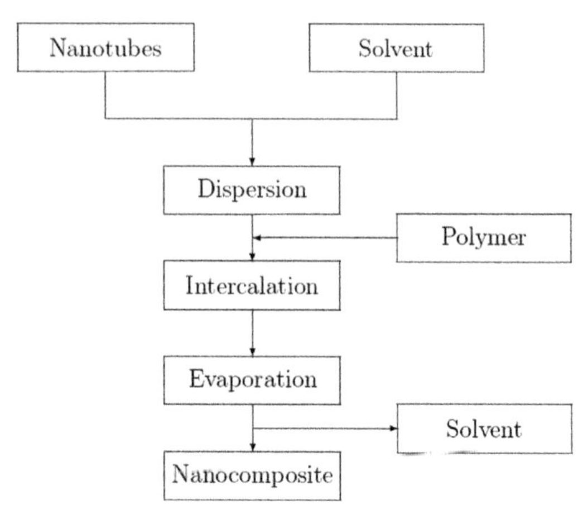

FIGURE 6.2 Flowchart showing the different stages of the solution processing of a polymeric composite.

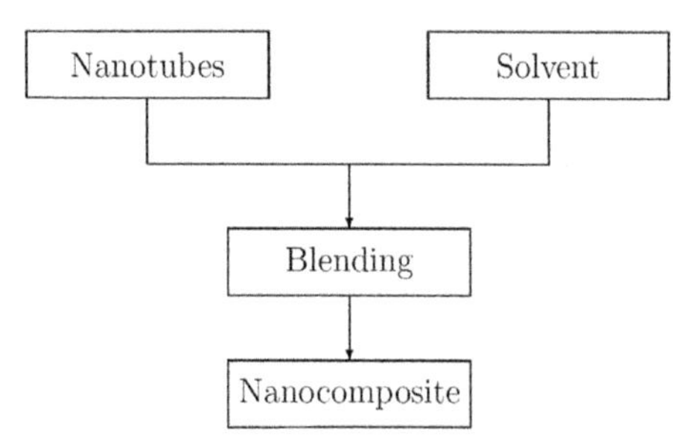

FIGURE 6.3 Flowchart showing the different stages of the melt processing of a polymer composite.

can be mixed with the molten material through mixing by shear [6]. Samples can be manufactured in high volume using the extrusion processing. Figure 6.3 shows the flowchart of the melt processing.

Some advantages of this technique are its speed and simplicity, not to mention its compatibility with standard industrial techniques [7,8].

Reaction processing or in situ polymerization is considered to be a very effective method to improve the dispersion of nanotubes and their interaction with the polymer matrix. In general, CNTs are first mixed with the monomers and then the polymer composite is obtained by polymerizing these monomers under certain conditions [5]. For thermosets such as epoxy or unsaturated polyester, a curing agent (hardener) or catalyst is added to initiate polymerization [6]. In the case of thermoplastics, the polymerization may be initiated by the addition of an

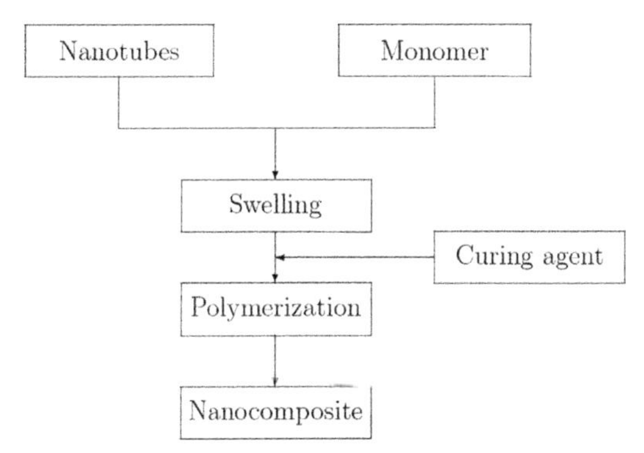

IGURE 6.4 Flowchart showing the different stages of the reaction processing of a olymeric composite.

nitiator or by increasing the temperature. Figure 6.4 shows the flowchart for the pro- essing reaction. This technique is particularly important for the preparation of nsoluble or unstable polymers, which cannot be prepared by the solution or melt rocessing.

.2 Technologies applied for the preparation of polymeric matrix nanocomposites

)epending on the preparation method used for the fabrication of polymer matrix omposites containing CNTs, different technologies are applicable independently r in combination for dispersing CNTs and subsequently obtaining the nanocompo- ite. Here we introduce some of the techniques used for preparing nanocomposites, rom the simplest such as magnetic or mechanical stirring to the more elaborate such s calendering and dual asymmetric centrifugation.

.2.1 Magnetic stirring

magnetic stirrer is a device widely used in laboratories and consists of a rotating nagnet or a stationary electromagnet that creates a rotating magnetic field. This de- ice is used to make a stir bar, immerse in a liquid, quickly spin, or stirring or mixing solution, for example. A magnetic stirring system generally includes a coupled eating system for heating the liquid (Figure 6.5).

Much of the current magnetic stirrers rotate the magnets by means of an electric notor. This type of equipment is one of the simplest to prepare mixtures. Magnetic tirrers are silent and provide the possibility of stirring closed systems without the eed for isolation, as in the case with mechanical agitators.

Because of their size, stir bars can be cleaned and sterilized more easily than ther devices such as stirring rods. However, the limited size of the stir bars enables

FIGURE 6.5 A stir bar mixing a solution in a magnetic stirrer with coupled heating.

Figure sourced from Keison website [9]

using this system only for volumes less than 4 L. In addition, viscous liquid or dense solutions are barely mixed using this method. In these cases, some kind of mechanical stirring is usually required.

A stir bar consists of a magnetic bar used to agitate a liquid mixture or solution (Figure 6.6). Because the glass does not affect a magnetic field significantly, and most of the chemical reactions are performed in glass vials or beakers, stirring bars function adequately in glassware commonly used in laboratories. Typically,

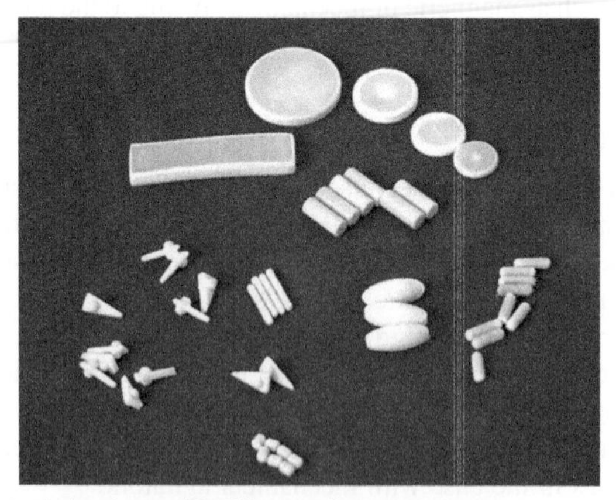

FIGURE 6.6 Typical stirring bars with different sizes and shapes.

Figure sourced from Lihe website [10]

stirring bars are coated with Teflon® or glass, so they are chemically inert and do not contaminate or react with the system in which they are immersed. Their shape may vary to increase efficiency during stirring. Their size varies from a few millimeters to a few centimeters.

The magnetic or mechanical stirring by itself is not effective for dispersing nanoparticles. This technique has been used for dispersing nanotubes; however, its use is usually coupled with other more powerful dispersion techniques such as sonication.

6.2.2 Mechanical stirring

This is probably the device best known for preparing mixtures. Basically, a motor is responsible for the rotation of the blades. The shape of the blades can be varied to increase the efficiency of stirring once the shear generated is a function of the aerodynamics of the blades. Figure 6.7 shows a mechanical stirrer and different types of blades.

Mechanical stirring is usually more efficient than magnetic stirring, but its use in dispersing nanotubes tends to occur in addition to dispersion techniques such as sonication, dual asymmetric centrifugation, and three-roll mill.

6.2.3 High-shear mixer

A high-shear mixer is used to disperse one phase (liquid or particulate) that would normally be immiscible in a continuous phase (liquid). A typical example of the

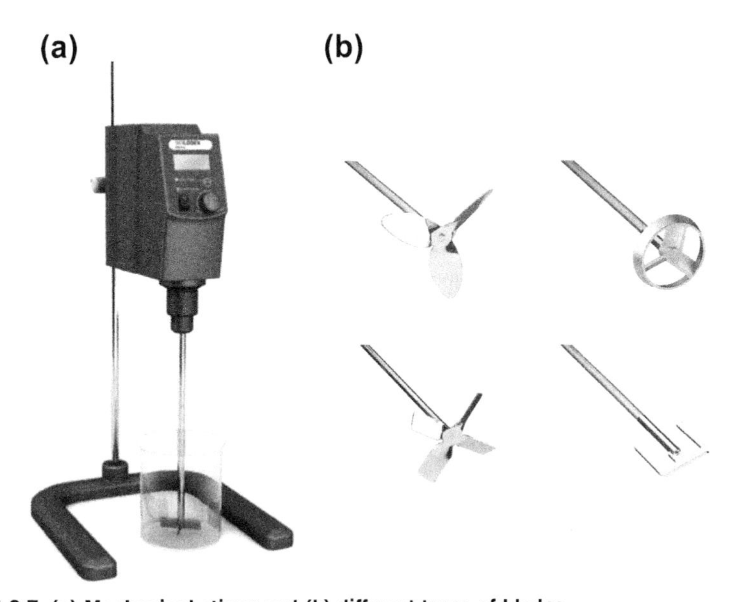

FIGURE 6.7 (a) Mechanical stirrer and (b) different types of blades.

Figure sourced from Spectra Services website [11]

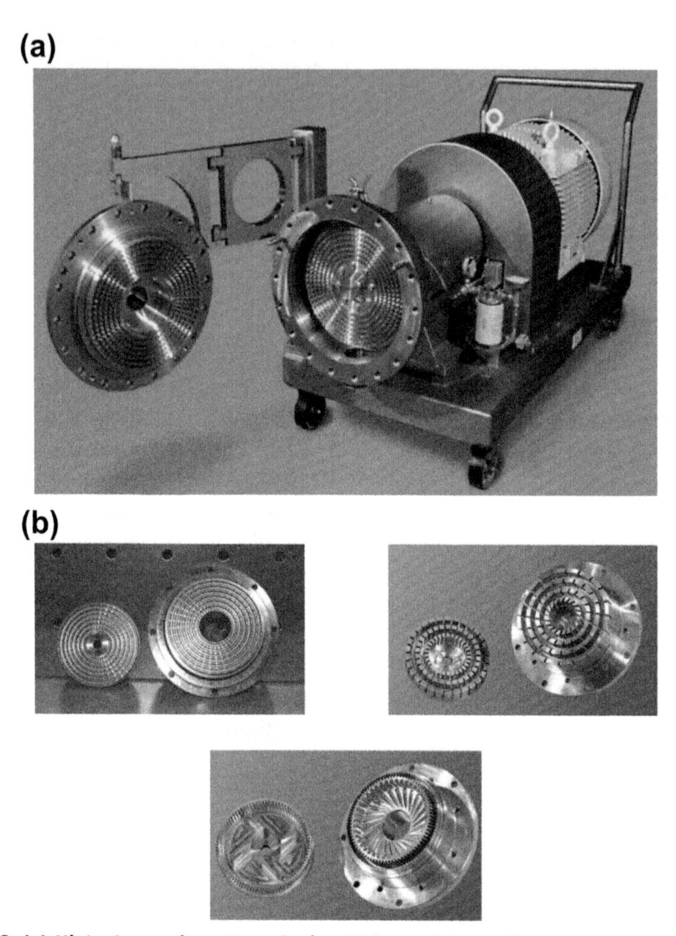

FIGURE 6.8 (a) High-shear mixer, Ross Series 700 and (b) the different types of rotors for creating high shear.

Figure sourced from Mixers.com [12]

application of high-shear mixing is the preparation of emulsions of oil in water. The principle of operating the equipment is simple: a motor rotates a rotor in at a high speed, creating high shear. Figure 6.8 shows a high-shear mixer and different kinds of rotors.

Large companies have been using high-shear mixers for dispersion of CNTs in thermosetting resins such as epoxy and polyurethane (PU) [13]. This equipment is a great option for the preparation of master batches containing CNTs for industrial applications.

EXAMPLE 6.1

Using high-shear mixer for the preparation of PU matrix composites reinforced with CNTs [13].

The Multi Wall Carbon Nanotubes (MWCNTs) used in this case study were C150P Baytubes® from Bayer MaterialScience. These CNTs have an average diameter of 13 nm and a length greater than 1 μm. To improve the quality of dispersion, a dispersing agent, polyvinyl butyral (supplied by Kuraray), was used. The PU system used consists of two parts: polyol and isocyanate. These two parts should be mixed to obtain the PU. Figures 6.9(a) and (b) show CNT suspensions prepared in isocyanate with and without the aid of dispersant agents, respectively. Figures 6.9(c) and (d) show CNTs suspended in polyol with and without dispersant, respectively. The suspension of CNTs in polyol is improved

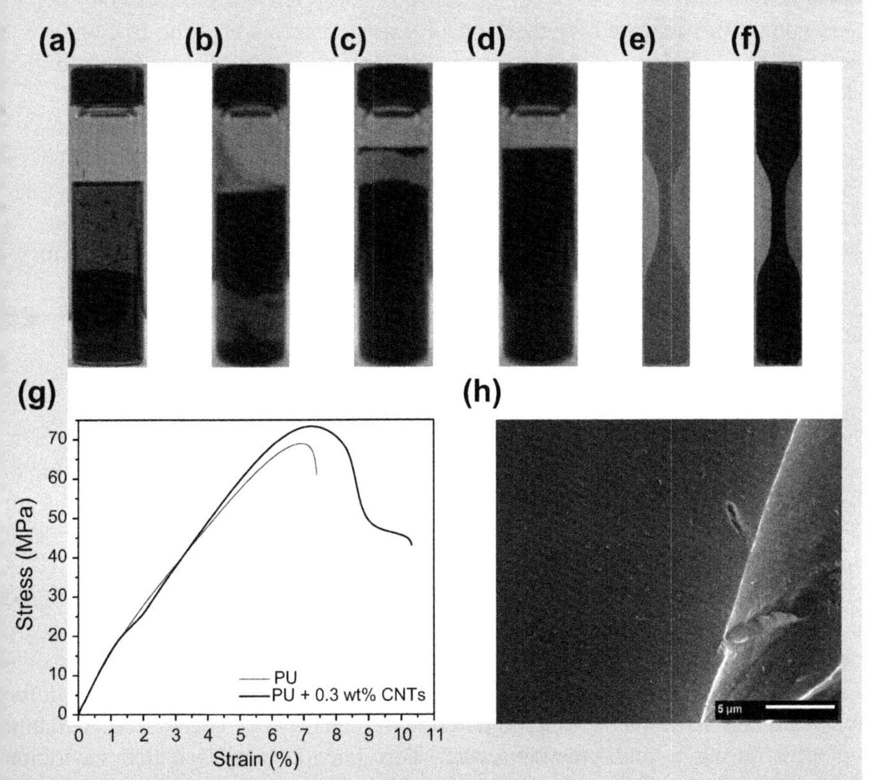

FIGURE 6.9 Suspensions of CNTs prepared in isocyanate (a) without and (b) with the use of a dispersant; in polyol (c) without and (d) with a dispersant; specimens for tensile test (e) PU and (f) PU + CNTs; (g) stress–strain curve obtained for PU and nanocomposites, and (h) distribution of CNTs in PU.

through the use of additives, whereas in the case of isocyanate there was no improvement. This preliminary result was crucial in deciding the preparation route for the composites.

After the dispersion studies, composites containing 0.3 wt% of CNTs were prepared using a high-shear mixer. During the dispersion of nanotubes in the polyol, bubbles are generated in the suspension of CNTs, and so a degassing process was used. Basically, the suspensions are placed under vacuum. Then, isocyanate, the second part of the system, was added to the mixture of CNTs and polyol. Finally, the samples were cured at room temperature for 18 h and postcured at 90 °C for an additional 6 h. Samples were also prepared without CNTs for reference.

Tensile tests were conducted on the universal testing machine Instron 1001, according to the standards of the American Society for Testing Materials (ASTM 638-03). Figures 6.9(e) and (f) show the reference samples (pure PU) and composites prepared for mechanical testing, respectively. The fracture surface of the samples was analyzed in a scanning electron microscope (JEOL JSM-6510LV) operating at a voltage of 30 kV.

Figure 6.9(g) shows the stress–strain curves obtained. The addition of CNTs improved the mechanical properties of the PU. In this case, the tenacity of PU has increased from 3.41 MJ/m^3 to 5.70 MJ/m^3 because of the addition of CNTs (an increase of 38%). This result is a consequence of a homogeneous dispersion of the CNTs (Figure 6.9(h)) and improved interaction between CNT and the polymer matrix.

6.2.4 Sonication

Sonication is the act of applying ultrasonic energy to agitate particles in a sample. In a laboratory, this is usually accomplished by use of an ultrasonic bath (Figure 6.10(a)) or ultrasonic probe (Figure 6.10(b)), colloquially known as a sonicator.

During the sonication of liquids, the sound waves that propagate into the liquid medium result in alternate cycles of high pressure (compression) and low pressure (rarefaction) that depend on frequency. During cycles of low pressure, ultrasonic waves of high intensity create small empty bubbles in the liquid. When the bubbles attain a volume at which they can no longer absorb energy, they collapse violently during a high-pressure cycle. This phenomenon is called cavitation (Figure 6.11(a)).

The implosion of cavitation bubbles results in micro-jets and microturbulences of up to 1000 km/h. This leads to mechanical stress on the electrostatic attractive forces between particles (van der Waals forces, for example). Large particles are subject to surface erosion due to cavitation collapse in the surrounding liquid or a reduction in size (Figure 6.12) resulting from fission through interparticle collision

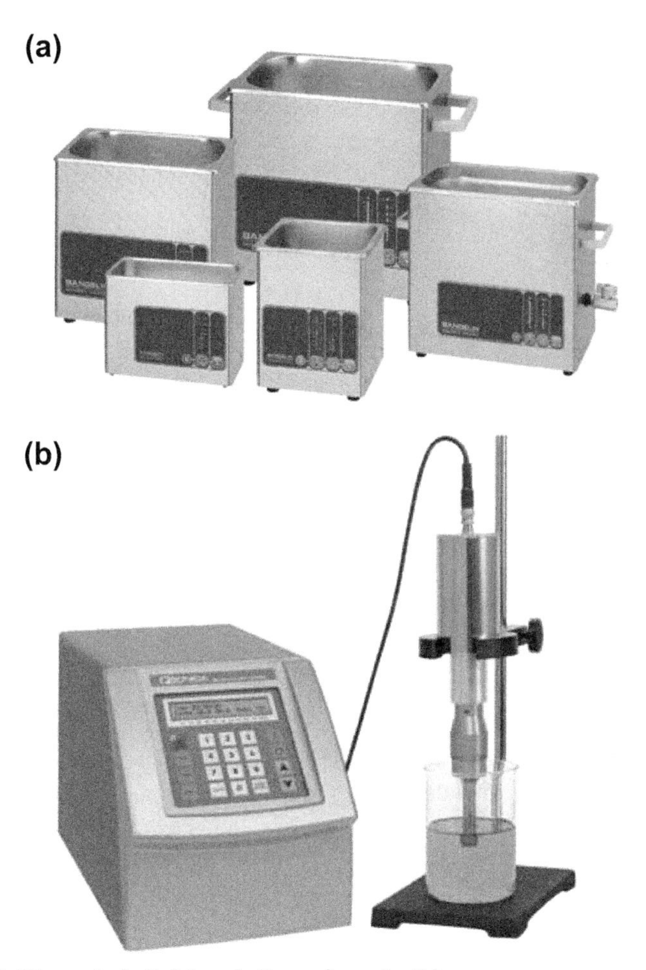

FIGURE 6.10 Ultrasonic bath (a) and ultrasonic probe (b).

Figure sourced from Sonicator.com [14]

or the collapse of cavitation bubbles formed on the surface. During the implosion, emperatures of 5000 K and pressures up to ~2000 atm can be reached locally.

The sonication technology can be used for:

- Quick dissolving, through breakdown of molecular interactions;
- Energy source for certain chemical reactions (sonochemistry);
- Removal of gases dissolved in liquids (degassing) by sonication of the liquid while it is under vacuum;
- Clean surfaces.

The probes used for sonication (Figure 6.11(b)) are made of titanium and machined to specific sizes and shapes so that when they reach the resonant frequency

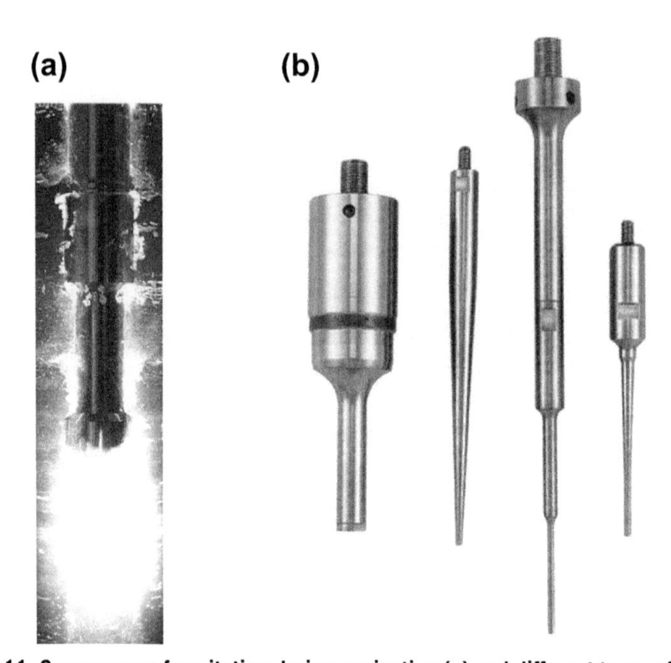

FIGURE 6.11 Occurrence of cavitation during sonication (a) and different types of probes (b).

Figures sourced from Sonicator.com and Hielscher.com [14,15]

they expand and contract longitudinally. This mechanical vibration is amplified and transmitted down the length of the probe. The temperature monitoring prevents overheating of the sample. The diameter of the probe defines the portion of the sample that can be effectively processed. Tips with smaller diameters provide high-intensity sonication but the energy is focused in a small, concentrated area. Tips with larger diameters can process large volumes but offer lower intensity. Over time, the probes erode and must be replaced.

Apparatus applied in laboratories can be used for volumes up to 2 L, whereas industrial sonicators are used in the process of development of batches of 0.5 up to 2000 L, or flows of 0.1 L up to 20 m^3/h. The sonication technology is extensively used to disperse nanoparticles such as CNTs in laboratories. However, this technique is not considered to be the most attractive for industrial applications. Note that excessive use of sonication can cause fracture of nanotubes, drastically reducing their aspect ratio or even destroying it partially.

6.2.5 Dual asymmetric centrifugation

A dual asymmetric centrifuge (DAC) is a type of centrifuge equipment in which, as usual, a container is rotated around the axis at a defined distance and velocity. The main difference of a DAC in relation to normal centrifugation is that the vessel containing the sample rotates around its own center (vertical axis) during the normal spin (Figure 6.13). This results in two overlapping movements of the sample in

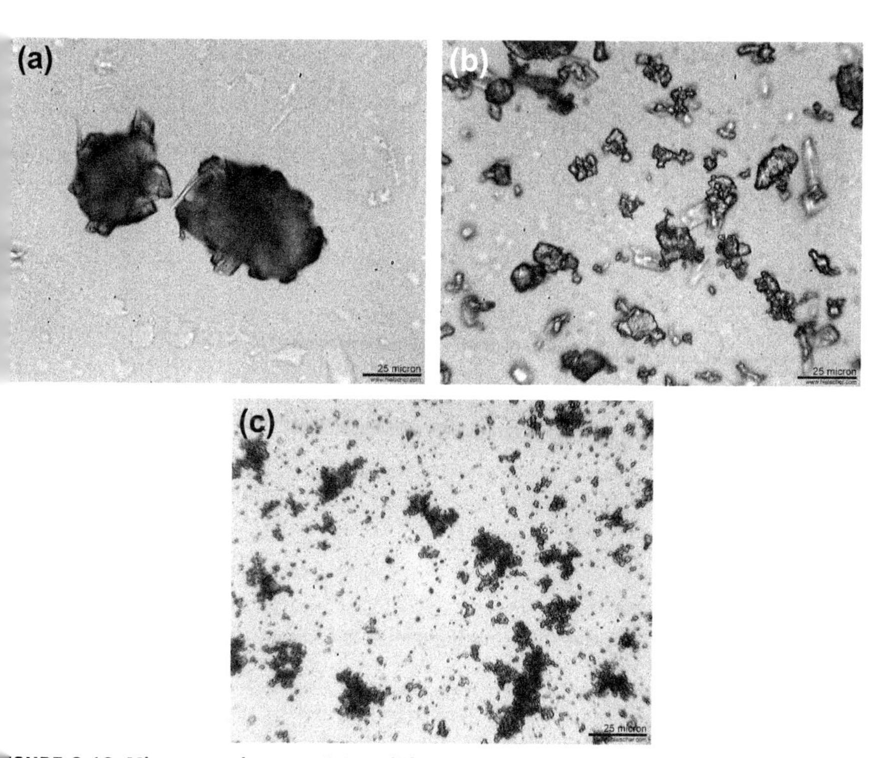

IGURE 6.12 Microscopy images of the milling process of sodium carbonate (Na_2CO_3) in sopropanol.

ote that the size of the particles decreases from (a) to (c).

Figure sourced from Hielscher.com [16]

ne container: while the main rotation "pushes" the sample in a direction outside inertia), the container rotation around its own center "pushes" the sample in the pposite direction resulting from adhesion between the material making up the sam- le and the container in rotation. This latter movement, the internal transport of the ample, is effective if there is sufficient adhesion of the sample in the material of the ontainer and the sample is viscous enough, because both influence the amount of nergy that can be transferred in the sample. This amount of movement results in high shear strength within the sample.

DAC has been used since the 1970s as a convenient technology for rapid mixing f viscous components. Because the DAC mixing is extremely fast, DAC technology also called SpeedMixer.

Figure 6.14 shows a schematic drawing of a DAC. The main arm of rotation of the AC makes an angle of ~40° with the plane of rotation. At this angle, the rotation rm forces the contents of the container into the corner formed between the bottom nd the wall of the container. The DAC shown in Figure 6.13 provides a maximum peed of 3540 rpm and allows the sample to an acceleration of 911 *g*.

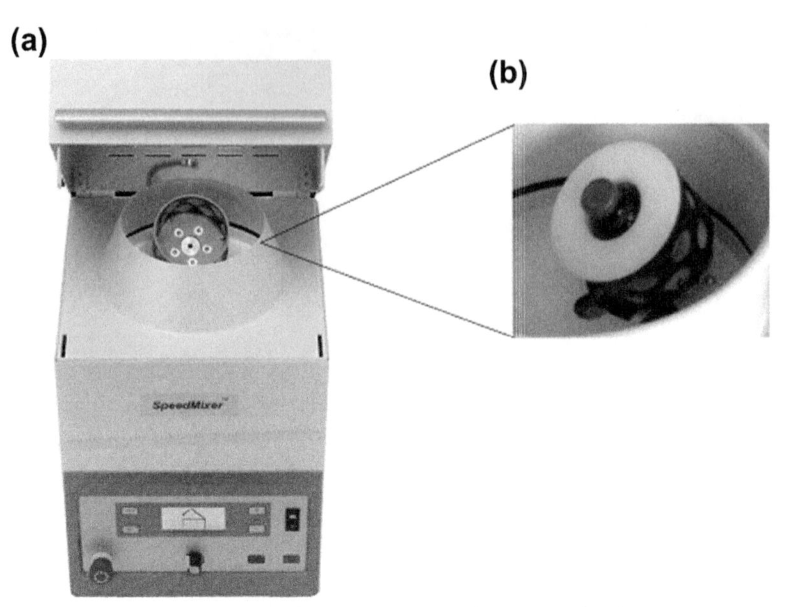

FIGURE 6.13 Typical dual asymmetric centrifuge used in laboratories for samples up to 150 *g*.

DAC with the lid open (a) and view of the rotation chamber (b).

Figures sourced from Speedmixer.com and Dual asymmetric centrifugation [17,18].

The DAC has been greatly used for the dispersion of CNTs in epoxy matrices.

6.2.6 Three-roll mill (calender)

A three-roller mill (or simply calender) is a machine that uses the shear force created by three horizontally positioned rolls rotating in opposite directions and different speeds relative to each other to mix, refine, disperse, or homogenize viscous materials. Figure 6.15 shows a calender typically used in laboratories.

The operation of a calender is simple (Figure 6.16). The material to be processed is placed between the feed roll and the center roll. Each adjacent roll rotates in progressively higher speeds. For example, if the feed roll (first roll) rotates at 30 rpm, the center roll should spin at 90 rpm and 270 rpm on the last roll. The material is transferred from the center roll to the end roll by adhesion. The dispersion is achieved from shear forces generated between the adjacent rolls. The processed material is then removed from the calender using a spatula.

The opening (gap) between the rollers can be mechanically or hydraulically adjusted and maintained. Typically the gap between the rollers is much greater than the particle size. In some cases, such as nanoparticles, the band gap is gradually decreased to reach the desired level of dispersion. Currently, gaps as small as 5 μm can be used.

This equipment is widely used for mixing paints, high-performance ceramics, cosmetics, carbon/graphite, pharmaceuticals, dental composites, pigmented coatings,

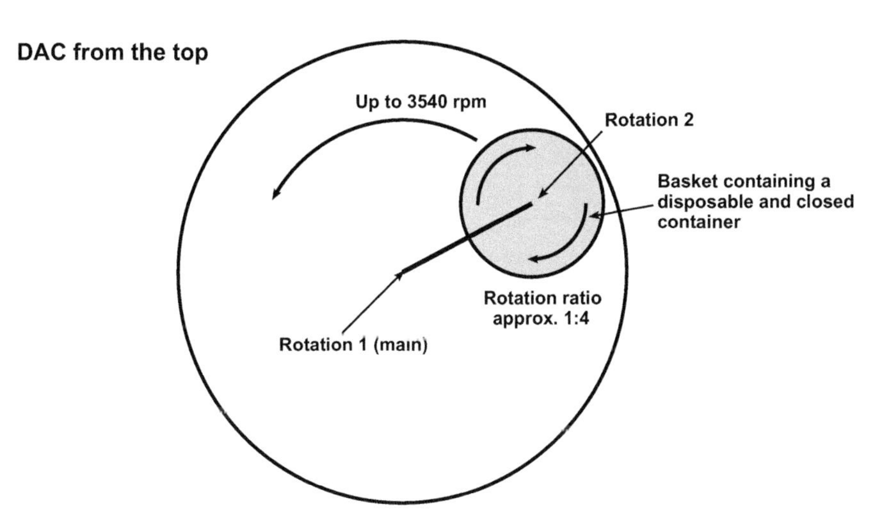

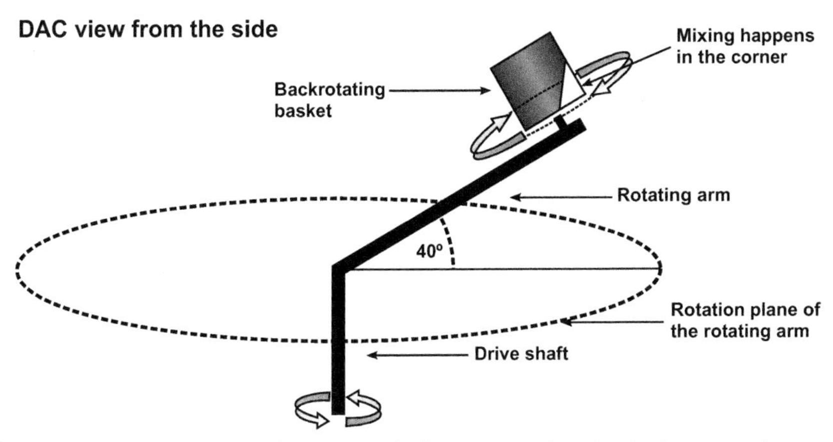

FIGURE 6.14 Schematic drawing of the principle of operation of a dual asymmetric centrifuge (DAC).

Figure sourced from Dual asymmetric centrifugation [18]

adhesives, and sealants and recently proved extremely efficient for the dispersion of CNTs [2,21]. In fact, the use of three-roll mills has been proven to be one of the most suitable for the dispersion of nanotubes in epoxy resins at an industrial scale and, as a result, companies use this equipment to prepare master batches.

6.2.7 Extrusion

Extrusion is the processing method most commonly used for rubbers and thermoplastics. As the name suggests, this process occurs in an extruder (Figure 6.17). Basically, the raw material in pellet form feeds the hopper mounted on the extruder

FIGURE 6.15 Three-roll mill used in laboratories.

Figure sourced from Exakt website [19]

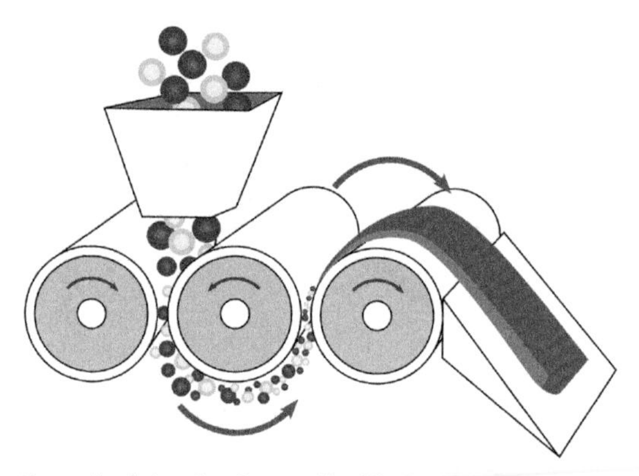

FIGURE 6.16 Operating principle of a three-roll mill.

Figure sourced from Exakt website [20]

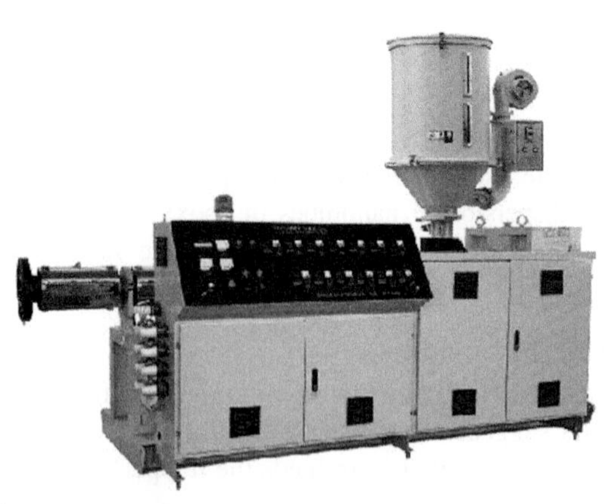

FIGURE 6.17 Single-screw extruder.

Figure sourced from Faygounian website [22]

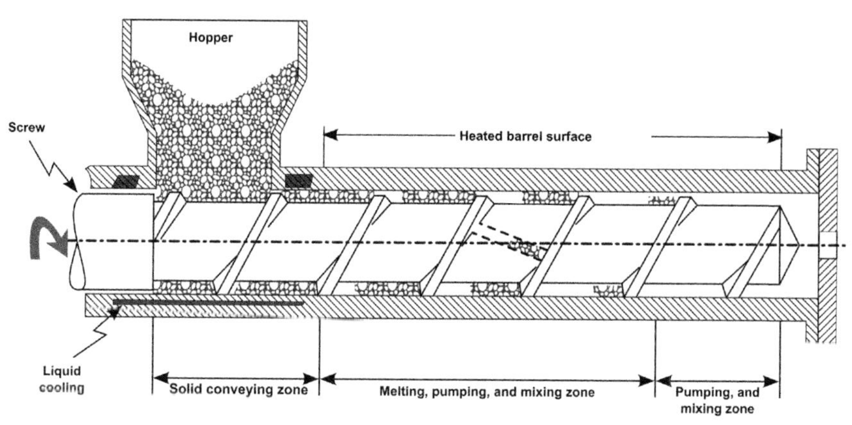

IGURE 6.18 Schematic representation of a single-screw extruder.

Figure sourced from AIPMA website [23]

arrel by simple gravity (Figure 6.18). Additives such as colorants, ultraviolet nhibitors, or even CNTs may be mixed with the raw material before it reaches he hopper.

The material enters through the hole in the rear of the barrel and makes contact vith the screw. The rotating screw forces the plastic through the barrel, which is usu-lly heated at the melting temperature of the polymer. In most cases, one heating rofile is adapted to the barrel where three or more zones of controlled heating tem-erature gradually increases from the rear to the front. This enables gradual melting f the polymer as it is pushed through the barrel and reduces the risk of overheating, vhich may cause degradation.

Most extruders also contain cooling sections (coolers) to maintain the tempera-ure below certain values. This is necessary because the actual friction (shear) that akes part in the cannon generates heat. After going through various sections of ne screw the molten plastic enters the matrix, which gives the final shape to the ma-erial. Cooling of the final piece is then required to enable solidification.

The extrusion technique is frequently used for the preparation of nanocomposites vith thermoplastic matrix because this shear force between screw and barrel of an xtruder is sufficient for dispersion of nanotubes. In fact, nanotubes can be broken p during extrusion. Moreover, the alignment of nanotubes in polymer fibers can e achieved by using extrusion.

'o learn more...

Ve encourage readers to consult the following references:

. Tadmor Z. Principles of polymer processing. 2nd ed. Wiley-Interscience; 2006, ISBN-13: 978-0471387701.

2. Koo JH. Polymer nanocomposites: processing, characterization, and applications. 1st ed. McGraw-Hill Professional; 2006, ISBN-13: 978-0071458215.
3. Leng J, Kin-tak Lau A. Multifunctional polymer nanocomposites. 1st ed. CRC Press; 2010, ISBN-13: 978-1439816820.

References

[1] Tugrul Seyhan A, Tanoglu M, Schulte K. Tensile mechanical behavior and fracture toughness of MWCNT and DWCNT modified vinyl-ester/polyester hybrid nanocomposites produced by 3-roll milling. Mater Sci Eng A 2009;523:85–92.

[2] Fiedler B, Gojny FH, Wichmann MHG, Nolte MCM, Schulte K. Fundamental aspects of nano-reinforced composites. Compos Sci Technol 2006;66:3115–25.

[3] Baek J-B, Lyons CB, Tan L-S. Grafting of vapor-grown carbon nanofibers via in situ polycondensation of 3-phenoxybenzoic acid in poly(phosphoric acid). Macromolecules 2004;37:8278–85.

[4] Loos MR. Development of functionalized polyoxadiazole nanocomposites. TuTech Innovation, Inc. 2010; ISBN 978-3941492257. p. 173.

[5] Coleman JN, Khan U, Blau WJ, Gun'ko YK. Small but strong: a review of the mechanical properties of carbon nanotube-polymer composites. Carbon 2006;44:1624–52.

[6] Du J-H, Bai J, Cheng H-M. The present status and key problems of carbon nanotube based polymer composites. eXPRESS Polym Lett 2007;1:253–73. http://dx.doi.org/10.3144/expresspolymlett.2007.39.

[7] Andrews R, Jacques D, Minot M, Rantell T. Fabrication of carbon multiwall nanotube/polymer composites by shear mixing. Macromol Mater Eng 2002;287:395–403. http://dx.doi.org/10.1002/1439-2054(20020601)287:6<395::AID-MAME395>3.0.CO;2-S.

[8] Breuer O, Sundararaj U. Big returns from small fibers: a review of polymer/carbon nanotubes composites. Polym Compos 2004;25(6):630–45.

[9] www.keison.co.uk/stuart_cr302.shtml.

[10] www.lihe-china.com/ptfe_coated_magnets.htm.

[11] http://spectraservices.com/84020301.html.

[12] www.mixers.com/whitepapers/guide_nano.pdf.

[13] Loos MR, Yang J, Feke DL, Manas-Zloczower I, Unal S, Younes U. Enhancement of fatigue life of polyurethane composites containing carbon nanotubes. Compos Part B Eng 2013:740–4.

[14] www.sonicator.com/sonicatorq500.aspx.

[15] www.hielscher.com/ultrasonics/water_disinfection.htm.

[16] www.hielscher.com/ultrasonics/mill_01.htm.

[17] www.speedmixer.com/.

[18] Massing U, Cicko S, Ziroli V. Dual asymmetric centrifugation (dac)—a new technique for liposome preparation. J Control Release 2008;125(1):16–24.

[19] www.exakt.de/superfine-models.30+m52087573ab0.0.html.

[20] www.exakt.de/exakt-three-roll-mil.38+m52087573ab0.0.html.

[21] Gojny FH, Wichmann MHG, Fiedler B, Schulte K. Influence of different carbon nanotubes on the mechanical properties of epoxy matrix composites – a comparative study. Compos Sci Technol 2005;65:2300–13.

[22] http://faygounion.en.made-in-china.com/.

[23] www.aipma.net/plasticprocess01.html.

CHAPTER

Applications of CNTs

7

CHAPTER OUTLINE

Because of their exceptional properties, carbon nanotubes (CNTs) are already used in various applications and have been gaining place in the market. Although the price of CNTs has declined in the past few years, it may take a while for products containing CNTs to become more economically viable. However, this does not

Carbon Nanotube Reinforced Composites. http://dx.doi.org/10.1016/B978-1-4557-3195-4.00007-2

prevent the imagination of engineers and scientists to create the most diverse applications. The figure shows a small-scale wind turbine that had the blades manufactured using CNTs [1].

7.1 Carbon nanotubes: present and future applications

The sports industry has found several applications for nanotubes. The company Nanoledge, Canada, uses carbon nanotubes (CNTs) to make tennis rackets stronger and lighter (Figure 7.1) [2]. The company Montreal from Finland uses multiwalled carbon nanotubes (MWCNTs) produced by Bayer MaterialScience, Germany, in the manufacturing of hockey sticks (Figure 7.2) [3]. Still in the area of sports, the company Zyvex has provided resins containing dispersed CNTs for the manufacture of Easton brand baseball bats (Figure 7.3) [4].

In the maritime sector, the company Synergy Yachts applied the resin Zyvex used in the manufacturing of a part of the sailboat Synergy 350RL (Figure 7.4) [5]. The entire hull is made with carbon fiber, whereas the mast contains dispersed CNTs. The application of dispersed CNTs in cars and airplanes is also now under study, and on a small scale is already a reality [1]. Nanotubes also have applications in transportation. The bike company BMC Switzerland manufactured the world's first bicycle containing dispersed CNTs and it was used in the Tour de France 2006 (Figure 7.5) [6]. The construction of this bike, with a framework of less than 1 kg of mass, was the result of a partnership between the BMC and the companies Zyvex and Easton.

FIGURE 7.1 Tennis racket made of carbon nanotubes.

Figure sourced from Nanooze website [2]

IGURE 7.2 Hockey sticks produced with multiwalled carbon nanotubes.

Figure sourced from Montreal Hockey website [3]

IGURE 7.3 Easton baseball bat manufactured with carbon nanotubes.

Figure sourced from Zyvextech website [4]

GURE 7.4 Sail boat built with carbon fibers and nanotubes to reinforce the mast.

Figure sourced from Zyvextech website [5]

FIGURE 7.5 Bicycle with frame made of carbon nanotubes. CNTs make the bike lighter and with a high mechanical resistance.

Figure sourced from Azo Nano website [6]

Perhaps, one of the most interesting applications of CNTs is their use in the manufacturing of shoes. It is already possible to buy the Adidas adizero with nanotubes (Figure 7.6) [7]. In this case, CNTs also made the shoes lighter and more rigid.

Recently, carbon nanotubes were used to manufacture the blades of a wind turbine on a small scale (Figure 7.7). The prototype, built by Marcio R. Loos et al., consists of glass fibers and polyurethane; the addition of CNTs improved the fatigue resistance of the blades [1]. In this case, block copolymers have been used to aid in the dispersion of nanotubes.

CNTs' conductivity has a great sensitivity to the surface adsorbates and permits their use as highly sensitive nano-sensors [8]. These properties make CNTs attractive for a wide range of electrochemical biosensors (Figure 7.8). Applications range

FIGURE 7.6 Adidas adizero shoes manufactured with carbon nanotubes.

Figure sourced from Adidas website [7]

FIGURE 7.7 Wind turbine blades made of carbon nanotubes.

Figure sourced from Google [1]

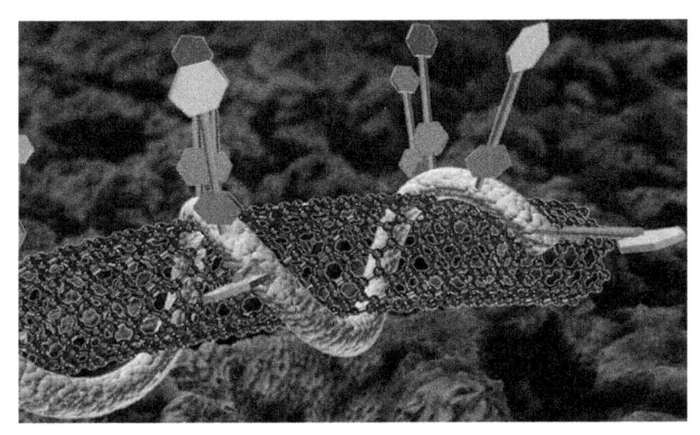

FIGURE 7.8 Hydrophilic microbiosensor based on DNA-modified single-walled carbon nanotube decorated with platinum black nanoparticles for attaching enzymes.

Figure sourced from Microbiosensors based on DNA modified single-walled carbon nanotube and Pt black nanocomposites [10]

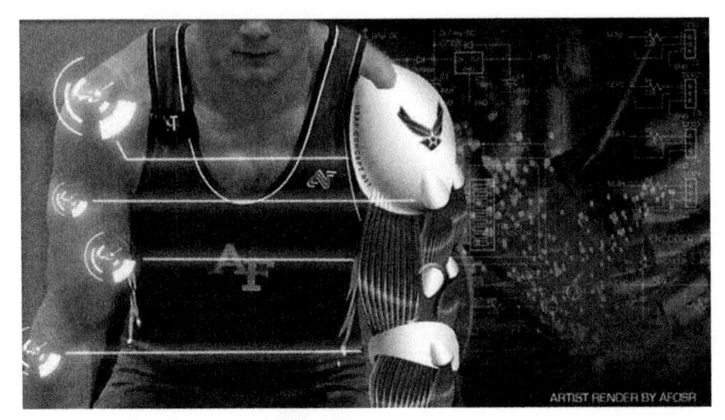

FIGURE 7.9 Artistic rendering of Air Force Research Laboratory–sponsored University of Texas at Dallas efforts to create a carbon nanotube–based artificial muscle for Air Force aerospace and space applications.

Figure sourced from Air Force Office of Scientific Research [12]

from amperometric enzyme electrodes to DNA hybridization biosensors. In addition, CNTs can also be modified with protein receptors that pick up even weak traces of specific chemicals [9]. Once the receptor interacts with the chemical, a current appears on the nanotube, enabling the identification of the target. This technology can be used to search for toxins, bombs, or even be used to test whether someone has skin cancer by checking for traces of chemicals common in these cases.

Researchers at the UT Dallas Alan G. MacDiarmid NanoTech Institute have created a CNT aerogel that expands and contracts as it converts electricity into chemical energy [11]. Because of their ability to retain their shape after being compressed many times, CNTs are suitable for artificial muscles (Figure 7.9). When in an aerogel form, the tubes grow denser under stress, in the same way as weight lifting does to natural muscles. Although natural muscles can contract at about 20% per second, the new artificial muscles can contract at about 30,000% per second.

Geckos can climb up smooth surfaces because of the tiny hairs on their feet, exploiting the electrostatic force between themselves and the wall (Figure 7.10). Scientists have developed a synthetic gecko tape by transferring micropatterned carbon nanotube arrays onto flexible polymer tape based on the hierarchical structure found on the foot of a gecko lizard [13]. The synthetic tape can support a shear stress nearly four times higher than the gecko can. The carbon nanotube-based tape offers an excellent synthetic option as a dry conductive reversible adhesive in microelectronics, robotics, and space applications.

CNTs are excellent field emitters and are great for conductive surfaces. In 2008, Samsung unveiled the world's first carbon nanotube color active matrix electrophoretic display (EPD) e-paper [15]. Because the EPD does not require backlighting, its energy consumption is minimum and it is visible under direct sunlight (Figure 7.11). In addition, the image on the display is retained without the need to constantly refresh. This technology is a low-energy display option for mobile devices.

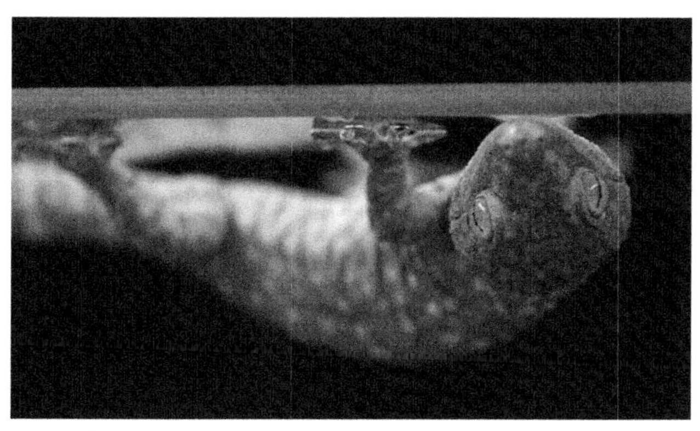

IGURE 7.10 Geckos' unique dry-adhesive toe pads allow them to cling, walk, and even leep upside-down on glass and other slippery surfaces.

Figure sourced from University of Akron's Research [14]

IGURE 7.11 Samsung display.

Figure sourced from Tech News Daily website [9]

A major goal in hard tissue engineering is to combine scaffold materials with ving cells to develop biologically active substitutes that can restore tissue functions 6]. CNTs have been used as part of a bottom-up approach in bone tissue engineerig (Figure 7.12). Additionally, their rod-like shape and nanoscale dimensionality iakes them a morphological biomimetic of the fibrillar proteins in the extracellular iatrix [17]. Although it is technically challenging to fabricate an effective scaffold ystem based on CNTs, the results are encouraging. Studies suggest that CNTs djoining bone induce little local inflammatory reaction, permit bone repair, show igh bone-tissue compatibility, become integrated into new bone, and accelerate one formation [18].

There are many types of drug delivery systems that are currently available [20]. he ability of CNTs to penetrate into the cells brings about the possibility of using

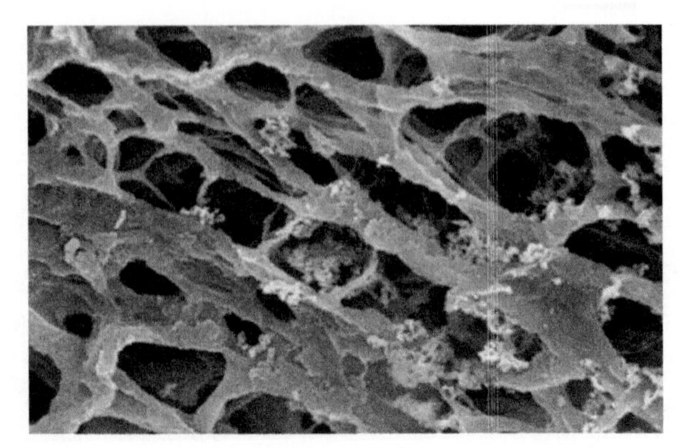

FIGURE 7.12 Scanning electron microscopy analysis of hydroxylapatite formation on collagen composite (4 wt% of CNTs) after 1 week.

Figure sourced from Nanostructured 3-D collagen/nanotube biocomposites for future bone regeneration scaffolds [19].

CNTs as an alternative and efficient tool for transporting and translocating small drug molecules [21,22]. Studies indicate that CNTs can be functionalized with bioactive peptides, proteins, nucleic acids, and drugs, and used to deliver their therapeutic molecules to cells and organs (Figure 7.13). Once functionalized, CNTs

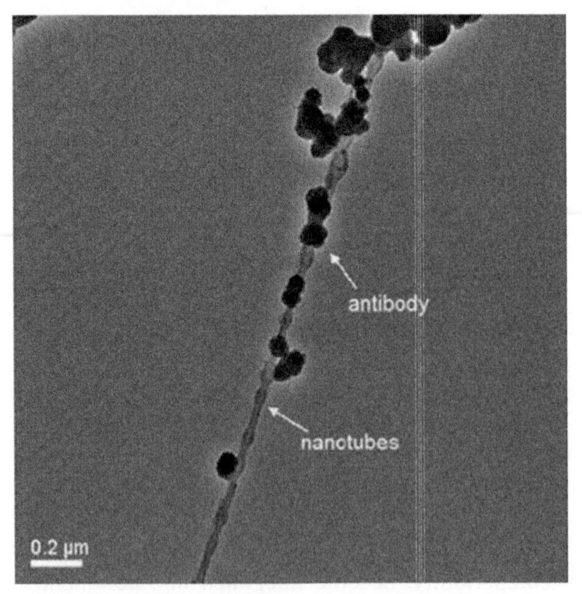

FIGURE 7.13 Transmission electron micrograph of the antibody functionalized single-walled carbon nanotubes (SWCNTs).

Figure sourced from Integrated Molecular Targeting... [23

display low toxicity and are not immunogenic. Results have shown that animals could recover their functions after a stroke because of the silencing of a gene following intracranial injection of small interfering RNA complexed to carbon nanotubes [9].

Because of their higher energy density, long-term operation stability, low level of heating, fast discharge/charge time, and safety, supercapacitors are attractive as an alternative power source [24]. CNTs have been shown to be suitable materials for polarizable electrodes as both single-walled carbon nanotubes (SWCNTs) and MWCNTs have been studied for electrochemical supercapacitor electrodes, as shown in Figure 7.14. In fact, CNT-based electrode materials, including CNTs, CNT/oxide composite, and CNT/polymer composite, satisfy the three basic requirements of high capacitance, low resistance, and stability for use as electrode in supercapacitors. The elevated surface area of nanotubes also enables the energy to be stored all over the tube, not only at the ends, like in a conventional capacitor [9,25].

Extensive research has been carried out toward the development of CNT-based organic solar cells (Figure 7.15) [27]. Many studies have focused on synergistically combine CNTs with complementary moieties, such as metal oxides, quantum dots, and/or polymers together, to create unique and novel heterostructures, combining the favorable optoelectronic properties of each individual component aiming to improve the overall efficiency of the resulting solar cell configurations. These systems are developed to substitute for semitransparent indium tin oxide in photovoltaic

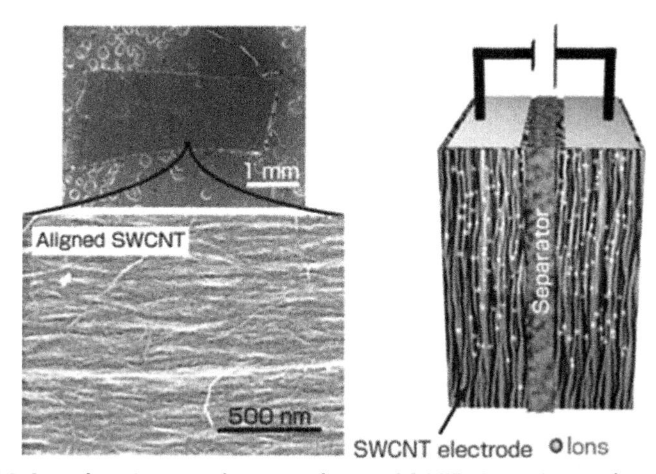

FIGURE 7.14 Scanning electron microscope image of SWNT electrode showing ordered pore structure and alignment (left) and typical cell assembly of a super capacitor electrode using SWCNTs (right).

Figure sourced from Extracting the full potential of single-walled carbon nanotubes as durable supercapacitor electrodes operable at 4 V with high power and energy density [26]

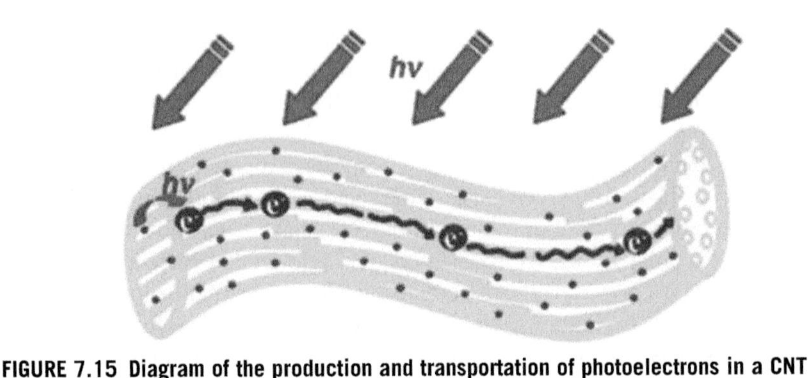

FIGURE 7.15 Diagram of the production and transportation of photoelectrons in a CNT composite fiber used for solar cell fabrication. The top arrows represent sunlight. The tube, dots, and spheres correspond to the CNT and photoelectrons, respectively.

Figure sourced from Flexible, light-weight, ultrastrong, and semiconductive carbon nanotube fibers for a highly efficient solar cell [30]

cells [28]. Power conversion efficiencies of up to 14% [29] have been achieved using CNT-based heterostructures as components of semiconducting-based solar cells. Although the results are encouraging, it should be noted that the efficiencies of CNT-based solar cells, as compared with conventional silicon-based solar cells, remains relatively low.

Because of its huge surface area of up 1000 m^2/g, size, and adsorption properties, carbon nanotubes are ideal for filtration of dissolved salts, toxic chemicals, and biological contaminants from water. Using CNT-based membranes, salt and other ionic compounds can be filtered out of seawater or brackish water [31]. This concept is depicted in Figure 7.16. All these can be achieved by spending substantially less energy than is needed to achieve similar results using conventional polymer-based membranes. This technology can help to bring potable water for millions of people around the world as well as help extract drinking water from the ocean.

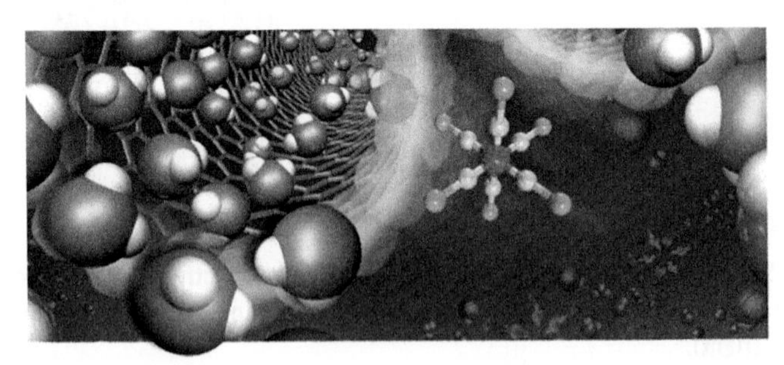

FIGURE 7.16 Inside CNT water molecules flow at an ultrafast rate. The image depicts an ion being rejected by a CNT pore.

Figure sourced from Science & Technology Review [31]

According to the Department of Energy, half of a fuel cell's price comes from he platinum catalyst used to speed up the reaction that produces energy [32]. latinum is a limited natural resource and its cost ranges from $800 to $2200 er ounce [33]. Carbon nanotubes are promising for application as support for catlytic materials to prepare novel highly efficient electrodes for fuel cells because of heir catalytic activity [33,34]. Figure 7.17 shows a carbon-nanotube-based atalyst.

A group of Chinese researchers discovered that very thin carbon nanotube films ould emit loud sounds after they are fed by sound frequency electric currents [36]. his finding allowed the manufacture of CNT thin film loudspeakers, which are a anometer thick and are transparent, flexible, stretchable, and magnet-free Figure 7.18). These new single-element thin film loudspeakers can be tailored into variety of insulating surfaces, such as room walls, magazines, pillars, eilings, flags, windows, and even clothes. The thin films could be used in devices uch as earphones and buzzers. The new speakers do not generate sound by vibrating e surrounding air molecules like conventional speakers do. They harness a phenomnon called the thermo-acoustic effect: when an electric current runs through the nanobe thin films, they heat and expand the air near them, creating sound waves.

CNTs make a major commercial success in lithium-ion batteries (Figure 7.19). hey have been found as outstanding conductive additives for these types of battees. Because of their elevated specific surface area and conductivity, the addition of NTs in the graphite anode can lead to graphite/CNT electrodes with enhanced yclability, rate capability, and safety. CNTs also aid electrical connectivity and echanical integrity [37]. For example, it was found that the addition of 10 wt%

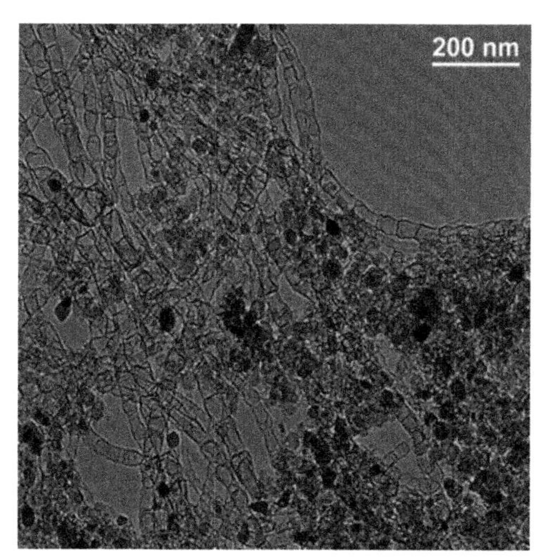

GURE 7.17 High-resolution microscopic image of a new type of carbon nanotube–based atalyst.

Figure sourced from Los Alamos National Laboratory [35]

FIGURE 7.18 Paper-thin cylinder composed of carbon nanotubes emits sound in all directions.

Figure sourced from Flexible, stretchable, transparent carbon nanotube thin film loudspeakers [36

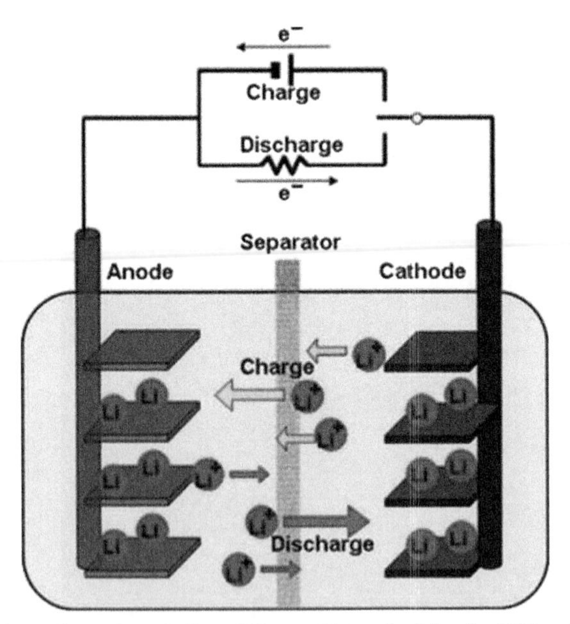

FIGURE 7.19 Schematic representation of the working principle of a lithium-ion battery.

Figure sourced from Carbon nanomaterials for advanced energy conversion and storage [39

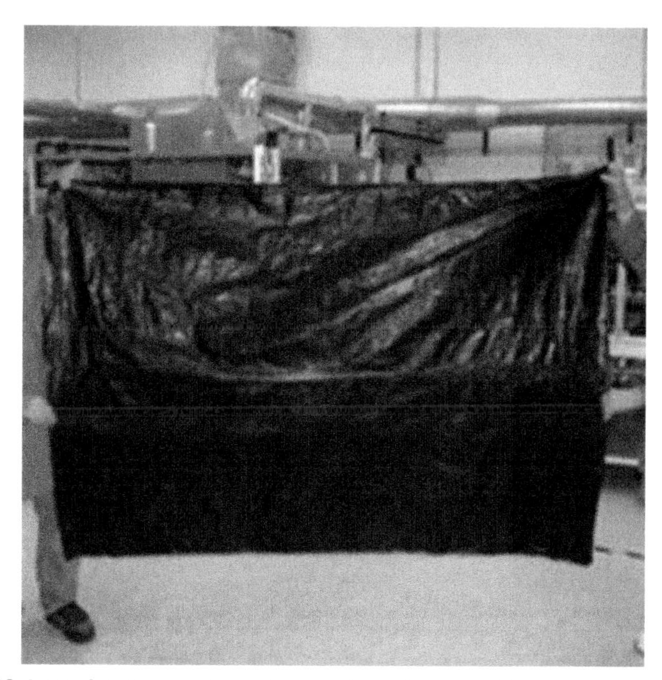

FIGURE 7.20 Large-format carbon nanotube textiles which could be used for electromagnetic interference shielding.

Figure sourced from Nanocomp Technologies [44]

of MWCNTs maintained the cyclic efficiency of a graphite anode at almost 100% up to 50 cycles [38]. Currently, more than 50% of notebooks and cell phones are already using batteries that contain CNTs.

Electromagnetic interference (EMI) is an undesirable offshoot of electronics. EMI shielding is necessary to protect the environment and workspace from radiation coming from electronic equipment and also to protect sensitive circuits [40]. Compared with conventional metal-based EMI shielding materials, conducting polymers composites reinforced with CNTs represent a novel class of materials that possess unique combination of properties useful for suppression of electromagnetic noises [41]. This class of materials has gained popularity because of their light weight, flexibility, resistance to corrosion, and processing advantages [40]. In fact, there are companies currently producing CNT-based lightweight electromagnetic shields such as shown in Figure 7.20 [42,43].

Because of their low electron scattering and their band gap, SWCNTs are also attractive for manufacturing transistors [45]. Printed CNT transistors on polymer film (Figure 7.21) or simply CNT thin-film transistors (TFTs) are attractive for driving organic light-emitting diode displays. They have shown higher mobility than amorphous silicon [46] and can be deposited by low temperature and

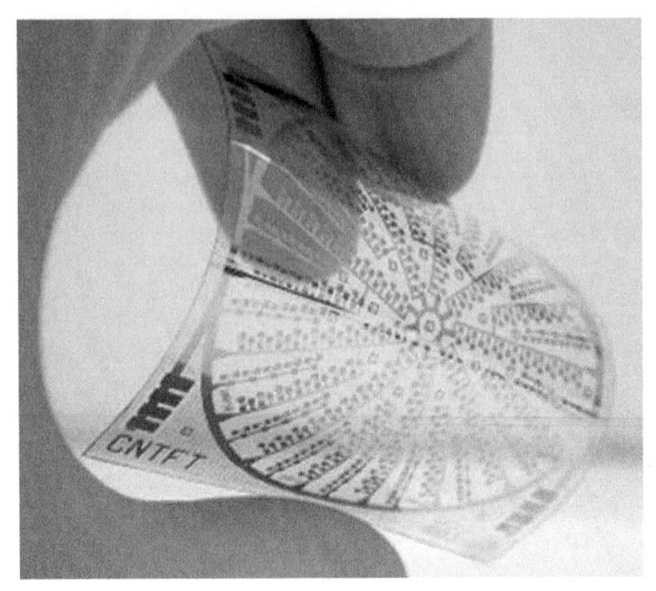

FIGURE 7.21 Printed CNT transistors on a flexible plastic substrate.

Figure sourced from University of Stuttgart [48]

nonvacuum methods, such as printing. Applications envisioned for TFTs include flexible electronics, such as displays, radiofrequency identification for packaging or on clothes, D flip-flop, and smart materials [47].

To learn more...

We encourage readers to consult the following references:

1. Qing Zhang, editor. Carbon nanotubes and their applications (Pan Stanford series on carbon-based nanomaterials); April 23, 2012, ISBN-13: 978-9814241908.
2. Morinobu Endo, Michael S Strano, Pulickel M Ajayan. Potential applications of carbon nanotubes. Carbon Nanotubes Topics in *Appl Phys* 2008; 111: 13–61, ISBN: 978-3-540-72,864-1.

References

[1] www.sites.google.com/site/cntcomposites/in-the-news.
[2] www.nanooze.org/sample-post-5/.
[3] www.montrealhockey-usa.com/index.htm.
[4] www.zyvextech.com/.
[5] www.zyvextech.com/news/2006/10/first-cnt-sailboat-mast-released-at-nanotx-06.

[6] www.azonano.com/article.aspx?articleid=1298.
[7] www.adidas.com/us/product/running-adizero-adios-2-shoes/QP620?cid=G95119.
[8] Wang Joseph. Carbon-nanotube based electrochemical biosensors: a review. Electroanalysis 2005;17(1):7–14. http://dx.doi.org/10.1002/elan.200403113/. Article first published online: November 19, 2004.
[9] www.technewsdaily.com/16158-10-uses-for-carbon-nanotubes.html.
[10] Shi Jin, Cha Tae-Gon, Claussen Jonathan C, Diggs Alfred R, Choibf Jong Hyun, Porterfield D Marshall. Microbiosensors based on DNA modified single-walled carbon nanotube and Pt black nanocomposites. Analyst 2011;136:4916–24. http://dx.doi.org/10.1039/C1AN15179G.
[11] www.utdallas.edu/news/2009/03/20-001.php.
[12] www.nanopatentsandinnovations.blogspot.com.br/2009/11/carbon-nanotube-artificial-muscles-for.html.
[13] Ge Lichui, Sethi Sunny, Ci Lijie, Ajayan Pulickel M, Dhinojwala Ali. Carbon nanotube-based synthetic gecko tapes. PNAS June 26, 2007;104(26):0792–10795. http://dx.doi.org/10.1073/pnas.0703505104.
[14] www.cleveland.com/science/index.ssf/2012/09/university_of_akrons_research.html.
[15] www.cleantechnica.com/2008/10/22/samsung-demonstrates-worlds-first-carbon-nanotube-based-display/#kS1oq7qGGHcReDuM.99.
[16] Han Zhao Jun, Rider Amanda E, Ishaq Musarat, Kumar Shailesh, Kondyurin Alexey, Bilek Marcela MM, et al. Carbon nanostructures for hard tissue engineering. RSC Adv 2013;3:11058–72. http://dx.doi.org/10.1039/C2RA23306A. First published online January 4, 2013.
[17] Newman Peter, Hons Andrew Minett, Ellis-Behnke Rutledge, Zreiqat Hala. Carbon nanotubes: their potential and pitfalls for bone tissue regeneration and engineering. Nanomed. Nanotechnol Biol Med, Available online: June 12, 2013. http://dx.doi.org/10.1016/j.nano.2013.06.001.
[18] Usui Yuki, Aoki Kaoru, Narita Nobuyo, Murakami Narumichi, Nakamura Isao, Nakamura Koichi, et al. Carbon nanotubes with high bone-tissue compatibility and bone-formation acceleration effects. Small February 1, 2008;4(2):240–6. http://dx.doi.org/10.1002/smll.200700670.
[19] da Silva Edelma E, Della Colleta Heloisa HM, Ferlauto Andre S, Moreira Roberto L, Resende Rodrigo R, Oliveira Sergio, et al. Nanostructured 3-D collagen/nanotube biocomposites for future bone regeneration scaffolds. Nano Res 2009;2:462–73.
[20] Bianco Alberto, Kostarelos Kostas, Prato Maurizio. Applications of carbon nanotubes in drug delivery. Curr Opin Chem Biol December 2005;9(6):674–9. http://dx.doi.org/10.1016/j.cbpa.2005.10.005.
[21] Pantarotto D, Briand JP, Prato M, Bianco A. Translocation of bioactive peptides across cell membranes by carbon nanotubes. Chem Commun (Camb); 2004:16–7. http://dx.doi.org/10.1039/b311254c.
[22] Shi Kam NW, Jessop TC, Wender PA, Dai H. Nanotube molecular transporters: internalization of carbon nanotube-protein conjugates into mammalian cells. J Am Chem Soc 2004;126:6850–1. http://dx.doi.org/10.1021/ja0486059.
[23] Shao N, Lu S, Wickstrom E, Panchapakesan B. Integrated molecular targeting of IGF1T and HER2 surface receptors and destruction of breast cancer cells using single wall carbon nanotubes. Nanotechnology 2007;18:1–9.
[24] Pan Hui, Li Jianyi, Feng YuanPing. Carbon nanotubes for supercapacitor. Nanoscale Res Lett 2010;5(3):654–68. http://dx.doi.org/10.1007/s11671-009-9508-2.

[25] http://japantechniche.com/2011/04/29/development-of-a-single-walled-nanotubes-super-capacitor-electrode-by-aist/.
[26] Izadi-Najafabadi Ali, Yasuda Satoshi, Kobashi Kazufumi, Yamada Takeo, Futaba don N, Hatori Hiroaki, et al. Extracting the full potential of single-walled carbon nanotubes as durable supercapacitor electrodes operable at 4 V with high power and energy density. Adv Mater September 15, 2010;22(35):E235–41. http://dx.doi.org/10.1002/adma.200904349.
[27] Wang Lei, Liu Haiqing, Konik Robert M, Misewich James A, Wong Stanislaus S. Carbon nanotube-based heterostructures for solar energy applications. Chem Soc Rev 2013;42:8134–56. http://dx.doi.org/10.1039/C3CS60088B.
[28] Campidelli S, Klumpp C, Bianco A, Guldi DM, Prato M. Functionalization of CNT: synthesis and applications in photovoltaics and biology. J Phys Org Chem 2006;19: 531–9. http://dx.doi.org/10.1002/poc.1052.
[29] Tune DD, Flavel BS, Krupke R, Shapter JG. Carbon nanotube-silicon solar cells. Adv Energy Mater 2012;2:1043–55. http://dx.doi.org/10.1002/aenm.201200249.
[30] Chen Tao, Wang Shutao, Yang Zhibin, Feng Quanyou, Sun Xuemei, Li Li, et al. Flexible, light-weight, ultrastrong, and semiconductive carbon nanotube fibers for a highly efficient solar cell. Angew Chem February 18, 2011;50(8):1815–9. http://dx.doi.org/10.1002/anie.201003870.
[31] www.str.llnl.gov/OctNov10/fornasiero.html.
[32] www.discovermagazine.com/2009/jul-aug/09-ways-carbon-nanotubes-just-might-rock-world#.UkAJLdJ6bVE.
[33] www.westfloridacomponents.com/blog/carbon-nanotubes-can-reduce-the-price-of-fuel-cells/.
[34] Khantimerov SM, Kukovitsky EF, Sainov NA, Suleimanov NM. Fuel cell electrodes based on carbon nanotube/metallic nanoparticles hybrids formed on porous stainless steel pellets. Int J Chem Eng 2013;2013. http://dx.doi.org/10.1155/2013/157098. Article ID 157098, 4 pp.
[35] www.kurzweilai.net/los-alamos-carbon-nanotube-catalyst-could-jumpstart-e-cars-green-energy.
[36] Xiao Lin, Chen Zhuo, Feng Chen, Liu Liang, Bai Zai-Qiao, Wang Yang, et al. Flexible stretchable, transparent carbon nanotube thin film loudspeakers. Nano Lett October 29 2008;8(12):4539–45. http://dx.doi.org/10.1021/nl802750z. Publication Date (Web).
[37] Sotowa Chiaki, Origi Gaku, Takeuchi Masataka, Nishimura Yoshiyuki, Takeuchi Kenji Jang In Young, et al. The reinforcing effect of combined carbon nanotubes and acetylene blacks on the positive electrode of lithium-ion batteries. ChemSusChem November 24, 2008;1(11):911–5. http://dx.doi.org/10.1002/cssc.200800170.
[38] Endo M, Hayashi T, Kim YA. Large-scale production of carbon nanotubes and their applications. Pure Appl Chem 2006;78:1703. http://dx.doi.org/10.1351/pac200678091703.
[39] Dai Liming, Chang Dong Wook, Baek Jong-Beom, Lu Wen. Carbon nanomaterials for advanced energy conversion and storage. Small April 23, 2012;8(8):1130–66. http://dx.doi.org/10.1002/smll.201101594.
[40] Li Ning, Huang Yi, Du Feng, He Xiaobo, Lin Xiao, Gao Hongjun, et al. Electromagnetic interference (EMI) shielding of single-walled carbon nanotube epoxy composites. Nano Lett 2006;6(6):1141–5. http://dx.doi.org/10.1021/nl0602589.
[41] Saini Parveen, Arora Manju. In: De Souza Gomes Ailton, editor. Microwave Absorption and EMI Shielding Behavior of Nanocomposites Based on Intrinsically Conducting

Polymers, Graphene and Carbon Nanotubes, New Polymers for Special Applications. ISBN: 978-953-51-0744-6. InTech, doi:10.5772/48779. Available from: www.intechopen.com/books/new-polymers-for-special-applications/microwave-absorption-and-emi-shielding-behavior-of-nanocomposites-based-on-intrinsically-conducting-; 2012.

[42] ANS Composites Synthetic fibers—www.appliednanostructuredsolutions.com/archives/4.

[43] Axson—www.axson-group.com.

[44] Nanocomp engineers demonstrate the company's large-format carbon nanotube textiles. www.scientificamerican.com/article.cfm?id=carbon-nanotube-emi-protection.

[45] De Volder Michael FL, Tawfick Sameh H, Baughman Ray H, Hart A John. Carbon nanotubes: present and future commercial applications. Science February 1, 2013; 339:535−9. http://dx.doi.org/10.1126/science.1222453.

[46] Dong-ming Sun, Timmermans Marina Y, Ying Tian, Nasibulin Albert G, Kauppinen Esko I, Shigeru Kishimoto, et al. Flexible high-performance carbon nanotube integrated circuits. Nat Nanotechnol 2011;6:156.

[47] Timmermans Marina Y, Estrada David, Nasibulin Albert G, Wood Joshua D, Behnam Ashkan, Sun Dong-ming, et al. Effect of carbon nanotube network morphology on thin film transistor performance. Nano Res 2012;5(5):307−19. http://dx.doi.org/10.1007/s12274-012-0211-8.

[48] www.igm.uni-stuttgart.de/forschung/arbeitsgebiete/cnt/index.en.html.

CHAPTER

s It Worth the Effort to Reinforce Polymers with Carbon Nanotubes?

Marcio R. Loos[1], Karl Schulte[2]

Department of Macromolecular Science and Engineering, Case Western Reserve University, Cleveland, OH, USA[1], Institute of Polymer & Composites, Hamburg University of Technology, Germany[2]

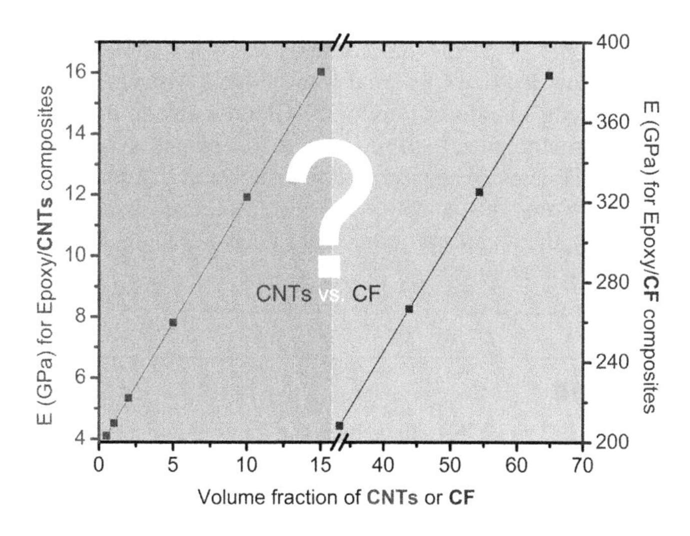

HAPTER OUTLINE

rbon Nanotube Reinforced Composites. http://dx.doi.org/10.1016/B978-1-4557-3195-4.00008-4

Carbon nanotubes (CNTs) show exceptional properties that render them attractive for incorporation in a new generation of high-performance engineering composites with tailored properties. Although a great deal of work has been done toward using CNTs as a reinforcing agent in polymer composites, the full potential of CNTs is yet to be reached. In this work, two case studies are proposed in order to analyze the effectiveness of CNTs and carbon fibers (CFs) as reinforcing agents. Micromechanics models for the stiffness and strength of hybrid composites, comprising CNTs and CFs, are derived by considering the concept of effective fibers. In addition, the 2009 prices of commercially available CNTs are reviewed. The strongest, the stiffest, and the cheapest CFs commercially available are compared with single-walled CNTs (SWCNTs) and multiwalled CNTs (MWCNTs). The simulated results from the micromechanics models show that the use of CFs makes the acquisition of composites with maximum tensile strengths of 4.18 GPa possible. Analysis of the cost versus property relation showed that CNTs are the most viable strengthening option for achieving composites with strengths of up to 11.61 GPa. It also revealed that CFs are the most viable stiffening option, making composites with Young's moduli of up to 383 GPa possible at the expense of the material's toughness. Moreover, to achieve CNTs true potential, several challenges have to be faced. CNTs have to be produced with higher purity, longer lengths, better integrity, in larger amounts, and at lower cost. Moreover, issues such as orientation of the CNTs, their concentration, interfacial adhesion, distribution, and dispersion have to be overcome.

8.1 Introduction

Since Iijima's report in 1991, CNTs have been the focus of considerable research [1] (Figure 8.1). During the past decade, a great deal of effort has been given toward maximizing the potential of CNTs as reinforcing agents in polymer matrix composites. Despite this effort, the full potential of CNT-reinforced composites has not been realized because of processing difficulties and load transfer limitations between the matrix and the nanotubes. Results from simulations predict that composites containing CNTs should have exceptional mechanical properties [3–5]. Table 8.1 shows numerous results reporting the effect of CNTs on tensile properties of various polymer matrices. Although great improvements have been achieved, the results for the elastic modulus and strength of polymer composites have usually been disappointing, particularly when compared with advanced composites reinforced with high-performance continuous fibers.

CFs refer to fibers that are at least 92 wt% carbon in composition [56]. The high moduli of CFs stem from the fact that the carbon layers tend to be parallel to the fiber axis. Moreover, the density of CFs is quite low, making the specific moduli (E/ρ) of high-modulus CFs exceptionally high. Carbon fiber–based composites, particularly those with polymeric matrices, have become the dominant advanced composite materials for aerospace, automobile, sporting goods, and other applications because of

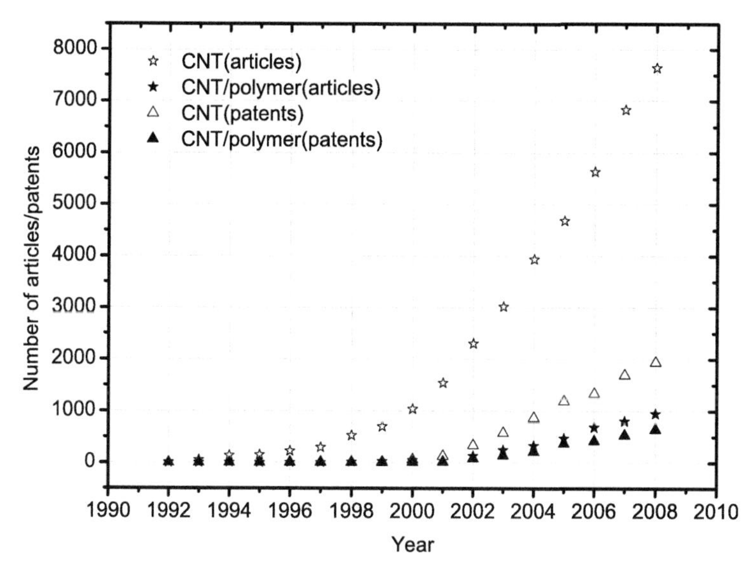

FIGURE 8.1 Number of published articles and patents on nanotubes or nanotube/polymer as a function of year.

CNT, carbon nanotubes.

Figure sourced from Is it worth the effort to reinforce polymers...? [2]

their high strength, high modulus, low density, and reasonable cost. As the price of CFs have decreased with time, their applications have continued to broaden, and now even include the construction industry, which uses CFs to reinforce concrete.

Commercial CFs are fabricated by using pitch or poly(acrylonitrile) (PAN) as the precursor. Among the existing high-performance CFs, those based on pitch can attain higher moduli than those based on PAN because pitch is more graphitizable than PAN. However, the fibers based on PAN can attain a higher tensile strength and greater elongation than those based on pitch. Although the tensile stress–strain curves of CFs are linear to fracture, the main drawback of CFs' mechanical properties is their low ductility, which is lower than those of glass, quartz, and Kevlar fibers. The ductility of high-modulus CFs is even lower than that of high-strength CFs. Commercially available CFs can have a Young's modulus as high as 900 GPa, a tensile strength of up to 6.4 GPa, and for CFs with low moduli, an elongation at break of 2.2%.

On the other hand, for CNTs, experimental results have shown Young's moduli of up to 1800 GPa, tensile strengths of up to 150 GPa, and elongations at break of up to 15% [3]. Figure 8.2 shows a comparison between the Young's modulus, tensile strength, and elongation at break of various CFs with SWCNTs and MWCNTs. The main advantages of CNTs over CFs are their extremely high tensile strength and elongation at break. Another key point for CNTs is the possibility to process composites using standard industrial techniques such as extrusion. Nevertheless, many

Table 8.1 Mechanical Properties Enhancement of Selected Nanocomposite Materials

Matrix or Polymer	Filler	Φ (wt%)	E_{matrix} (GPa)	E_{max} (GPa)	E % Increase	σ_{matrix} (MPa)	σ_{Max} (MPa)	σ % Increase	Year and Reference
ABPBO	MWCNTs	5	1.8	2.6	+44	37	67	+81	2008 [6]
Epoxy	MWCNTs	2	1.18	1.39	+18	52	62	+19	2002 [7]
Epoxy	DWCNTs	0.1	3.29	3.50	+6	–	–	–	2004 [8]
Epoxy	MWCNTs	8	1.05	0.750	−29	43	70	+63	2007 [9]
Epoxy	SWCNTs	0.1	1.10	1.72	+56	–	–	–	2008 [10]
Epoxy	MWCNTs	1	1.21	1.61	+33	26	58	+123	2008 [11]
Epoxy	MWCNTs	0.7	0.52	0.95	+83	–	–	–	2008 [12]
Epoxy	MWCNTs	0.1				47	65	+38	2008 [12]
Epoxy	MWCNTs	5	1.9	2.9	+53	46	52	+13	2008 [13]
Epoxy	MWCNTs	1	1.97	1.77	−10	47.3	47.9	+1	2008 [14]
Epoxy	SWCNTs	1.0	2.76	3.49	+26	64.1	74.7	+17	2008 [15]
Epoxy	MWCNTs	2	2.6	2.9	+12	–	–	–	2009 [16]
Epoxy	SWCNTs	1	3.4	4.2	+24	–	–	–	2009 [17]
Epoxy	MWCNTs	16.5	2.5	20.4	+716	89.1	231.5	+160	2009 [18]
iPP	MWCNTs	2.5	0.60	1.42	+137	28.7	34.5	+20	2008 [19]
LDPE	MWCNTs	10	0.24	0.44	+83	10.7	15.6	+46	2008 [20]
Nylon 610	MWCNTs	1.5	0.9	2.4	+167	35.9	51.4	+43	2009 [21]
PA6	MWCNTs	1	3.3	4.7	+42	60.4	71.5	+18	2008 [22]
PA1010	MWCNTs	30	1.02	1.91	+87	–	–	–	2006 [23]
PBO	MWCNTs	5	3.6	5.2	+44	68	119	+75	2008 [24]

PBO	MWCNTs	0.54	66.6	99.8	+50	1.18	1.51	+28	2008 [25]
PC	MWCNTs	0.5	1.48	2.16	+46	41.4	61.0	+47	2004 [26]
PE	MWCNTs	8	~1.3	~1.7	+31	–	–	–	2009 [27]
PE	SWCNTs	0.5	0.81	0.8	−1	34.1	33.3	−2	2004 [28]
PEI	MWCNTs	1	1.47	2.05	+39	77.5	95.0	+23	2007 [29]
PEN	MWCNTs	0.5	1.68	1.98	+18	66.3	87.8	+32	2008 [30]
PEO	SWCNTs	1	0.06	0.15	+150	–	–	–	2002 [31]
PI	SWCNTs	1	2.2	3.2	+45	105	105	0	2004 [32]
PI	MWCNTs	5	–	–		92	133	+45	2006 [33]
PI	MWCNTs	6.98	2.3	3.7	+31	102	134	+31	2007 [34]
PI	MWCNTs	5	0.91	1.21	+33	–		–	2007 [35]
PI	MWCNTs	14.3	2.84	3.90	+37	115.6	95.2	−18	2004 [26]
PLLA	MWCNTs	3	0.21	0.32	+52	–	–	–	2008 [36]
PolyENB	MWCNTs	1.6	1.89	2.02	+7	52.3	52.1	0	2009 [37]
PMMA	SWCNTs	5	~3.1	~5	+61	–	–	–	2000 [38]
PP	MWCNTs	1	4.4	5.7	+30	500	520	+4	2008 [39]
PP	MWCNTs	5	1.28	2.15	+68	28.2	35.25	+25	2009 [40]
PP	MWCNTs	1.5				25.16	60.74	+141	2008 [41]
PP	MWCNTs	2.0	0.773	1.684	+118	–	–	–	2008 [41]
PP	MWCNTs	0.3	1.570	2.107	+34	30.71	53.98	+76	2008 [24]
PP	MWCNTs	5	4.6	7.1	+54	490	570	+16	2002 [42]
PP	SWCNTs	0.8	4.2	4	−5	430	420	−2	2003 [43]
pPEK	MWCNTs	17	4.0	6.7	+68	–	–	–	2009 [44]
PS	MWCNTs	1	~1.19	~1.69	+42	~12.8	~16	+25	2000 [45]

Continued

Table 8.1 Mechanical Properties Enhancement of Selected Nanocomposite Materials—cont'd

Matrix or Polymer	Filler	Φ (wt%)	E_{matrix} (GPa)	E_{max} (GPa)	E % Increase	σ_{matrix} (MPa)	σ_{Max} (MPa)	σ % Increase	Year and Reference
PS	MWCNTs	2.5[a]	2.02	2.72	+35	–	–	–	2008 [46]
PU	MWCNTs	2.5	40	150	+275	–	–	–	2007 [47]
PU	MWCNTs	10	0.235	0.444	+89	10.7	15.6	+46	2007 [48]
PVA	SWCNTs	5	~4	~6.2	+55	–	–	–	2004 [49]
PVA	MWCNTs	9.1	5.6	25.3	+352	15.7	42.3	+169	2005 [50]
PVA	MWCNTs	5	1.95	4.68	+140	72.6	119.4	+64	2007 [51]
PVOH	MWCNTs	1	1.8	10.4	+478	–	–	–	2007 [52]
SAN	MWCNTs	1	1.48	2.23	+51	27.8	55.3	+99	2005 [53]
TPU	SWCNTs	0.5	7.7	14.5	+88	12.4	13.3	+7	2006 [54]
UHMWPE	MWCNTs	5	~122.6	~136.8	+12	~3510	~4170	+19	2006 [55]

Φ, Loading; E, Young's modulus; σ, ultimate tensile strength or tensile strength at yield; ABPBO, poly(2,5-benzoxazole); iPP, isotatic polypropylene; LDPE, low-density polyethylene; PA-6, polyamide-6; PA1010, polyamide-1010; PBO, poly(1,4-phenylene-cis-benzobisoxazole); PC, polycarbonate; PE, polyethylene; PEI, polyether imide; PEN, poly(ethylene naphthalene); PEO, polyethylene oxide; PI, polyimide; PLLA, poly(lactic acid); PolyENB, poly(5-ethylidene-2-norbornene); PMMA, polymethyl methacrylate; PP, polypropylene; pPEK, para-polyetherketone; *PS, polystyrene; PU, polyurethane; PVA, polyvinyl acetate; PVOH, poly(vinyl alcohol); SAN, styrene-acrylonitrile; TPU, thermoplastic polyurethane; UHMWPE, ultrahigh-molecular-weight polyethylene; SWCNT, single-walled CNT; MWCNT, multiwalled CNT.*

[a] *vol%.*

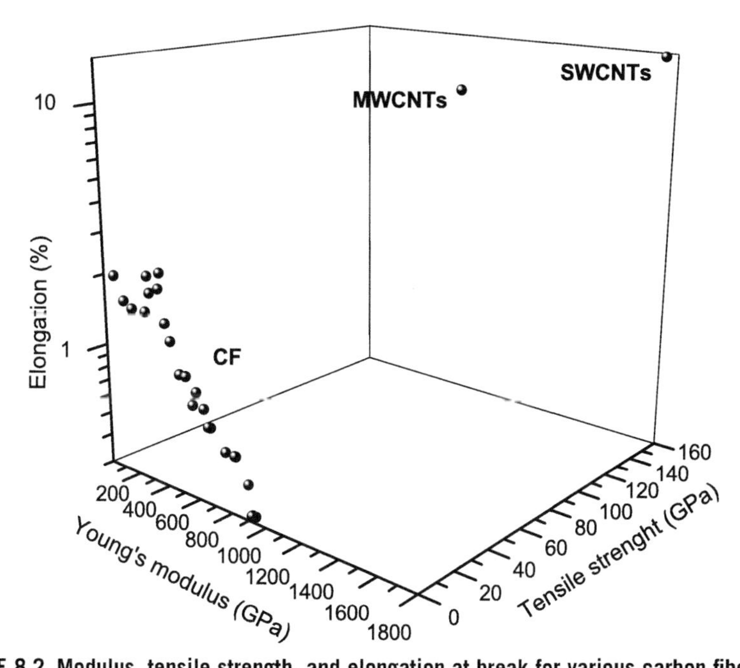

FIGURE 8.2 Modulus, tensile strength, and elongation at break for various carbon fibers (CFs) compared with single-walled CNTs (SWCNTs) and multiwalled CNTs (MWCNTs).

Figure sourced from Is it worth the effort to reinforce polymers...? [2]

issues have to be overcome to fully realize the potential of CNTs as reinforcing agents. The main challenges in this area are the purification and functionalization, the ability to disperse CNTs within the matrix, interactions between the CNTs and the host matrix, and alignment within the host matrix. In addition to these challenges, there is no known preparation method that gives CNTs with uniform proportions.

Surface modification of CNTs is the most common strategy used to increase nanotube–polymer interactions, decrease filler self-aggregation, and thus improve load transfer. Even by making use of such strategies or by aligning the CNTs, in many cases the improvements in tensile strength and modulus are coupled with a reduction in strain at break, indicating a decreased ductility.

After examining the full potential of CNTs as a mechanical reinforcing agent, the next issue is cost. Table 8.2 presents the 2009 prices of various commercially available CNTs. The prices are a function of diameter, length, purity, and method of manufacturing. For SWCNTs, double-walled CNTs (DWCNTs), and MWCNTs, the prices per gram are in the range of \$32–2500, \$21–1600, and \$0.5–136, respectively. For CFs, the price per gram can vary between \$0.037 and \$1.8.

Considering both the mechanical properties and the cost of CFs and CNTs, the following questions arise: Is it worth the effort to reinforce polymers with CNTs? Will CNTs replace CFs as a reinforcing agent? These two questions cannot be fully answered simply by comparing the mechanical properties and price of CFs and

Table 8.2 Price and Some Properties of Different Types of Commercially Available Carbon Nanotubes (CNTs)

Product	Supplier/Country	Purity (wt% or vol% ≥)	*d* (nm)	*l* (μm)	Cost ($/g)	Production Method
SWCNTs	M K Impex/Canada	90% CNTs	1–2	5–30	32	CVD
SWCNTs	MER/USA	12% SWCNTs	1.2–1.4	10–50	35	CVD
SWCNTs	MicrotechNano/USA	90% CNTs 50% SWCNTs	<2	5–15	39	CVD
SWCNTs	NTP Shenzhen/China	90% CNTs 50% SWCNTs	<2	5–15	50	CVD
SWCNTs	CSI/USA	40–60% C	1.4	1.5	50	Arc discharge
SWCNTs	NanoCarbLab/Russia	40% CNTs	1.2–1.4	1–5	60	Arc discharge
SWCNTs	CheapTubes/USA	90% CNTs 5% MWCNTs	1–2	5–30	65	CVD
SWCNTs	Alpha nano/China	90% C	1–2	5–20	65	CVD
SWCNTs	Heji/China	90% C	1–2	10–20	72	CVD
SWCNTs	BuckyUSA/USA	90% C	0.7–2.5	0.5–10	77	CVD
SWCNTs	Arry/Germany	90% C	1–2	5–20	90	CVD
SWCNTs	Nanoamor/USA	90% SWCNTs	1–2	5–30	100	CVD
SWCNTs	Nanothinx/Greece	85% CNTs	0.8–1.4	≥5	120	CVD
SWCNTs	Helix/USA	90% C	1.3	0.5–40	124	CVD
SWCNTs	Your-tool/Germany	90% CNT 60% SWCNTs	2	20	129	–
SWCNTs	Chengdu/China	90% C	1–2	–	180	CVD
SWCNTs	Nanoshel/USA	98% CNTs 70% SWCNTs	0.7–2	3–8	180	Arc discharge
SWCNTs	SES research/USA	90% CNTs 50% SWCNTs	<2	1–5	199	CVD
SWCNTs	Nanomaterial store/USA	90% SWCNTs	1–2	5–30	280	CVD

SWCNTs	NanoCarbLab/Russia	70–80% CNTs	1.2–1.4	1–5	380	Arc discharge
SWCNTs	CSI/USA	90% C	1.4	0.5–15	400	Arc discharge
SWCNTs	Nanocyl/Belgium	70% C	2	–	447	CVD
SWCNTs	Carbolex/USA	70–90% SWCNTs	1.4	2–5	800	–
SWCNTs	CNI technology/USA	95% CNTs	0.8–1.2	0.1–1	2000	HiPCo
SWCNTs	NanoLab/USA	90% SWCNTs	1–1.5	>10	2500	CVD
DWCNTs	Heji/China	60% C	1.3–5	–	21	CVD
DWCNTs	MicrotechNano/USA	90% CNTs 50% DWCNTs	<5	5–15	30	CVD
DWCNTs	Chengdu/China	60% C	2–4	–	30	CVD
DWCNTs	NTP Shenzhen/China	90% CNTs 50% DWCNTs	<5	5–15	50	CVD
DWCNTs	Nanoamor/USA	50% DWCNTs	<5	5–15	54	CVD
DWCNTs	Arry/Germany	80% C	1.3–3	5–15	83	CVD
DWCNTs	Helix/USA	90% C	4	0.5–40	124	CVD
DWCNTs	Your-tool/Germany	90% CNTs 60% DWCNTs	3	20	129	–
DWCNTs	Xintek/USA	85% CNTs	3	2–6	160	CVD
DWCNTs	Nanocyl/Belgium	90% C	3.5	1–10	179	CVD
DWCNTs	NanoCarbLab/Russia	20–30% DWCNTs	–	–	250	Arc discharge
DWCNTs	Nanomaterial store/USA	60% DWCNTs	<3	<20	310	CVD
DWCNTs	NanoLab/USA	95% CNTs	3–5	1–5	350	CVD
DWCNTs	SES research/USA	90% CNTs 50% DWCNTs	<5	5–15	385	CVD
DWCNTs	NanoCarbLab/Russia	90% DWCNTs	–	–	1500	Arc discharge
DWCNTs	Xintek/USA	98% CNTs	2	2–6	1600	CVD
MWCNTs	Alpha nano/China	95% C	20–30	5–20	0.5	CVD
MWCNTs	Bayer/Germany	95% C	13	>1	0.60	CVD

Continued

Table 8.2 Price and Some Properties of Different Types of Commercially Available Carbon Nanotubes (CNTs)—cont'd

Product	Supplier/Country	Purity (wt% or vol% ≥)	*d* (nm)	*l* (μm)	Cost ($/g)	Production Method
MWCNTs	Bayer/Germany	99% C	13–16	>1	0.75	CVD
MWCNTs	Chengdu/China	95% C	<8	–	1.5	CVD
MWCNTs	CheapTubes/USA	95% MWCNTs	<8	10–30	2.5	CVD
MWCNTs	Nanoamor/USA	95% CNTs	8–15	10–50	4	CVD
MWCNTs	M K Impex/Canada	95% CNTs	<8	10–30	12	CVD
MWCNTs	Heji/China	95% C	<8	0.5–200	12	CVD
MWCNTs	Helix/USA	95% C	<10	0.5–40	13	CVD
MWCNTs	Your-tool/Germany	95% CNTs	<10	5–15	14	–
MWCNTs	Arry/Germany	95% C	5–10	5–20	14	CVD
MWCNTs	Nanothinx/Greece	98% CNTs	25–40	≥10	23	CVD
MWCNTs	NTP Shenzhen/China	85% CNTs	<10	5-15	25	CVD
MWCNTs	MicrotechNano/USA	95% CNTs	~5	5–15	29	CVD
MWCNTs	MER/USA	90% CNTs	25–45	30	35	CVD
MWCNTs	Nanomaterial store/USA	95% CNTs	<8	10–30	35	CVD
MWCNTs	Future carbon/Germany	98% C	15	10-50	37	CVD
MWCNTs	Catalytic materials/USA	99% CNTs	10	–	40	CVD
MWCNTs	Nanoshel/USA	90% CNTs 70% MWCNTs	4–12	3–10	45	Arc discharge
MWCNTs	Nanocyl/Belgium	95% C	9.5	1.5	45	CVD
MWCNTs	NanoLab/USA	95% CNTs	10–20	1–5	75	CVD
MWCNTs	BuckyUSA/USA	95% C	5–15	1–10	80	CVD
MWCNTs	Xintek/USA	90% CNTs	8	10–20	120	CVD
MWCNTs	SES research/USA	95% CNTs	<10	1–2	136	CVD

d, average diameter; l, average length. The values showed above are for reference only. The cost may change according to the amount of CNTs purchased. Here the cost was valued considering CNTs amounts from 1 to 1000 g. The purity of CNTs is determined by using different methods such as TEM, Raman, TGA, or EA;

CNTs. It is necessary to analyze the cost versus property relation of composites reinforced with these fillers.

To answer these questions, this study involves deriving an equation based on the Halpin–Tsai model, and, using the concept of effective fiber, predicting the elastic modulus and tensile strength of epoxy matrix composites reinforced with CNTs, CFs, or a combination of the two materials. Based on the simulation results, the feasibility of the composite materials, that is to say, of the CNTs and CFs as a reinforcement agent, is then ranked. To be as realistic as possible, we chose for our simulations the strongest, the stiffest, and the cheapest PAN-based CFs commercially available and compared them with MWCNTs.

8.2 Theories

8.2.1 Micromechanics modeling of fiber composites

Many micromechanics models are available to predict the tensile properties of fiber composites [4,5,57,58]. The Halpin–Tsai equations are a set of empirical relationships that enable the property of a composite material to be expressed in terms of the properties of the matrix and reinforcing phases considering their proportions and geometry. Despite of their empirical nature, these equations are quite popular in materials science. For the longitudinal Young's modulus of aligned fiber composites, the Halpin–Tsai equation states [4,8,59]

$$\frac{E_{11}}{E_m} = \frac{1 + \zeta\eta V_f}{1 - \eta V_f} \tag{8.1}$$

$$\eta = \frac{(E_f/E_m) - 1}{(E_f/E_m) + \zeta} \tag{8.2}$$

$$\zeta = 2\frac{l}{d} \tag{8.3}$$

where E_{11}, E_m, and E_f are the composite, matrix, and fiber elastic moduli, respectively; V_f is the fiber volume fraction; and ζ is a shape parameter dependent upon reinforcement geometry and orientation. The constant η takes into account the modulus of the matrix and fiber.

Thostenson and Chou [4] modified the Halpin–Tsai equation toward its applicability to nanotube reinforced composites. Considering that the outer wall of the nanotubes act as an effective solid fiber, with the same deformation behavior, diameter (d) and length (l) of the nanotube, the parameter η can be rewritten as

$$\eta' = \frac{(E_{NT}/E_m) - (d/4t)}{(E_{NT}/E_m) + (l/2t)} \tag{8.4}$$

where E_{NT} is the nanotube elastic modulus, l and d are the length and average outer diameter of the nanotube, and t is the thickness of graphite layer (0.34 nm).

CNTs can be used as a single reinforcing phase (binary systems) or as an additional reinforcing phase in conjunction with CFs in hybrid composites (ternary systems). The former case has been widely considered in the literature experimentally as well as theoretically. Nevertheless, the latter case has been studied experimentally in some detail, whereas the modeling of such hybrid composites has been less considered.

We will now derive an equation based on the Halpin–Tsai model and in the concept of effective fiber [4] to predict the elastic properties of hybrids composites containing CNTs and CFs. As we will demonstrate, the equations predict the same values as Eqns (8.1)–(8.4) for the cases of composites containing only CFs or only CNTs.

For a hybrid composite containing CNTs and CFs, the fiber volume fraction is expressed as

$$V_f = V_{NT} + V_{CF} \tag{8.5}$$

Substituting the equation above into Eqn (8.1) yields

$$\frac{E_{11}}{E_m} = \frac{1 + \zeta\eta V_{CF} + \zeta'\eta' V_{NT}}{1 - \eta V_{CF} - \eta' V_{NT}} \tag{8.6}$$

By substituting Eqn (8.2) and Eqn (8.4) into Eqn (8.6) and considering the limiting value of ζ when continuous aligned CFs are used $\zeta = \infty$

$$\lim_{\zeta \to \infty} \frac{E_{11}}{E_m} \tag{8.7}$$

We get

$$E_{11} = \frac{E_{NT}[1 + \zeta' V_{NT} + (E_{CF}V_{CF}/E_m) - V_{CF}] + E_{CF}V_{CF}l/2t + E_m l/2t(1 - V_{CF}) - E_m\zeta' V_{NT}d/4t}{(E_{NT}/E_m)(1 - V_{NT}) + l/2t + V_{NT}d/4t} \tag{8.8}$$

$$\zeta' = 2\frac{l}{d} \tag{8.9}$$

This equation gives the Young's modulus in the direction of the alignment of composites containing aligned CNTs, CFs, or a hybrid composite containing both of them. A similar calculation can be used to derive an equation for the tensile strength of hybrids composites containing CNTs and CFs. The theoretical tensile strength parallel to the fiber in the composites can be approximated with reasonable accuracy by

$$\sigma_c = \sigma_f V_f + \sigma_m (1 - V_f) \tag{8.10}$$

where σ_c, σ_f, and σ_m are the composite, fiber, and matrix strengths, respectively. For composites containing CNTs, where $l > l_c$ the composite strength, can be described by [5]

$$\sigma_c = (\eta_s \sigma_{NT} - \sigma_m) V_{NT} + \sigma_m \tag{8.11}$$

$$\eta_s = 1 - \frac{l_c}{2l} \tag{8.12}$$

/here η_s is the strength efficiency factor and l_c is the critical length. This equation oes not consider that the polymer–nanotube interaction result in the formation of n interfacial polymer region with properties different of the bulk polymer [60]. 'or a hollow cylinder, the critical length is given by [61]

$$l_c = \frac{\sigma_{NT} d}{2\tau}\left(1 - \frac{d_i^2}{d^2}\right) \tag{8.13}$$

/here τ is the interfacial shear strength between the hollow tube and the surrounding olymer. For MWCNTs (or DWCNTs), two cases can be considered: (1) only the utmost wall carries the load and (2) all walls carry the load. In the former case ne effective tensile strength of the carbon nanotube, according to the "effective ber" definition [4] can be expressed as

$$\sigma_{eff} = \frac{4t}{d}\sigma_{NT} \tag{8.14}$$

/hereas in the latter case, we have

$$\sigma_{eff} = \frac{\left(d^2 - d_i^2\right)}{d^2}\sigma_{NT} \tag{8.15}$$

The properties of the nanotubes are extremely dependent on the interaction be-veen the outmost layers and the internal layers. Because, in the case of strength, the nterwall sliding may occur before ultimate fracture of the nanotubes, and the inter-al layers of DWCNTs and MWCNTs do not contribute fully to the nanotube rength, it is reasonable that the value obtained using Eqn (8.14) or at least values etween this and the one given by Eqn (8.15) should be considered for modeling.

Substituting Eqn (8.5) into Eqn (8.11) and assuming the strength efficiency ictor to be equal to 1, for continuous aligned fibers in the direction of alignment, 'e have

$$\sigma_c = \eta_s \sigma_{NT} V_{NT} + \sigma_{CF} V_{CF} + \sigma_m [1 - (V_{NT} + V_{CF})] \tag{8.16}$$

If $V_{NT} = 0$, then Eqn (8.15) reduces to the Eqn (8.10), whereas for $V_{CF} = 0$, qn (8.16) becomes Eqn (8.11). This equation gives the tensile strength in the direc-on of the alignment of composites containing CNTs, CFs, or hybrid composite con-iining both of them.

The density of hybrid composites can be calculated using a simple rule of iixture as

$$\rho_c = \rho_{NT} V_{NT} + \rho_{CF} V_{CF} + \rho_m [1 - (V_{NT} + V_{CF})] \tag{8.17}$$

'here ρ_c, ρ_{NT}, ρ_m, and ρ_{CF} are the composite, CNT, matrix, and CF densities, respec-vely. In addition, the final materials cost of the composites is defined as

$$C_c = W_m C_m + W_{NT} C_{NT} + W_{CF} C_{CF} \tag{8.18}$$

where C_c, C_{NT}, C_m, and C_{CF} are the composite, CNT, matrix, and CF cost per kilogram, respectively. W_{NT}, W_m, and W_{CF} are the nanotube, matrix, and CFs weight fraction, respectively. Because of several challenges in processing of CNT-reinforced composites, especially with aligned CNTs, the manufacturing costs involved in the preparation of composites are not accounted for.

8.2.2 Cost versus property analysis

The following simple case studies are used to rank the feasibility of CNTs and CFs as reinforcing agents. Because cost is a crucial factor in material selection process, the following method compares the material cost for equivalent material property requirements [62]. The property could be any property critical to the application. Therefore, this method determines the weight required by different materials to meet the desired property. The weight determines the cost of the material. For instance, for structural applications, the volume of the material required to carry the load is determined and then, by multiplying by the density, the weight is obtained. If tensile strength is the determining factor for the selection of a material, then the weight required by different materials of choice are determined to have the same tensile strength values.

8.2.2.1 Case study I—beam designed to carry an axial load

Let us suppose two materials, I and II, are selected for comparison purposes for the fabrication of a beam. If the member is designed to carry an axial load P, then the cross-sectional area required by beam I and II will be

$$A_I = \frac{P}{\sigma_I} \tag{8.19}$$

$$A_{II} = \frac{P}{\sigma_{II}} \tag{8.20}$$

where σ_I and σ_{II} are the tensile strengths of materials I and II, respectively.

If the member is a solid circular rod, then the ratio of the diameters will be

$$\frac{D_I}{D_{II}} = \left(\frac{\sigma_{II}}{\sigma_I}\right)^{1/2} \tag{8.21}$$

For rods of equal lengths, the weight ratio is

$$\frac{W_I}{W_{II}} = \frac{\rho_I \sigma_{II}}{\rho_{II} \sigma_I} \tag{8.22}$$

If the cost per unit weight of materials I and II is C_I and C_{II}, respectively, then the ratio of total material costs T_I and T_{II} will be

$$\frac{T_I}{T_{II}} = \frac{C_I \rho_I \sigma_{II}}{C_{II} \rho_{II} \sigma_I} \tag{8.23}$$

The ratio of total material costs for different materials enable us to rank or compare the feasibility of these materials. For example, Eqn (8.23) gives values <1 when material I is more feasible for an application than material II. Of course, when material II is more feasible than material I Eqn (8.23) gives values >1.

8.2.2.2 Case study II—beam designed for bending stiffness

Similar to tensile strength, modulus can be a determining factor for the selection of a material. Let us suppose a beam is designed for bending stiffness [62]. For instance, deflection (δ) of a simply supported beam loaded at the center is given by

$$\delta = \frac{PL^3}{48EI} \tag{8.24}$$

where P is the load applied, L is the total length of the beam, E is the Young's modulus of the composite, and I is the moment of inertia of the cross-section. The stiffness of the beam may be given by

$$\frac{P}{\delta} = \frac{48EI}{L^3} \tag{8.25}$$

For the same length of beam, stiffness is proportional to EI. Therefore, for two materials, systems I and II,

$$E_I I_I = E_{II} I_{II} \tag{8.26}$$

For a rectangular beam shape, with width b and thickness h, the moment of inertia is given by

$$I = \frac{bh^3}{12} \tag{8.27}$$

Considering two beams of the same width, from Eqns (8.26) and (8.27), we have

$$\frac{h_{II}}{h_I} = \left(\frac{E_I}{E_{II}}\right)^{1/3} \tag{8.28}$$

The relative weight can be written as

$$\frac{W_I}{W_{II}} = \frac{\rho_I}{\rho_{II}} \left(\frac{E_{II}}{E_I}\right)^{1/3} \tag{8.29}$$

Thus, the ratio of total material cost of material I (T_I) and II (T_{II}) can be compared as

$$\frac{T_I}{T_{II}} = \frac{C_I \rho_I}{C_{II} \rho_{II}} \left(\frac{E_{II}}{E_I}\right)^{1/3} \tag{8.30}$$

8.2.3 Results and discussion

To analyze the feasibility of CNTs as a reinforcement agent in relation to commercially available CFs, we propose different model composites comprising single-phase reinforced epoxy composites as well as hybrid composites. Therefore, to be as realistic as possible, we choose for our simulations:

1. The stiffest commercial carbon fiber from Toray Industries, M60J, which has a Young's modulus of 588 GPa. This CF also happens to be the most expensive.
2. The strongest carbon fiber, T1000GB, which has an ultimate tensile strength of 6.4 GPa.
3. The cheapest carbon fiber, T700SC.
4. MWCNTs, which are assumed to have a Young's modulus of 950 GPa and tensile strength of 120 GPa.
5. The epoxy resin/hardener combination Aradur® MY720/Aradur® 9664-1, commercialized by Huntsman Advanced Materials. The MY720 system is a high-performance multifunctional resin frequently used in the preparation of prepregs for aircraft structures.

The properties of all the materials considered in our model composites can be seen in Table 8.3. We have chosen MWCNTs instead of SWCNTs or DWCNTs in our simulations because they can be produced in higher amounts and are the cheapest CNTs commercially available. We also considered only PAN-based CFs because of their higher ductility when compared with pitch-based CFs. Considering the theoretical shear strength $\tau = 50$ MPa [61] as a conservative value in Eqn (8.13), the critical length of MWCNTs described in Table 8.3 was calculated to be 8.52 μm. It is worth noting that the interfacial shear strength of high-modulus CFs in epoxy usually ranges between 50 and 100 MPa. The effective modulus of the CNTs (i.e., the modulus considering that the outer wall of the nanotubes act as an effective solid fiber) was determined to be 882 GPa.

Table 8.3 Different Properties of the Materials Composing the Composite Models Considered in the Simulations [3,61–64]

Material	ρ (g/cm^3)	E (GPa)	σ_M (GPa)	d	l (μm)	Cost ($/g)
MY720	1.25	3.7	0.060	–	–	0.044
T1000GB	1.80	294	6.4	5 μm	–	0.27
M60J	1.93	588	3.9	5 μm	–	1.8
T700SC	1.80	230	4.9	5 μm	–	0.037
MWCNTs	2.09	950	120	Ext: 15 nm Int: 4 nm	50	10

ρ, density; E, Young modulus; σ_M, tensile strength; d, diameter; l, length. T1000GB, M60J, and T700SC: catalog for TORAYCA, Toray Industries Inc. (Toray), high-performance carbon fiber Torayca, 2009.

In addition, in this study, we considered the tensile strength of the MWCNTs as ne arithmetic mean of the two values obtained by using Eqn (8.14), 11 GPa, and qn (8.15), 111 GPa. Thus a final value of 61 GPa was assumed during calculations. ven though MWCNTs can be purchased from $0.60/g, we assumed here a price of 10/g. We believe this price, or at least a value near this, better represents the CNTs escribed in Table 8.3, still with a purity of >99% MWCNTs rather than >99%. Ioreover, we did not take into account any waviness effect of the CNTs and ssumed that they have a narrow distribution of diameter and length.

Table 8.4 gives the properties as calculated according to Eqns (8.8), 3.16)–(8.18), whereas Figures 8.3 and 8.4 are plots of the cost of different compos-e models as a function of their specific elastic modulus and tensile strength, respec-vely. The specific stiffness is a parameter that can be used to measure technical erformance of one material relative to another. This parameter is of extreme impor-ance in engineering applications because both high stiffness and low density present ajor selection criteria [65]. The analyses of the results show that hybrid composites omprising both CNTs and CFs can be the most expensive ones, followed by the NT-reinforced composites. Carbon fiber–reinforced composites are the cheapest nes. In addition, the highest specific stiffness is obtained by using CFs followed y hybrid composites (Figure 8.3). Moreover, CFs lead to the lowest specific trength between all composites (Figure 8.4). The highest specific strengths are ffered by both CNT-reinforced composites and hybrid composites.

It is also worth noting that the real advantage of CNTs over CFs, considering the roperties calculated here, is their tensile strength. Also, as the elongation at break of ne CFs is below that of most available polymers, an increase in elongation of com-osites will not be expected. Nevertheless, CNTs have an elongation at a break of up) 15% [3], which could possibly lead to a higher improvement of the elongation and lso ductility of composites.

To truly analyze the feasibility of the simulated composite models, we further nalyze the cost versus property relation of the two case studies proposed previously Section 8.2.2). Using Eqn (8.23), considering tensile strength as the determining ctor for the selection, we ranked the feasibility of the composite models showed Table 8.4. The results are depicted in Figure 8.5. The lower the ratio T_I/T_{II} in qn (8.23), the more feasible the composite is for the considered application. The F T700SC is the most viable option to tensile strengths of up to 3.21 GPa. How-ver, for tensile strengths higher than this value, up to 11.61 GPa, hybrid composites ontaining CNTs and CFs are the most viable option. The highest tensile strength an only be achieved with hybrid composites (50 vol% T1000GB + 15 vol% IWCNTs). Note that the CFs considered here does not offer the possibility of btaining composites with tensile strength higher than 4.18 GPa. Figure 8.5 also ows that CNTs are more viable than the stiffest CFs, M60J. Thus, CNTs are the ost viable strengthening option.

Using Eqn (8.30), considering stiffness as the determining factor for the election, we again ranked the feasibility of the composite materials showed in able 8.4. The results are depicted in Figure 8.6. For Young's moduli of up to

Table 8.4 Mechanical Properties, Density, and Price of Different Composite Materials

MWCNTs (vol%)	T1000GB (vol%)	M60J (vol%)	T700SC (vol%)	E_{11} (GPa0	σ_M (GPa)	ρ (g/cm^3)	E_{11}/ρ (GPa/g/cm^3)	σ_M/ρ (GPa/g/cm^3)	Cost ($/kg)
0.5	–	–	–	4.11	0.34	1.25	3.28	0.27	127
1	–	–	–	4.52	0.62	1.26	3.59	0.49	209
2	–	–	–	5.34	1.18	1.27	4.22	0.93	373
5	–	–	–	7.81	2.85	1.29	6.04	2.21	849
10	–	–	–	11.92	5.65	1.33	8.93	4.24	1604
15	–	–	–	16.03	8.44	1.38	11.65	6.14	2312
–	35	–	–	105.30	2.28	1.44	73.00	1.58	143
–	45	–	–	134.34	2.91	1.50	89.71	1.95	166
–	55	–	–	163.36	3.55	1.55	105.23	2.28	188
–	65	–	–	192.40	4.18	1.61	119.69	2.60	208
–	–	35	–	208.20	1.40	1.49	139.92	0.94	841
–	–	45	–	266.64	1.79	1.56	171.36	1.15	1024
–	–	55	–	325.06	2.17	1.62	200.16	1.34	1192
–	–	65	–	383.50	2.56	1.69	226.65	1.51	1346
–	–	–	35	82.90	1.75	1.44	57.47	1.22	41
–	–	–	45	105.54	2.24	1.50	70.47	1.49	40

–	–	–	55	128.16	2.72	1.55	82.55	1.75	40
–	–	–	65	150.80	3.21	1.61	93.81	1.99	39
0.5	64.5	–	–	191.36	4.43	1.61	118.79	2.75	272
1	64	–	–	190.32	4.68	1.61	118.04	2.90	335
2	63	–	–	188.24	5.17	1.62	116.55	3.20	461
5	60	–	–	182.02	6.66	1.62	112.09	4.10	835
10	55	–	–	171.64	9.14	1.64	104.77	5.58	1451
15	50	–	–	161.25	11.61	1.65	97.58	7.03	2056
0.5	–	64.5	–	380.99	2.82	1.69	225.06	1.66	1397
1	–	64	–	378.49	3.08	1.69	223.48	1.82	1448
2	–	63	–	373.48	3.60	1.70	220.31	2.12	1549
5	–	60	–	358.44	5.16	1.70	210.85	3.03	1852
10	–	55	–	333.39	7.76	1.71	195.19	4.54	2354
15	–	50	–	308.33	10.36	1.72	179.68	6.04	2850
0.5	–	–	64.5	150.08	3.46	1.61	93.28	2.15	104
1	–	–	64	149.36	3.72	1.61	92.75	2.31	168
2	–	–	63	147.92	4.23	1.61	91.69	2.62	297
5	–	–	60	143.61	5.76	1.62	88.54	3.55	681
10	–	–	55	136.43	8.31	1.64	83.36	5.08	1311
15	–	–	50	129.24	10.86	1.65	78.28	6.58	1931

E_{11}, Young's modulus in the longitudinal direction; σ_M, ultimate tensile strength; ρ, density.

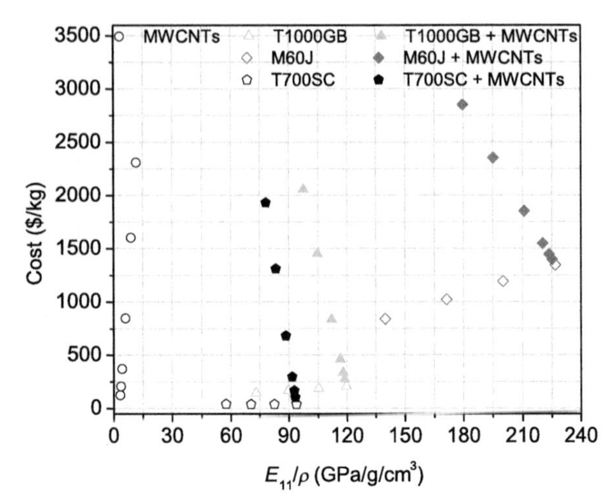

FIGURE 8.3 Cost of different composite materials as a function of their specific stiffness.

MWCNTs, multiwalled CNTs.

Figure sourced from Is it worth the effort to reinforce polymers...? [2]

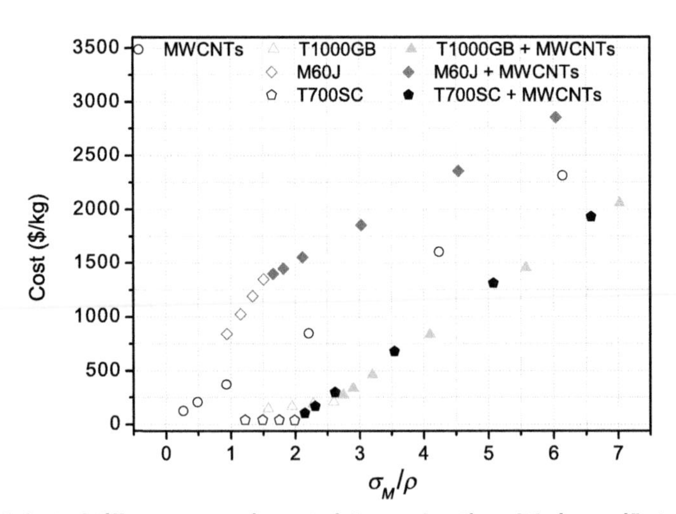

FIGURE 8.4 Cost of different composite materials as a function of their specific tensile strength.

MWCNTs, multiwalled CNTs.

Figure sourced from Is it worth the effort to reinforce polymers...? [2]

150 GPa, the CFs T700SC are the most viable option. However, for intermediate and high moduli, the CFs T1000GB and M60J become the most viable options, respectively. Despite MWCNTs being the stiffest option of all the fillers considered during calculations, their cost makes them the least viable option.

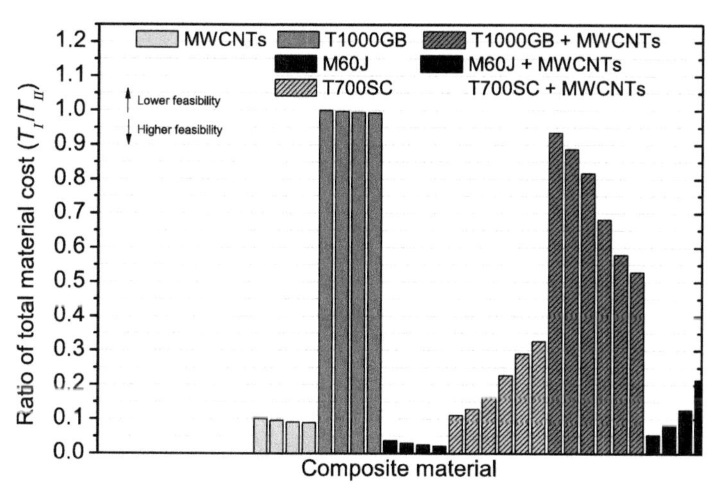

IGURE 8.5 Cost versus property analysis, according to Table 8.4, considering tensile :rength as a determining factor.

IWCNTs, multiwalled CNTs.

Figure sourced from Is it worth the effort to reinforce polymers...? [2]

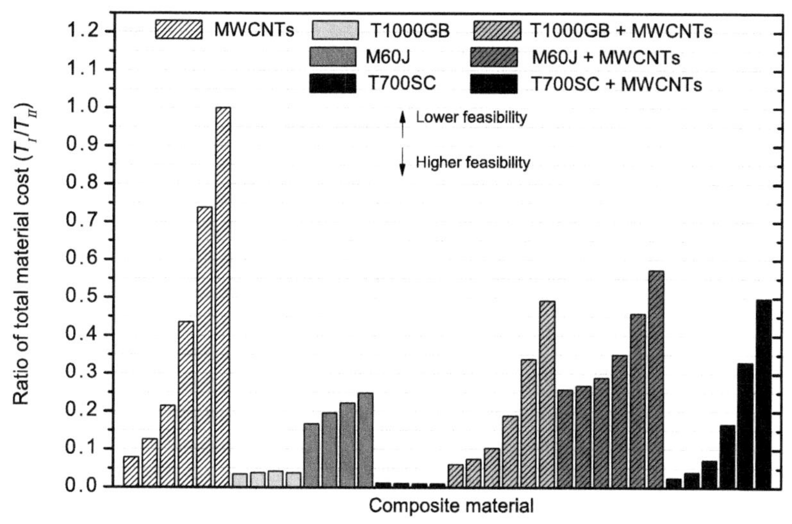

GURE 8.6 Cost versus property analysis, according to Table 8.4, considering stiffness as a etermining factor.

WCNTs, multiwalled CNTs.

Figure sourced from Is it worth the effort to reinforce polymers...? [2]

Because their strength is around 20 times higher than that of the strongest carbon ber and their elongation is seven times higher, CNTs should be considered as reinrcing filler for polymer nanocomposites. However, to achieve their true potential, everal challenges have to be faced. CNTs have to be produced with higher purity, integrity, in larger amounts, and at lower price. They have to have longer lengths. Concerning the composites' technology, issues such as orientation of the CNTs, their concentration, interfacial adhesion, distribution, and dispersion have to be overcome [66].

The global market for chemical fibers and yarns is steadily increasing with growing world population [67,68]. With that, demand for new industrial fibers with special or improved properties emerges. Among other things, stronger and/or electrical conductive fibers are required for applications as reinforcement fibers, smart clothing, electromagnetic shields, or armors [69]. An alternative route toward CNT-based materials with superior properties consists of using CNTs fibers. For production of fibers with diameters ranging between 10 and 100 μm, however, only nanoscaled fillers as CNTs can be used. The superior mechanical and physical properties of individual CNTs provide the input for researchers in developing high-performance continuous fibers based upon CNTs. Although the prices for high-purity nanotubes are still too high for commercial success, a breakthrough is expected within the next few years.

8.3 CONCLUSION

CNT-reinforced polymer composites are an emerging class of high-performance materials with unique and promising properties. The combined use of CNTs with more economic CFs is a way to reduce the cost of advanced composites. In this study, a micromechanical approach used for modeling fiber composites was modified considering the effective fiber concept toward its applicability to hybrid composites. Two case studies were proposed in order to rank the feasibility of CNTs and CFs as reinforcing agents. The results from calculations showed that the use of CFs makes the acquisition of composites with a maximum tensile strength of 4.18 GPa possible. In addition, the analyses of the cost versus property relation showed that CNTs are the most viable strengthening option to achieve composites with strengths of up to 11.61 GPa. Considering the actual CNT prices, CFs came out to be the most viable stiffening option, enabling composites with a Young's modulus of up to 383 GPa but with a loss of toughness. The discussion also showed that, to achieve the true potential of nanotubes, several challenges have to be faced. CNTs have to be produced with higher purity, integrity, in larger amounts, and at a lower price. They have to have longer lengths. Concerning the composites technology, issues such as orientation of the CNTs, their concentration, interfacial adhesion, distribution, and dispersion have to be overcome.

Acknowledgments

Author M. R. Loos wishes to thank Michelle K. Sing for her help in revising and editing this work.

References

[1] Iijima S. Helical microtubules of graphitic carbon. Nature 1991;354(6348):56–8.
[2] Loos MR, Schulte K. Is it worth the effort to reinforce polymers with carbon Nanotubes? Macromol Theory Simul 2011;20:350–62.
[3] Andrews R, Weisenberger MC, Qian D, Meier MS, Cassity K. Carbon nanotubes, multi-walled. In: Lukehart Charles M, editor. Nanomaterials: inorganic and bioinorganic perspectives, vol. 1. United Kingdom: Wiley; 2009.
[4] Thostenson ET, Chou T-W. On the elastic properties of carbon nanotube-based composites: modeling and characterization. Phys D Appl Phys 2003;36(5):573.
[5] Coleman JN, Khan U, Blau WJ, Gun'ko YK. Small but strong: a review of the mechanical properties of carbon nanotube-polymer composites. Carbon 2006;44(9):1624–52.
[6] Zhou C, Wang S, Zhuang Q, Han Z. Enhanced conductivity of polybenzazoles doped by carboxylated multi-walled carbon nanotubes. Carbon 2008;46(9):1232–40.
[7] Ying Z, Du JH, Bai S, Li F, Liu C, Cheng HM. Mechanical properties of surfactant-coating carbon nanofiber/epoxy composite. Int J Nanosci 2002;1(5–6):425–30.
[8] Gojny FH, Wichmann MHG, Köpke U, Fiedler B, Schulte K. Carbon nanotube-reinforced epoxy-composites: enhanced stiffness and fracture toughness at low nanotube content. Compos Sci Technol 2004;64(15):2363–71.
[9] Guo P, Chen X, Gao X, Song H, Shen H. Fabrication and mechanical properties of well-dispersed multiwalled carbon nanotubes/epoxy composites. Compos Sci Technol 2007;67(15–16):3331–7.
[10] Valentin L, Puglia D, Carniato F, Boccaleri E, Marchese L, Kenny JM. Use of plasma fluorinated single-walled carbon nanotubes for the preparation of nanocomposites with epoxy matrix. Compos Sci Technol 2008;68(3–4):1008–14.
[11] Kim JY, Han SI, Hong S. Effect of modified carbon nanotube on the properties of aromatic polyester nanocomposites. Polymer 2008;49:3335–45.
[12] Zou W, Du Z-J, Liu Y-X, Yang X, Li H-Q, Zhang C. Functionalization of MWNTs using polyacryloyl chloride and the properties of CNT-epoxy nanocomposites. Compos Sci Technol 2008;68:3259–64.
[13] Yeh M-K, Hsieh T-H. Fabrication and mechanical properties of multi-walled carbon nanotubes/epoxy nanocomposites. Mater Sci Eng A 2008;483–484:289–92.
[14] Hernández-Pérez A, Avilés F, May-Pat A, Valadez-González A, Herrera-Franco PJ, Bartolo-Pérez P. Effective properties of multiwalled carbon nanotube/epoxy composites using two different tubes. Compos Sci Technol 2008;68:1422–31.
[15] Sun L, Warren GL, O'Reilly JY, Everett WN, Lee SM, Davis D, et al. Mechanical properties of surface-functionalized Epoxy/SWCNT nanocomposites. Carbon 2008;46:320–8.
[16] Fu S-Y, Chen Z-K, Hong S, Han CC. The reduction of carbon nanotube (CNT) length during the manufacture of CNT/polymer composites and a method to simultaneously determine the resulting CNT and interfacial strengths. Carbon 2009;47(14):3192–200.
[17] Hubert P, Ashrafi B, Adhikari K, Meredith J, Vengallatore S, Guan J, et al. Synthesis and characterization of carbon nanotube reinforced epoxy – viscosity and elastic properties correlation. Compos Sci Technol 2009;69:2274–80.
[18] Cheng QF, Wang JP, Wen JJ, Liu CH, Jiang KL, Li QQ, et al. Carbon nanotube/epoxy composites fabricated by resin transfer molding. Carbon 2010;48(1):260–6.

[19] Bikiaris D, Vassiliou A, Chrissafis K, Paraskevopoulos KM, Jannakoudakis A, Docoslis A. Effect of acid treated multi-walled carbon nanotubes on the mechanical, permeability, thermal properties and thermo-oxidative stability of isotactic polypropylene. Polym Degrad Stab 2008;93:952–67.
[20] Fragneaud B, Masenelli-Varlot K, Gonzalez-Montiel A, Terrones M, Cavaillé JY. Mechanical behavior of polystyrene grafted carbon nanotubes/polystyrene nanocomposites. Compos Sci Technol 2008;68:3265–71.
[21] Kang M, Myung SJ, Jin H-J. Nylon 610 and carbon nanotube composite by in situ interfacial polymerization. Polymer 2006;47(11):3961–6.
[22] Meng H, Sui GX, Fang PF, Yang R. Effects of acid- and diamine-modified MWNTs on the mechanical properties and crystallization behavior of polyamide 6. Polymer 2008;49:610–20.
[23] Zeng H, Gao C, Wang Y, Watts PCP, Kong H, Cui X, et al. In situ polymerization approach to multiwalled carbon nanotubes reinforced nylon 1010 composites: mechanical properties and crystallization behavior. Polymer 2006;47:113–22.
[24] Zhou C, Wang S, Zhuang Q, Han Z. Enhanced conductivity of polybenzazoles doped by carboxylated multi-walled carbon nanotubes. Carbon 2008;46(9):1232–40.
[25] Zhou C, Wang S, Zhang Y, Zhuang Q, Han Z. In-situ preparation and continuous fiber spinning of poly(p-phenylene benzobisoxazole) composites with oligo-hydroxylamide-functionalized multi-walled carbon nanotubes. Polymer 2008;49(10):2520–30.
[26] Ogasawara T, Ishida Y, Ishikawa T, Yokota R. Characterization of multi-walled carbon nanotube/phenylethynyl terminated polyimide composites. Compos Part A 2004;35(1):67–74.
[27] Adhikari AR, Chipara M, Lozano K. Processing effects on the thermo-physical properties of carbon nanotube polyethylene composite. Mater Sci Eng A 2009;526(1–2):123–7.
[28] Tong X, Liu C, Cheng HM, Zhao HC, Yang F, Zhang XQ. Surface modification of single-walled carbon nanotubes with polyethylene via in situ Ziegler-Natta polymerization. J Appl Polym Sci 2004;92:3697.
[29] Liu T, Tong Y, Zhang W-D. Preparation and characterization of carbon nanotubes/polyetherimide nanocomposite films. Compos Sci Technol 2007;67:406–12.
[30] Kim JY, Han SI, Hong S. Effect of modified carbon nanotube on the properties of aromatic polyester nanocomposites. Polymer 2008;49(15):3335–45.
[31] Geng HZ, Rosen R, Zheng B, Shimoda H, Fleming L, Liu J, et al. Fabrication and properties of composites of poly(ethylene oxide) and functionalized carbon nanotubes. Adv Mater 2002;14(19):1387–90.
[32] Siochi EJ, Working DC, Park C, Lillehei PT, Rouse JH, Topping CC, et al. Melt processing of SWCNT-polyimide nanocomposite fibers. Compos Part B 2004;35(5):439–46.
[33] Zhu B-K, Xie S-H, Xu Z-K, Xu Y-Y. Preparation and properties of the polyimide/multi-walled carbon nanotubes (MWNTs) nanocomposites. Compos Sci Technol 2006;66(3–4):548–54.
[34] Yuen S-M, Ma C-C, Lin Y-Y, Kuan H-C. Preparation, morphology and properties of acid and amine modified multiwalled carbon nanotube/polyimide composite. Compos Sci Technol 2007;67:2564–73.
[35] So HH, Cho JW, Sahoo NG. Effect of carbon nanotubes on mechanical and electrical properties of polyimide/carbon nanotubes nanocomposites. Eur Polym J 2007;43:3750–6.

36] Feng J, Sui J, Cai W, Wan J, Chakoli AN, Gao Z. Preparation and characterization of magnetic multi-walled carbon nanotubes-poly(1-lactide) composite. Mater Sci Eng B 2008;150(3):208–12.

37] Jeong W, Kessler MR. Effect of functionalized MWCNTs on the thermo-mechanical properties of poly(5-ethylidene-2-norbornene) composites produced by ring-opening metathesis polymerization. Carbon 2009;47:2406–12.

38] Haggenmueller R, Gommans HH, Rinzler AG, Fischer JE, Winey KI. Aligned single-wall carbon nanotubes in composites by melt processing methods. Chem Phys Lett 2000;330:219–25.

39] Lee G-W, Jagannathan S, Chae HG, Minus ML, Kumar S. Carbon nanotube dispersion and exfoliation in polypropylene and structure and properties of the resulting composites. Polymer 2008;49:1831–40.

40] Prashantha K, Soulestin J, Lacrampe MF, Krawczak P, Dupin G, Claes M. Masterbatch-based multi-walled carbon nanotube filled polypropylene nanocomposites: assessment of rheological and mechanical properties. Compos Sci Technol 2009;69:1756–63.

41] Yang B-X, Shi J-H, Pramoda KP, Goh SH. Enhancement of the mechanical properties of polypropylene using polypropylene-grafted multiwalled carbon nanotubes. Compos Sci Technol 2008;68:2490–7.

42] Kumar S, Doshi H, Srinivasarao M, Park JO, Schiraldi DA. Fibers from polypropylene/nano carbon fiber composites. Polymer 2002;43(5):1701–3.

43] Bhattacharyya AR, Sreekumar TV, Liu T, Kumar S, Ericson LM, Hauge RH, et al. Crystallization and orientation studies in polypropylene/single wall carbon nanotube composite. Polymer 2003;44:2373.

44] Bangarusampath DS, Ruckdäschel H, Altstädt V, Sandler JKW, Garray D, Shaffer MSP. Rheological and electrical percolation in melt-processed poly(ether ether ketone)/multi-wall carbon nanotube composites. Chem Phys Lett 2009;482(1–3):105–9.

45] Qian D, Dickey EC, Andrews R, Rantell T. Load transfer and deformation mechanisms in carbon nanotube-polystyrene composites. Appl Phys Lett 2000;76:2868–70.

46] Fragneaud B, Masenelli-Varlot K, Gonzalez-Montiel A, Terrones M, Cavaillé JY. Efficient coating of N-doped carbon nanotubes with polystyrene using atomic transfer radical polymerization. Compos Sci Technol 2008;68(15–16):3265–71.

47] Sahoo NG, Jung YC, Yoo HJ, Cho JW. Influence of carbon nanotubes and polypyrrole on the thermal, mechanical and electroactive shape-memory properties of polyurethane nanocomposites. Compos Sci Technol 2007;67:1920–9.

48] Guo P, Chen X, Gao X, Song H, Shen H. Fabrication and mechanical properties of well-dispersed multiwalled carbon nanotubes/epoxy composites. Compos Sci Technol 2007;67:3331–7.

49] Paiva MC, Zhou B, Fernando KAS, Lin Y, Kennedy JM, Sun YP. Mechanical and morphological characterization of polymer–carbon nanocomposites from functionalized carbon nanotubes. Carbon 2004;42:2849–54.

50] Chen W, Tao X, Xue P, Cheng X. Enhanced mechanical properties and morphological characterizations of poly(vinyl alcohol)–carbon nanotube composite films. Appl Surf Sci 2005;252:1404–9.

51] Mi Y, Zhang X, Zhou S, Cheng J, Liu F, Zhu H, et al. Morphological and mechanical properties of bile salt modified multi-walled carbon nanotube/poly(vinyl-alcohol) nanocomposites. Compos Part A 2007;38:2041–6.

[52] Ryan KP, Cadek M, Nicolosi V, Blond D, Ruether M, Armstrong G, et al. Carbon nanotubes for reinforcement of plastics? A case study with poly(vinyl alcohol). Compos Sci Technol 2007;67:1640–9.
[53] Wang M, Pramoda KP, Goh SH. Enhancement of the mechanical properties of poly(styrene-co-acrylonitrile) with poly(methyl methacrylate)-grafted multiwalled carbon nanotubes. Polymer 2005;46:11510–6.
[54] Chen W, Tao X. Production and characterization of polymer nanocomposite with aligned single wall carbon nanotubes. Appl Surf Sci 2006;252(10):3547–52.
[55] Ruan S, Gao P, Yu TX. Ultrastrong gelspun UHMWPE fibers reinforced using multiwalled carbon nanotubes. Polymer 2006;47:1604–11.
[56] Fietzer E, Künkele F. Today's carbon fibres-a new energy-saving and environment-friendly all-round material review. High Temp High Press 1990;22(3):239–66.
[57] Halpin JC, Kardos JL. The Halpin-Tsai equations: a review. Polym Eng Sci 1976;16(5): 344–52.
[58] Tucker III CL, Liang E. Stiffness predictions for unidirectional short-fiber composites: review and evaluation. Compos Sci Technol 1999;59:655–71.
[59] Peeterbroeck S, Breugelmans L, Alexandre M, Nagy JB, Viville P, Lazzaroni R. The influence of the matrix polarity on the morphology and properties of ethylene vinyl acetate copolymers-carbon nanotube nanocomposites. Compos Sci Technol 2007;67: 1659–65.
[60] Cooper CA, Cohen SR, Barber AH, Wagner HD. Detachment of nanotubes from a polymer matrix. Appl Phys Lett 2002;81:3873–5.
[61] Wagner HD. Nanotube–polymer adhesion: a mechanics approach. Chem Phys Lett 2002;361:57–61.
[62] Mazumdar S. Composites manufacturing: materials, product, and process engineering. 1st ed. Boca Raton: CRC Publisher; 2001. ISBN: 9780849305856.
[63] Stano K, Koziol K, Pick M, Motta M, Moisala A, Vilatela J, et al. Direct spinning of carbon nanotube fibres from liquid feedstock. Int J Mater Form 2008;1:59–62.
[64] Popov VN. Carbon nanotubes: properties and application. Mater Sci Eng 2004;43: 61–102.
[65] Xiao T, Ren Y, Liao K, Wu P, Li F, Cheng HM. Determination of tensile strength distribution of nanotubes from testing of nanotube bundles. Compos Sci Technol 2008;68:2937–42.
[66] Esawi AMK, Farag MM. Carbon nanotube reinforced composites: … current challenges. Mater Des 2007;28:2394–401.
[67] Pötschke P, Brünig H, Janke A, Fischer D, Jehnichen D. Orientation of multiwalled carbon nanotubes in composites with polycarbonate by melt spinning. Polymer 2005; 46:10355–63.
[68] Marcin A. Modification of fiberforming polymers by additives. Prog Polym Sci 2002; 27:853–913.
[69] Dalton AB, Collins S, Razal J, Munoz E, Ebron VH, Kim BG, et al. Continuous carbon nanotube composite fibers: properties, potential applications, and problems. J Mater Chem 2004;14:1–3.

CHAPTER

Reinforcement Efficiency of Carbon Nanotubes—Myth and Reality

9

Marcio R. Loos, Ica Manas-Zloczower

Department of Macromolecular Science and Engineering, Case Western Reserve University, Cleveland, OH, USA

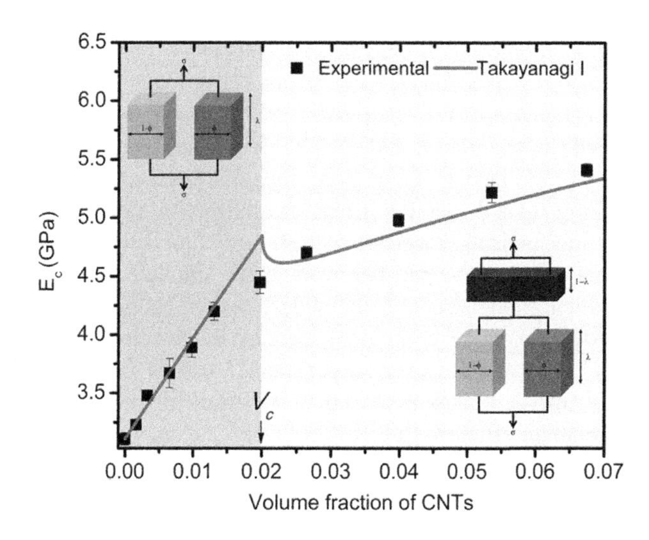

CHAPTER OUTLINE

Carbon Nanotube Reinforced Composites. http://dx.doi.org/10.1016/B978-1-4557-3195-4.00009-6

Different micromechanical models for the prediction of mechanical properties of carbon nanotube/polymer composites, taking into consideration filler percolation throughout the matrix, are considered. We demonstrate that the critical filler volume fraction where a percolating network of carbon nanotubes is forming marks a "turning point" in the reinforcement efficiency. Expectations for the reinforcing effect of carbon nanotubes at concentrations above a percolating threshold with the current technology are mostly unrealistic.

9.1 Introduction

Due to their outstanding electrical, thermal, and mechanical properties, carbon nanotubes (CNTs) have many potential applications and are considered to be ideally suited for the next generation of composite materials [1]. Probably the most widely used application of CNTs is as reinforcing agents for polymers. The key issues for transferring the CNT properties to composites are (1) homogeneous dispersion and distribution of CNTs in the matrix and (2) interaction between CNTs and the surrounding medium [2,3]. Due to their large specific surface area, CNTs tend to agglomerate, making their dispersion a real challenge. As a consequence, the reported results on mechanical properties of polymer composites are still far from satisfactory and are for the most part below expected simulated values. Commonly used micromechanical models such as the widely known Halpin-Tsai model or the simple rule of mixtures predict a continuous increase of the composite modulus with the addition of CNTs [4]. However, experimental results have shown that the enhancement of properties takes place up to a certain filler concentration, after which the reinforcement efficiency decreases [5,6]. One such example is shown in Figure 9.1 which compares experimental data for the elastic modulus of an epoxy system at various loadings of multi walled carbon nanotubes (MWCNTs) with predicted values using the Halpin-Tsai model and the rule of mixtures. There is an obvious "turning point" at a volume fraction of CNTs of 0.00521 which is not predicted by the models.

There are numerous examples in literature which suggest that above the critical concentration of CNTs, the properties either drop or the reinforcement efficiency decreases considerably [7—10]. It is well known that nanotubes tend to aggregate and above a critical concentration will form a 3D network inside the polymeric matrix. The volume fraction of CNTs at which this 3D network forms is called the percolation threshold (V_c). The "turning point" in the reinforcement efficiency usually takes place at volume fractions coincident with the percolation threshold. To the best of our knowledge, there are no micromechanical models accounting for the turnaround in mechanical properties of carbon nanotube reinforced composites at concentrations above a threshold value.

In this study, we derive and compare different micromechanical models that account for the presence of dispersed and agglomerated CNTs in the polymeric matrix and consider the effect of a percolation threshold on the elastic modulus of

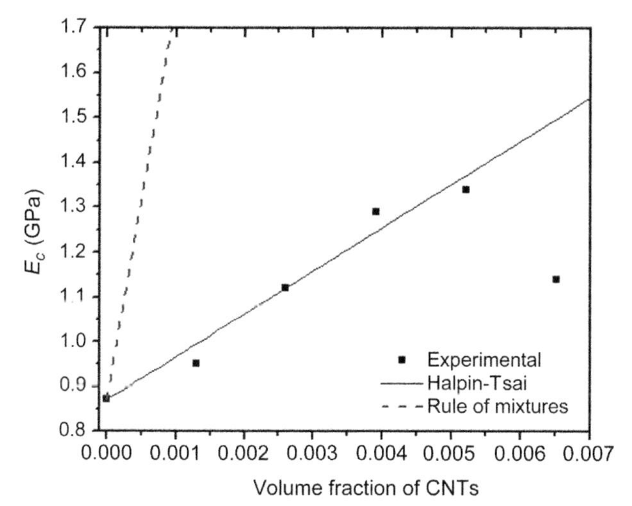

GURE 9.1 Theoretical Halpin-Tsai and rule of mixtures models and experimental data for the poxy composites [7] at various nanotube loadings. Predictions are made considering = 10 μm, $d = 50$ nm, $E_{NT} = 900$ GPa.

he composite. The suitability of these models is verified by comparing simulated alues with experimental data from literature. The results show that some models re able to predict mechanical properties over a wide range of testing conditions. Overall, the findings of this study strongly suggest that despite the exceptional properties of CNTs, their efficient use may be bound to limited concentrations.

.2 Models development

When using CNTs into a polymeric matrix, it is reasonable to assume that some of hese nanotubes will form agglomerates. At low concentrations of filler, and considering adequate dispersion procedures, most of the nanotubes will be well dispersed nto the matrix (Figure 9.2(a)). However, even at such low concentrations, some of he nanotubes will be agglomerated. At a critical volume fraction of the filler, the nanotubes will form a 3D network. We label this critical concentration the percolaon threshold, V_c. At this point the concentration of nanotubes in the system enables he formation of a 3D network (Figure 9.2(b)). At filler concentrations exceeding the ercolation threshold, the system will primarily engage all filler into the 3D network, lthough there still may be some well dispersed or agglomerated nanotubes present n the matrix (Figure 9.2(c)).

One can conceptually formulate a simple mechanical model composed of three onstituents/phases: the matrix, the well-dispersed CNTs, and the agglomerated nanotubes. All these constituents have different mechanical properties and they will contribute to the overall mechanical behavior of the composite system. For

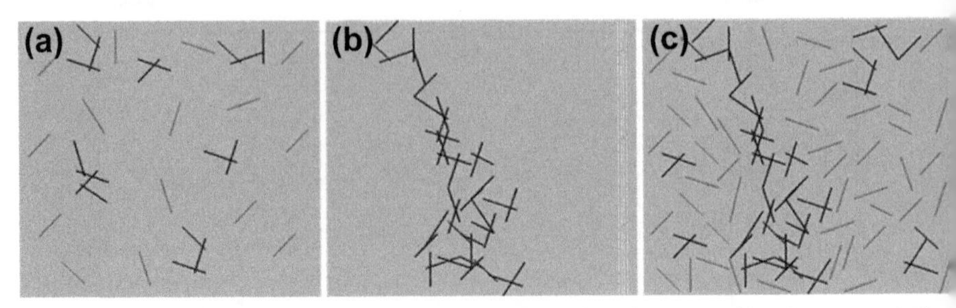

FIGURE 9.2 Schematic representation of a polymer matrix reinforced with a low amount of carbon nanotubes below the percolation threshold (a), the same matrix with a critical volume fraction of the filler (b), and concentrations exceeding the percolation threshold (c).

simplicity, we will consider that all three constituents exhibit a linear elastic behavior and are isotropic. The well-dispersed CNTs encompass all non-entangled units characterized by an average aspect ratio η. Agglomerated CNTs consist of the percolated network (at overall concentration in the system exceeding the threshold value) and all other entangled nanotubes present in the system. The models presented here do not account for nanotube/matrix debonding, slippage at the interface matrix/nanotube, or any voids or cracks present in the system. No residual stresses are present in the system.

9.2.1 Three-phase model in series and parallel

The composite system contains three phases, namely the matrix of volume fraction V_m and tensile modulus E_m, the dispersed CNTs of volume fraction V_{NT}^{dis} and modulus E_{NT}^{dis}, and the agglomerated CNTs of volume fraction V_{NT}^{agg} and modulus E_{NT}^{agg}. The three phases can be arranged in series or in parallel as shown schematically in Figure 9.3(a) and 9.3(b).

The mechanical properties of the agglomerated nanotubes will differ significantly from the ones of the well-dispersed ones. In fact, the properties of large-diameter CNT bundles are dominated by shear slippage of individual nanotubes within the bundle [11]. This inter-tube slippage within bundles lowers their intrinsic mechanical properties to a great extent and may partially explain the decrease in the modulus of a polymer composite at high volume fraction of CNTs. Experimental measurements on films of CNTs show values of elastic modulus orders of magnitude lower than the ones reported for individual CNTs [12]. Moreover, CNTs are not linked by strong hydrogen bonds, as is the case with other nano-fillers such as cellulose whiskers, and consequently, the formation of a percolating network can be detrimental for the mechanical properties of the composite system. We will explore this concept in predicting the decrease in mechanical properties of the system at nanotube concentrations above a critical value.

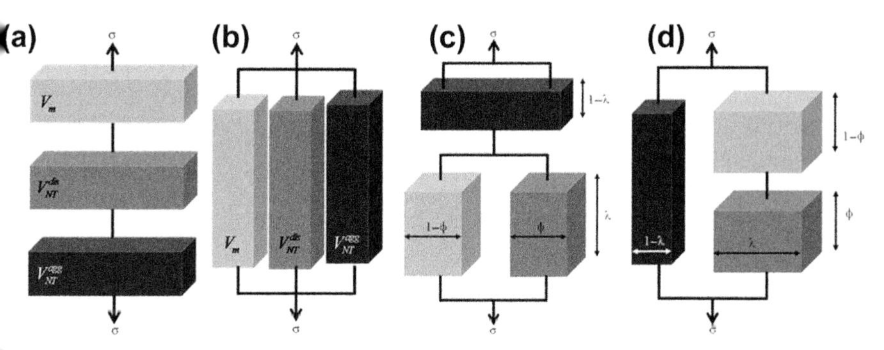

FIGURE 9.3 Schematic representation of different models. Three-phase composite system in series (a) and in parallel (b). Diagram of the Takayanagi model I (c) and II (d) with a percolation concept. ϕ is a function of the volume fraction of the parallel element and λ of the series element. Models I and II are also known as the series-parallel and parallel-series model, respectively.

For the case of a series arrangement (Figure 9.3(a)), the stress is identical in all the three phases, whereas the strain is additive and the composite modulus can be calculated from:

$$E_c = \frac{E_{NT}^{agg} E_{NT}^{dis} E_m}{V_{NT}^{agg} E_{NT}^{dis} E_m + V_{NT}^{dis} E_{NT}^{agg} E_m + \left(1 - V_{NT}^{agg} - V_{NT}^{dis}\right) E_{NT}^{agg} E_{NT}^{dis}} \tag{9.1}$$

For the case of a parallel arrangement of the three phases (Figure 9.3(b)), the strain is identical in all three phases, whereas the stress is additive. In this case the composite modulus is given by:

$$E_c = V_{NT}^{agg} E_{NT}^{agg} + V_{NT}^{dis} E_{NT}^{dis} + \left(1 - V_{NT}^{agg} - V_{NT}^{dis}\right) E_m \tag{9.2}$$

Equations (9.1) and (9.2) will predict a lower and upper bound of the modulus in the composite system, respectively.

One can differentiate between the well-dispersed and the agglomerated CNTs present in the system using the percolation theory. Chatterjee employed the concept of a "switching function," f_s, to emulate the percolation behavior in the vicinity of the percolation threshold [13]. The switching function is defined as:

$$\begin{aligned} f_s(V_{NT}) &= 0 \quad V_{NT} < V_c \\ f_s(V_{NT}) &= 1 - e^{\left(-A\left(V_{NT}/V_c\right) - 1\right)^{0.474}} \quad V_{NT} \geq V_c \end{aligned} \tag{9.3}$$

where A is an adjustable parameter that modulates the width of the transition. In other words, the domain within which the percolation probability exhibits power-law behavior can be appropriately confined to the vicinity of the threshold by a suitable choice of the parameter A [13]. For filler volume fractions exceeding the percolation threshold, a fraction $V_{NT}^{agg} = f_s V_{NT}$ of the filler particles in the system belongs to the percolating network, whereas the remaining rods $V_{NT}^{dis} = V_{NT}(1 - f_s)$ are treated as being uniformly dispersed within the matrix material.

The percolation threshold, V_c, depends on the carbon nanotube aspect ratio. Based on the excluded volume concept, Celzard et al. [14] and Balberg et al. [15] estimated the percolation threshold to be:

$$\phi_c = \frac{0.5}{\eta} \tag{9.4}$$

where $\eta = l/d$ is the aspect ratio of the CNTs, and l and d are the average length and diameter of the nanotubes, respectively. In systems with large distributions of aspect ratios, the percolation threshold can be lower than the value predicted by Eqn (9.4).

Whereas the inclusion of the percolation concept in the three-phase model in series and parallel can give some guidelines in terms of predicting the mechanical behavior of carbon nanotube filled systems, in order to better account for the effect of a percolating network on the mechanical properties, a more advanced model is required. Such a model may be the Takayanagi model which is described in the next section.

9.2.2 Takayanagi model

Takayanagi and coauthors developed a two-phase mechanical model to predict the modulus of a crystalline polymer, taking into consideration the crystalline and non-crystalline phases of a polymer and the fact that different parts of the crystalline phase can undergo different deformation under the application of stress [16]. A schematics for the series-parallel (model I) and parallel-series (model II) models is shown in Figure 9.3(c) and 9.3(d). The Takayanagi model was used successfully to analyze tensile properties for polymer blends and for composite systems with diverse filler morphologies [17,18]. Ouali and coauthors extended the parallel-series model of Takayanagi by introducing a percolating concept [19]. This enabled the authors to apply the model to predict mechanical properties for polymer composites reinforced with cellulose nano-whiskers and polymer blends [17]. Parameters λ and ϕ or their combinations reflect volume fractions of the system components. So far, the series-parallel model of Takayanagi has not been extended to include a percolating concept.

Following the concept put forward by Ouali et al. [19], we used the percolation concept in the series-parallel and parallel-series models of Takayanagi to predict mechanical properties of our three-phase system consisting of the matrix, the agglomerated carbon nanotube phase, and the well-dispersed carbon nanotube phase. Schematics of our models are shown in Figure 9.3(c) and 9.3(d). In our models, $1 - \lambda$ is the volume fraction of the percolating phase, mostly composed of agglomerated nanotubes, and $\lambda\phi$ is the volume fraction of the dispersed CNTs. It follows that the volume fraction of the CNTs is given by:

$$V_{NT} = 1 - \lambda + \lambda\phi \tag{9.5}$$

The volume fraction of the percolating phase varies between 0, at carbon nano-
ube concentrations below a percolation threshold, and 1 when the volume fraction
f CNTs approaches 1. Using the percolation theory, one can relate the volume frac-
ion of the percolating phase to the overall carbon nanotube concentration in the sys-
em through:

$$\begin{aligned} 1-\lambda &= V_{NT}\left(\frac{V_{NT}-V_c}{1-V_c}\right)^b \quad V_{NT} \geq V_c \\ 1-\lambda &= 0 \quad V_{NT} < V_c \\ 1-\lambda &= 1 \quad V_{NT} = 1 \end{aligned} \tag{9.6}$$

In Eqn (9.6), V_c is the critical volume fraction to reach a geometrical percolation
f nanotubes, and b is the percolation exponent believed to be 0.4 in a 3D geomet-
ical system [20].

In addition, the amount of well-dispersed and agglomerated CNTs present in the
ystem can be related to the overall CNT concentration by applying the concept of a
witching function (f_s) described in Eqn (9.3).

Considering model I (Figure 9.3(c)), the elastic modulus of the composite E_c can
e written as:

$$E_c = \frac{(1-V_{NT})E_m E_{NT}^{agg} + (V_{NT}+\lambda-1)E_{NT}^{dis}E_{NT}^{agg}}{(1-\lambda)(1-V_{NT})E_m + (1-\lambda)(\lambda+V_{NT}-1)E_{NT}^{dis} + \lambda^2 E_{NT}^{agg}} \tag{9.7}$$

here E_{NT}^{agg} and E_{NT}^{dis} are the Young's moduli of the percolated carbon nanotube
etwork and of the dispersed nanotubes, respectively, and E_m is the Young's modulus
f the matrix phase. Below the percolation threshold ($1-\lambda=0$), Eqn (9.7) simply
duces to the rule of mixtures:

$$E_c = (1-V_{NT})E_m + V_{NT}E_{NT}^{dis} \tag{9.8}$$

For model II (Figure 9.3(d)), the modulus of the composite is given by the
quation:

$$E_c = \frac{(1-\lambda)(1-V_{NT})E_{NT}^{dis}E_{NT}^{agg} + (1-\lambda)(V_{NT}+\lambda-1)E_m E_{NT}^{agg} + \lambda^2 E_m E_{NT}^{dis}}{(1-V_{NT})E_{NT}^{dis} + (V_{NT}+\lambda-1)E_m} \tag{9.9}$$

Below the percolation threshold ($1-\lambda=0$), Eqn (9.9) reduces to the series
odel (lower bound)

$$E_c = \frac{E_m E_{NT}^{dis}}{(1-V_{NT})E_{NT}^{dis} + V_{NT}E_m} \tag{9.10}$$

In some studies using the Takayanagi equations with the percolation concept, it
as been suggested that the formation of a percolating network of fillers is the mech-
nism responsible for improving the composite mechanical properties [21]. In such
ases, fillers like, for example, cellulose whiskers are linked by strong hydrogen

bonds, and model II with percolation captures well the system mechanical behavior. However, in the case of CNTs that form a geometrical percolating network without strong hydrogen bonds, the network formation can be detrimental to the system mechanical properties. Such behavior is better described using the Takayanagi model I with the percolation concept.

9.3 Application

9.3.1 Model comparison

In this section we compare the various models described for predicting the elastic modulus of polymer composites containing CNTs. We consider a composite system with CNTs of average aspect ratio of 333, rendering a percolation threshold of 0.0015. We employ Chatterjee's switching function with $A = 1.2$ to describe percolation behavior. The modulus of the agglomerated phase is considered to be $E_{NT}^{agg} = 50$ GPa, whereas the Young's modulus of the matrix and dispersed CNTs are assumed to be 3 and 900 GPa [12], respectively. Figure 9.4 compares the moduli calculated from the Takayanagi models I and II with the three-phase model in series and parallel.

Takayanagi model I predicts an increase in the composite modulus with the amount of CNTs in the system up to the percolation threshold ($V_c = 0.0015$). Once a percolating network is formed, the modulus drops continuously when further increasing the CNT amount. As Figure 9.4 illustrates, the reinforcement effect is significantly higher before reaching the percolation threshold but deteriorates

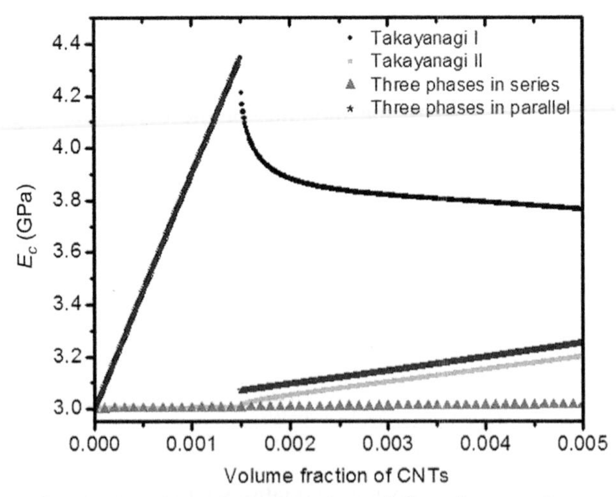

FIGURE 9.4 Elastic modulus of a carbon nanotubes-reinforced composite according to the Takayanagi models I and II and the three-phase model in series and parallel. $E_{NT}^{agg} = 50$ GPa, $E_{NT}^{dis} = 900$ GPa, $E_m = 3$ GPa, $l = 5$ μm, $d = 15$ nm, $A = 1.2$.

considerably afterward. The results obtained for the three-phase model in parallel are identical to the Takayanagi model I up to the threshold concentration, not surprisingly, since both models consider the matrix and dispersed CNTs to be arranged in series below the percolation threshold. Once the CNT network is formed, the reinforcement efficiency decreases instantly.

The three-phase series model gives a lower bound for the composite modulus considering that all three phases experience the same stress and the strain is additive. Takayanagi model II closely follows this lower bound up to the percolation threshold, followed by a slight increase in the composite modulus once the CNT network is formed. The three-phase parallel model is based on the assumption that all three phases can deform to the same extent upon the application of stress, which will render an upper bound to the composite modulus. However, a more realistic picture can be obtained if one differentiates between the deformation of the dispersed and percolated CNTs, which is achieved by employing Takayanagi model I.

9.3.2 Fitting experimental data

In order to analyze the suitability of the various mechanical models for predicting system properties, experimental data from literature are compared with model predictions. Figure 9.5(a) shows the experimental data obtained by Wu et al. for silane-modified MWCNT/epoxy composites [7]. The MWCNTs have been treated for 24 h in a mixture of HNO_3 and H_2SO_4, which may explain the rather low aspect ratio reported by the authors. We have employed Takayanagi I model with Chatterjee's percolation function to fit the experimental data. Our fitting parameters are the modulus of the dispersed CNT and the modulus of the percolated network, as well as the parameter A of the switching function. The model fits the experimental data very well. We also show the three-phase parallel model with Chatterjee's percolation function employing the same material parameters. Noteworthy are the rather low values for the CNT modulus, which may be due to their rather low aspect ratio, low interfacial shear strength, integrity, possible waviness [22] as well as sliding between tube interfaces in MWCNTs [23].

Omidi et al. prepared MWCNT/epoxy composites using a high power tip sonication method for the CNT dispersion and reported experimental values for the composites' moduli for CNT concentrations as high as 10 wt% [24]. In Figure 9.5(b) we compare Omidi's experimental results with model predictions. Note that Takayanagi model I captures well the change in CNT reinforcing efficiency once the percolation concentration is reached. The three-phase parallel model does not fit the experimental data well.

Martone et al. reported results for epoxy/MWCNT composites prepared using a high-energy sonicator [5]. The experimental data for the bending modulus as well as the simulated values are presented in Figure 9.5(c). Noteworthy, the calculated percolation threshold of 0.0005 coincides with the experimental value obtained by the authors via electrical conductivity measurements. In addition, morphological studies revealed also the formation of a percolating network.

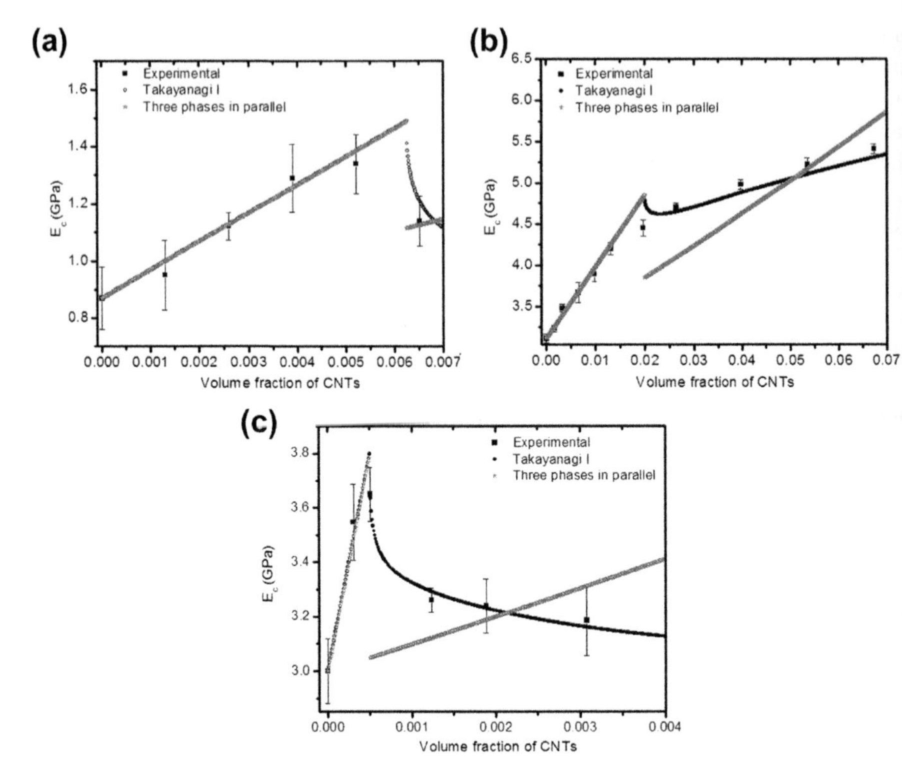

FIGURE 9.5 Comparison of experimental and simulated elastic modulus of MWCNT/epoxy composites. (a) Reported [7] MWCNTs and matrix material parameters: $l = 4\ \mu m$, $d = 50$ nm, $E_m = 0.87$ GPa; fitting parameters: $E_{NT}^{dis} = 100$ GPa, $E_{NT}^{agg} = 40$ GPa, $A = 2.9$. (b) Reported [24] MWCNTs and matrix material parameters: $l = 1\ \mu m$, $d = 40$ nm, $E_m = 3.11$ GPa; fitting parameters: $E_{NT}^{dis} = 90$ GPa, $E_{NT}^{agg} = 40$ GPa, $A = 0.75$. (c) Reported [5] MWCNTs and matrix material parameters: $l = 10\ \mu m$, $d = 10$ nm, $E_m = 3$ GPa; fitting parameters: $E_{NT}^{dis} = 1600$ GPa, $E_{NT}^{agg} = 100$ GPa, $A = 1.6$.

Figure 9.6(a) shows the modulus for ultrahigh-molecular-weight polyethylene/MWCNT composites prepared by Ko et al. [25]. The CNTs were treated in concentrated H_2SO_4 and functionalized. The results show that up to the critical percolation threshold, the CNTs reinforce the polymer matrix, but at higher volume fractions of CNTs the modulus drops significantly, even below the modulus of the pristine resin. The Takayanagi model I can fit these unexpected results very well.

Phenolic-based composites reinforced with MWCNTs have been prepared by Yeh et al. [26]. The CNTs were dispersed in the phenolic resin via sonication. The simulated modulus using Takayanagi model I satisfactorily matches the experimental data (Figure 9.6(b)).

Guojian et al. reported the elastic modulus for poly(vinyl chloride)/MWCNT composites [27]. Figure 9.6(c) shows these results as well as simulated values using

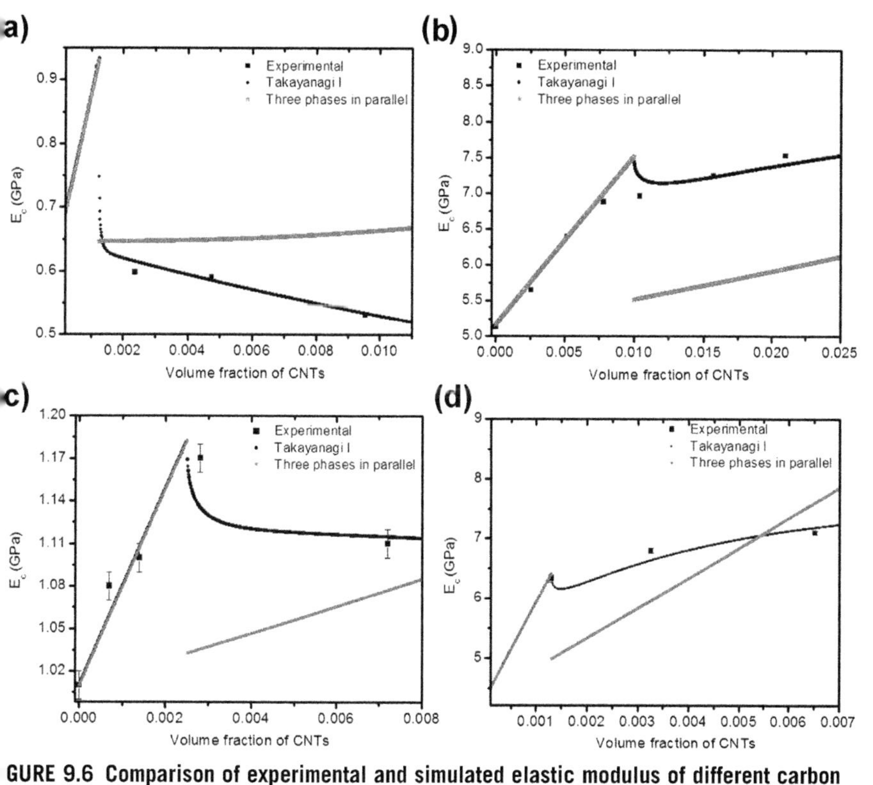

GURE 9.6 Comparison of experimental and simulated elastic modulus of different carbon anotube-based composites. (a) multi walled carbon nanotube/ultrahigh-molecular-weight olyethylene composites; reported [25] multi walled carbon nanotubes and matrix material arameters: $l = 10\ \mu m$, $d = 25$ nm, $E_m = 0.65$ GPa; fitting parameters: $E_{NT}^{dis} = 230$ GPa, $^{agg}_{NT} = 0.028$ GPa, $A = 9$. (b) multi walled carbon nanotube/phenolic composites; reported 6] multi-walled carbon nanotubes and matrix material parameters: $l = 2\ \mu m$, $d = 40$ nm, $_m = 5.17$ GPa; fitting parameters: $E_{NT}^{dis} = 240$ GPa, $E_{NT}^{agg} = 40$ GPa, $A = 0.8$. (c) multi-alled carbon nanotube/poly(vinyl chloride) composites; reported [27] MWCNTs and matrix aterial parameters: $l = 7\ \mu m$, $d = 35$ nm, $E_m = 1.01$ GPa; fitting parameters: $^{dis}_{NT} = 70$ GPa, $E_{NT}^{agg} = 10$ GPa, $A = 1.2$. (d) CNT/poly(vinyl)acrylonitrile composites; ported [28] carbon nanotubes and matrix material parameters: $l = 10\ \mu m$, $d = 26$ nm, $_m = 4.34$ GPa; fitting parameters: $E_{NT}^{dis} = 1600$ GPa, $E_{NT}^{agg} = 500$ GPa, $A = 0.68$.

akayanagi I model and the three-phase parallel model. Above the percolation reshold value, there is a decrease in the modulus, which is well captured by the akayanagi model I.

Hou et al. reported results for reinforced poly(vinyl)acrylonitrile using few-alled CNTs [28]. The CNTs were treated in HNO_3 and dispersed using a tip onicator. The experimental results (Figure 9.6(d)) show a reduction in the reinforcing effect of the CNTs after the formation of a percolating network throughout the matrix. This effect is well captured by the simulated curve using Takayanagi model I.

The suitability of the Takayanagi model I in predicting the elastic modulus of CNT composites is well demonstrated in the examples presented above. We have shown also the simplified three-phase model system to reveal the differences between these two "upper bound" models. All the experimental data and the fitted curves presented in this paper point to the unrealistic expectations for the reinforcing effect of CNTs at concentrations above a percolating threshold. The critical volume fraction where a percolating network of CNTs is developed marks a "turning point" in the reinforcement. This critical volume fraction depends on a number of factors such as the quality of CNTs, aspect ratio polydispersity, CNT waviness effect, and CNT functionalization. Another important factor in attempting to predict mechanical properties of CNT-reinforced composites is the interaction between the filler and the matrix. The models presented in this work do not account for nanotube/matrix debonding or slippage at the interface matrix/nanotube. In view of this, one may look at these models as predictive of an upper bound in the mechanical properties of composite materials. This reinforces the conclusion that despite the exceptional properties of CNTs, their efficient use is bound to limited concentrations.

9.4 Conclusion

Different micromechanical models that account for the presence of dispersed and agglomerated CNTs in a polymeric matrix and consider the effect of filler percolation on the elastic modulus of composites have been presented. The experimental data and the fitted curves have shown that significant enhancement of mechanical properties takes place up to a certain filler concentration, after which the reinforcement efficiency decreases dramatically. This "turning point" in the reinforcement efficiency usually takes place at volume fractions coincident with the percolation threshold. This behavior is well captured by the Takayanagi model I. The findings of this study point out that despite the exceptional properties of CNTs, their efficient use as reinforcing agents is bound to limited concentrations. Moreover, these results suggest that expectations for exceptional reinforcement efficiency of CNTs at concentrations above a percolating threshold with the current technology are unrealistic. In order to use CNTs at their full potential, issues such as interfacial shear strength, purity, integrity, orientation, dispersion, and distribution/alignment will have to be carefully examined and addressed.

Acknowledgments

This material is based upon work supported by the Department of Energy and Bayer MaterialScience LLC under Award Number DE-EE0001361.

Disclaimer

This report was prepared as an account of work sponsored by an agency of the United States Government. Neither the United States Government nor any agency thereof, nor any of their employees, makes any warranty, express or implied, or assumes any legal liability or responsibility for the accuracy, completeness, or usefulness of any information, apparatus, product, or process disclosed, or represents that its use would not infringe privately owned rights. Reference herein to any specific commercial product, process, or service by trade name, trademark, manufacturer, or otherwise does not necessarily constitute or imply its endorsement, recommendation, or favoring by the United States Government or any agency thereof. The views and opinions of authors expressed herein do not necessarily state or reflect those of the United States Government or any agency thereof.

References

[1] Loos MR, Schulte K, Abetz V. Fast and highly efficient one-pot synthesis of polyoxadiazole/carbon nanotubes nanocomposites in mild acid. Polym Int 2010;60:517–28.

[2] Gomes D, Loos MR, Wichmann MHG, De la Veja A, Schulte K. Sulfonated polyoxadiazole composites containing carbon nanotubes prepared via in-situ polymerization. Compos Sci Technol 2009;69:220–7.

[3] Loos MR, Schulte K. Is it Worth the effort to reinforce polymers with carbon nanotubes? Macromolecular theory and simulations. Macromol Theor Simul 2011;20:350–62.

[4] Loos MR, Schulte K, Abetz V. Fast and highly efficient one-pot synthesis of polyoxadiazole/carbon nanotubes nanocomposites in mild acid. Polymer international. J Polym Sci Part A: Polym Chem 2010;48:517–28.

[5] Martone A, Formicola C, Giordano M, Zarrelli M. Reinforcement efficiency of multiwalled carbon nanotube/epoxy nano composites. Compos Sci Technol 2010;70:1154–60.

[6] O'Connor I, Hayden H, O'Connor S, Coleman JN, Gun'ko YK. Polymer reinforcement with kevlar-coated carbon nanotubes. J Phys Chem C 2009;113:20184–92.

[7] Wu S-Y, Yuen S-M, Ma C-CM, Chiang C-L, Huang Y-L, Wu H-H, et al. Preparation, morphology, and properties of silane-modified MWCNT/epoxy composites. J Appl Polym Sci 2010;115:3481.

[8] Jeong W, Kessler MR. Effect of functionalized MWCNTs on the thermo-mechanical properties of poly(5-ethylidene-2-norbornene)composites produced by ring-opening metathesis polymerization. Carbon 2009;47:2406–12.

[9] Kanagaraj S, Varanda FR, Zhil'tsova TV, Oliveira MSA, Simões JAO. Mechanical properties of high density polyethylene/carbon nanotube composites. Compos Sci Technol 2007;67:3071–7.

[10] Thostenson ET, Chou T-W. On the elastic properties of carbon nanotube-based composites: modelling and characterization. J Phys D Appl Phys 2003;36:573.

[11] Coleman N, Khan U, Blau WJ, Gun'ko YK. Small but strong: a review of the mechanical properties of carbon nanotube-polymer composites. Carbon 2006;44:1624–52.

[12] Zhang X, Sreekumar TV, Liu T, Kumar SJ. Properties and structure of nitric acid oxidized single wall carbon nanotube films. Phys Chem B 2004;108:16435–40.

[13] Chatterjee P. A model for the elastic moduli of three-dimensional fiber networks and nanocomposites. J Appl Phys 2006;100(5):054302–8.

[14] Celzard A, McRae E, Deleuze C, Dufort M, Furdin G, Marêché F. Critical concentration in percolating systems containing a high-aspect-ratio filler. J Phys Rev B 1996;53(10):6209–14.
[15] Balberg I, Anderson CH, Alexander S, Wagner N. Excluded volume and its relation to the onset of percolation. Phys Rev B 1984;30:3933–43.
[16] Takayanagi M, Uemura S, Minami SJ. Application of equivalent model method to dynamic rheo-optical properties of crystalline polymer. J Polym Sci C 1964;5:113–22.
[17] Favier V, Chanzy H, Cavaille J. Polymer nanocomposites reinforced by cellulose whiskers. Macromolecules 1996;28:6365–7.
[18] Favier V, Cavaille JY, Canova GR, Shrivastava SC. Mechanical percolation in cellulose whisker nanocomposites. Polym Eng Sci 1997;37:1732–9.
[19] Ouali N, Cavaillé JY, Pérez J. Elastic, viscoelastic and plastic behavior of multiphase polymer blends. Plast Rubber Comp Process Appl 1991;16:55–60.
[20] Gennes De PG. Scaling concepts. In: Polymer physics. New York: Cornell University Press; 1979, ISBN-10: 080141203X, ISBN-13: 978–0801412035.
[21] Ljungberg N, Bonini C, Bortolussi F, Boisson C, Heux L, Cavaille JY. New nanocomposite materials reinforced with cellulose whiskers in ataractic polypropylene: effect of surface and dispersion characteristics. Biomacromolecules 2005;6:2732–9.
[22] Fisher FT, Bradshaw RD, Brinson LC. Fiber waviness in nanotube-reinforced polymer composites—I: modulus predictions using effective nanotube properties. Comp Sci Technol 2003;63(11):1689–703.
[23] Ogasawara T, Ishida Y, Ishikawa T, Yokota R. Characterization of multi-walled carbon nanotube/phenylethynyl terminated polyimide composites. Compos Part A 2004;35:67–74.
[24] Omidi M, Rokni DT, Milani AS, Seethaler RJ, Arasteh R. Prediction of the mechanical characteristics of multi-walled carbon nanotube/epoxy composites using a new form of the rule of mixtures. Carbon 2010;48:3218–28.
[25] Ko J-H, Chang J-H. Properties of ultrahigh-molecular-weight polyethylene nanocomposite films containing different functionalized multiwalled carbon nanotubes. Polym Eng Sci 2009;49:2168–78. http://dx.doi.org/10.1002/pen.21462.
[26] Yeh M-K, Tai N-H, Liu J-H. Mechanical behavior of phenolic-based composites reinforced with multi-walled carbon nanotubes. Carbon 2006;44:1–9.
[27] Guojian W, Lijuan W, Mei Z, Zhengmian C. Reinforcemente and toughening of poly(vinyl chloride) with poly(caprolactone) grafted carbon nanotube. Compos Part A 2009;40:1476–81.
[28] Hou Y, Tang J, Zhang H, Qian C, Feng Y, Liu J. Functionalized few-walled carbon nanotubes for mechanical reinforcement of polymeric composites. ACS Nano 2009;3:1057–62.

Richard Feynman's Talk

here's plenty of room at the bottom

his transcript of the classic talk that Richard Feynman gave on December 29, 1959, t the annual meeting of the American Physical Society at the California Institute f Technology (Caltech) was first published in Caltech Engineering and Science, olume 23:5, February 1960, pp 22–36. It has been made available on the web t www.zyvex.com/nanotech/feynman.html with their kind permission.

I imagine experimental physicists must often look with envy at men like amerlingh Onnes, who discovered a field like low temperature, which seems to e bottomless and in which one can go down and down. Such a man is then a leader nd has some temporary monopoly in a scientific adventure. Percy Bridgman, in esigning a way to obtain higher pressures, opened up another new field and was ble to move into it and lead us all along. The development of even higher vacuum as a continuing development of the same kind.

I would like to describe a field, in which little has been done, but in which an normous amount can be done in principle. This field is not quite the same as the thers in that it will not tell us much of fundamental physics (in the sense of, What are the strange particles?") but it is more like solid-state physics in the sense at it might tell us much of great interest about the strange phenomena that occur in omplex situations. Furthermore, a point that is most important is that it would have n enormous number of technical applications.

What I want to talk about is the problem of manipulating and controlling things n a small scale.

As soon as I mention this, people tell me about miniaturization, and how far it has rogressed today. They tell me about electric motors that are the size of the nail on our small finger. And there is a device on the market, they tell me, by which you can rite the Lord's Prayer on the head of a pin. But that is nothing; that is the most rimitive, halting step in the direction I intend to discuss. It is a staggeringly small orld that is below. In the year 2000, when they look back at this age, they will onder why it was not until the year 1960 that anybody began to move seriously this direction.

Why cannot we write the entire 24 volumes of the Encyclopedia Britannica on the ead of a pin?

Let us see what would be involved. The head of a pin is a sixteenth of an inch cross. If you magnify it by 25,000 diameters, the area of the head of the pin is en equal to the area of all the pages of the Encyclopedia Britannica. Therefore, l it is necessary to do is to reduce in size all the writing in the Encyclopedia by

25,000 times. Is that possible? The resolving power of the eye is about 1/120 of an inch—that is roughly the diameter of one of the little dots on the fine half-tone reproductions in the Encyclopedia. This, when you demagnify it by 25,000 times, is still 80 Å in diameter—32 atoms across, in an ordinary metal. In other words, one of those dots still would contain in its area 1000 atoms. So, each dot can easily be adjusted in size as required by photoengraving, and there is no question that there is enough room on the head of a pin to put all of the Encyclopedia Britannica.

Furthermore, it can be read if it is so written. Let us imagine that it is written in raised letters of metal; that is, where the black is in the Encyclopedia, we have raised letters of metal that are actually 1/25,000 of their ordinary size. How would we read it?

If we had something written in such a way, we could read it using techniques in common use today. (They will undoubtedly find a better way when we do actually have it written, but to make my point conservatively I shall just take techniques we know today.) We would press the metal into a plastic material and make a mold of it, then peel the plastic off very carefully, evaporate silica into the plastic to get a very thin film, then shadow it by evaporating gold at an angle against the silica so that all the little letters will appear clearly, dissolve the plastic away from the silica film, and then look through it with an electron microscope!

There is no question that if the thing were reduced by 25,000 times in the form of raised letters on the pin, it would be easy for us to read it today. Furthermore, there is no question that we would find it easy to make copies of the master; we would just need to press the same metal plate again into plastic and we would have another copy.

How do we write small?

The next question is: How do we *write* it? We have no standard technique to do this now. But let me argue that it is not as difficult as it first appears to be. We can reverse the lenses of the electron microscope in order to demagnify as well as magnify. A source of ions, sent through the microscope lenses in reverse, could be focused to a very small spot. We could write with that spot like we write in a TV cathode ray oscilloscope, by going across in lines, and having an adjustment which determines the amount of material which is going to be deposited as we scan in lines.

This method might be very slow because of space charge limitations. There will be more rapid methods. We could first make, perhaps by some photo process, a screen which has holes in it in the form of the letters. Then we would strike an arc behind the holes and draw metallic ions through the holes; then we could again use our system of lenses and make a small image in the form of ions, which would deposit the metal on the pin.

A simpler way might be this (though I am not sure it would work): We take light and, through an optical microscope running backwards, we focus it onto a very small photoelectric screen. Then electrons come away from the screen where the light is shining. These electrons are focused down in size by the electron microscope lenses to impinge directly upon the surface of the metal. Will such a beam etch away the metal if it is run long enough? I do not know. If it does not work for a metal surface, it must be possible to find some surface with which to coat the original pin so that, where the electrons bombard, a change is made which we could recognize later.

There is no intensity problem in these devices—not what you are used to in magnification, where you have to take a few electrons and spread them over a bigger and bigger screen; it is just the opposite. The light which we get from a page is concentrated onto a very small area, so it is very intense. The few electrons which come from the photoelectric screen are demagnified down to a very tiny area so that, again, they are very intense. I do not know why this has not been done yet!

That is the Encyclopedia Britannica on the head of a pin, but let us consider all the books in the world. The Library of Congress has approximately 9 million volumes; the British Museum Library has 5 million volumes; there are also 5 million volumes in the National Library in France. Undoubtedly there are duplications, so let us say that there are some 24 million volumes of interest in the world.

What would happen if I print all this down at the scale we have been discussing? How much space would it take? It would take, of course, the area of about a million pinheads because instead of there being just the 24 volumes of the Encyclopedia, there are 24 million volumes. The million pinheads can be put in a square of a thousand pins on a side, or an area of about 3 square yards. That is to say, the silica replica with the paper-thin backing of plastic, with which we have made the copies, with all this information, is on an area of approximately the size of 35 pages of the Encyclopedia. That is about half as many pages as there are in this magazine. All of the information which all of mankind has ever recorded in books can be carried around in a pamphlet in your hand—and not written in code, but as a simple reproduction of the original pictures, engravings, and everything else on a small scale without loss of resolution.

What would our librarian at Caltech say, as she runs all over from one building to another, if I tell her that, 10 years from now, all of the information that she is struggling to keep track of—120,000 volumes, stacked from the floor to the ceiling, drawers full of cards, storage rooms full of the older books—can be kept on just one library card! When the University of Brazil, for example, finds that their library is burned, we can send them a copy of every book in our library by striking off a copy from the master plate in a few hours and mailing it in an envelope no bigger or heavier than any other ordinary air mail letter.

Now, the name of this talk is "There is *Plenty* of Room at the Bottom"—not just "There is Room at the Bottom." What I have demonstrated is that there *is* room—that you can decrease the size of things in a practical way. I now want to show that there is *plenty* of room. I will not discuss now how we are going to do it, but only what is possible in principle—in other words, what is possible according to the laws of physics. I am not inventing antigravity, which is possible someday only if the laws are not what we think. I am telling you what could be done if the laws *are* what we think; we are not doing it simply because we have not yet gotten around to it.

Information on a small scale

Suppose that, instead of trying to reproduce the pictures and all the information directly in its present form, we write only the information content in a code of dots and dashes, or something like that, to represent the various letters. Each letter represents six or seven "bits" of information; that is, you need only about six or seven dots or dashes for each letter. Now, instead of writing everything, as I did before, on the *surface* of the head of a pin, I am going to use the interior of the material as well.

Let us represent a dot by a small spot of one metal, the next dash by an adjacent spot of another metal, and so on. Suppose, to be conservative, that a bit of information is going to require a little cube of atoms $5 \times 5 \times 5$—that is 125 atoms. Perhaps we need a hundred and some odd atoms to make sure that the information is not lost through diffusion, or through some other process.

I have estimated how many letters there are in the Encyclopedia, and I have assumed that each of my 24 million books is as big as an Encyclopedia volume, and have calculated, then, how many bits of information there are (10^{15}). For each bit I allow 100 atoms. And it turns out that all of the information that man has carefully accumulated in all the books in the world can be written in this form in a cube of material one two-hundredth of an inch wide—which is the barest piece of dust that can be made out by the human eye. So there is *plenty* of room at the bottom! Do not tell me about microfilm!

This fact—that enormous amounts of information can be carried in an exceedingly small space—is, of course, well known to the biologists, and resolves the mystery which existed before we understood all this clearly, of how it could be that, in the tiniest cell, all of the information for the organization of a complex creature such as ourselves can be stored. All this information—whether we have brown eyes, or whether we think at all, or that in the embryo the jawbone should first develop with a little hole in the side so that later a nerve can grow through it—all this information is contained in a very tiny fraction of the cell in the form of long-chain DNA molecules in which approximately 50 atoms are used for 1bit of information about the cell.

Better electron microscopes

If I have written in a code, with $5 \times 5 \times 5$ atoms to a bit, the question is: How could I read it today? The electron microscope is not quite good enough, with the greatest care and effort, it can only resolve about 10 Å. I would like to try and impress upon you while I am talking about all of these things on a small scale, the importance of improving the electron microscope by a hundred times. It is not impossible; it is not against the laws of diffraction of the electron. The wave length of the electron in uch a microscope is only 1/20 of an angstrom. So it should be possible to see the ndividual atoms. What good would it be to see individual atoms distinctly?

We have friends in other fields—in biology, for instance. We physicists often ook at them and say, "You know the reason you fellows are making so little progess?" (Actually I do not know any field where they are making more rapid progress han they are in biology today.) "You should use more mathematics, like we do." 'hey could answer us—but they're polite, so I'll answer for them: "What *you* should o in order for *us* to make more rapid progress is to make the electron microscope 00 times better."

What are the most central and fundamental problems of biology today? They are uestions like: What is the sequence of bases in the DNA? What happens when you ave a mutation? How is the base order in the DNA connected to the order of amino cids in the protein? What is the structure of the RNA; is it single-chain or doublehain, and how is it related in its order of bases to the DNA? What is the organization f the microsomes? How are proteins synthesized? Where does the RNA go? How oes it sit? Where do the proteins sit? Where do the amino acids go in? In photosynesis, where is the chlorophyll; how is it arranged; where are the carotenoids ıvolved in this thing? What is the system of the conversion of light into chemical nergy?

It is very easy to answer many of these fundamental biological questions; you ıst *look at the thing*! You will see the order of bases in the chain; you will see e structure of the microsome. Unfortunately, the present microscope sees at a scale hich is just a bit too crude. Make the microscope 100 times more powerful, and ıany problems of biology would be made very much easier. I exaggerate, of course, ut the biologists would surely be very thankful to you—and they would prefer that) the criticism that they should use more mathematics.

The theory of chemical processes today is based on theoretical physics. In this ense, physics supplies the foundation of chemistry. But chemistry also has analysis. you have a strange substance and you want to know what it is, you go through a ng and complicated process of chemical analysis. You can analyze almost anying today, so I am a little late with my idea. But if the physicists wanted to, ey could also dig under the chemists in the problem of chemical analysis. It would e very easy to make an analysis of any complicated chemical substance; all one ould have to do would be to look at it and see where the atoms are. The only ouble is that the electron microscope is 100 times too poor. (Later, I would like) ask the question: Can the physicists do something about the third problem of ıemistry—namely, synthesis? Is there a *physical* way to synthesize any chemical ıbstance?)

The reason the electron microscope is so poor is that the *f*-value of the lenses is nly 1 part to 1000; you do not have a big enough numerical aperture. And I know at there are theorems which prove that it is impossible, with axially symmetrical ationary field lenses, to produce an *f*-value any bigger than so-and-so; and therefore e resolving power at the present time is at its theoretical maximum. But in every eorem there are assumptions. Why must the field be axially symmetrical? Why must the field be stationary? Cannot we have pulsed electron beams in fields moving up along with the electrons? Must the field be symmetrical? I put this out as a challenge: Is there no way to make the electron microscope more powerful?

The marvelous biological system

The biological example of writing information on a small scale has inspired me to think of something that should be possible. Biology is not simply writing information; it is *doing something* about it. A biological system can be exceedingly small. Many of the cells are very tiny, but they are very active; they manufacture various substances; they walk around; they wiggle; and they do all kinds of marvelous things—all on a very small scale. Also, they store information. Consider the possibility that we too can make a thing very small which does what we want—that we can manufacture an object that maneuvers at that level!

There may even be an economic point to this business of making things very small. Let me remind you of some of the problems of computing machines. In computers we have to store an enormous amount of information. The kind of writing that I was mentioning before, in which I had everything down as a distribution of metal, is permanent. Much more interesting to a computer is a way of writing, erasing, and writing something else. (This is usually because we do not want to waste the material on which we have just written. Yet if we could write it in a very small space, it would not make any difference; it could just be thrown away after it was read. It does not cost very much for the material.)

Miniaturizing the computer

I do not know how to do this on a small scale in a practical way, but I do know that computing machines are very large; they fill rooms. Why cannot we make them very small, make them of little wires, little elements—and by little, I mean *little*. For instance, the wires should be 10 or 100 atoms in diameter, and the circuits should be a few thousand angstroms across. Everybody who has analyzed the logical theory of computers has come to the conclusion that the possibilities of computers are very interesting—if they could be made to be more complicated by several orders of magnitude. If they had millions of times as many elements, they could make judgments. They would have time to calculate what is the best way to make the calculation that they are about to make. They could select the method of analysis which, from their experience, is better than the one that we would give to them. And in many other ways, they would have new qualitative features.

If I look at your face I immediately recognize that I have seen it before. (Actually, my friends will say I have chosen an unfortunate example here for the subject of this illustration. At least I recognize that it is a *man* and not an *apple*.) Yet there is no machine which, with that speed, can take a picture of a face and say even that it is a man; and much less that it is the same man that you showed it before—unless it is exactly the same picture. If the face is changed; if I am closer to the face; if I am further from the face; if the light changes—I recognize it anyway. Now, this little computer I carry in my head is easily able to do that. The computers that we build are not able to do that. The number of elements in this bone box of mine are enormously greater than the number of elements in our "wonderful" computers. But our mechanical computers are too big; the elements in this box are microscopic. I want to make some that are *sub*-microscopic.

If we wanted to make a computer that had all these marvelous extra qualitative abilities, we would have to make it, perhaps, the size of the Pentagon. This has several disadvantages. First, it requires too much material; there may not be enough germanium in the world for all the transistors which would have to be put into this enormous thing. There is also the problem of heat generation and power consumption; TVA would be needed to run the computer. But an even more practical difficulty is that the computer would be limited to a certain speed. Because of its large size, there is finite time required to get the information from one place to another. The information cannot go any faster than the speed of light—so, ultimately, when our computers get faster and faster and more and more elaborate, we will have to make them smaller and smaller.

But there is plenty of room to make them smaller. There is nothing that I can see in the physical laws that says the computer elements cannot be made enormously smaller than they are now. In fact, there may be certain advantages.

Miniaturization by evaporation

How can we make such a device? What kind of manufacturing processes would we use? One possibility we might consider, since we have talked about writing by putting atoms down in a certain arrangement, would be to evaporate the material, then evaporate the insulator next to it. Then, for the next layer, evaporate another position of a wire, another insulator, and so on. So, you simply evaporate until you have a block of stuff which has the elements—coils and condensers, transistors and so on—of exceedingly fine dimensions.

But I would like to discuss, just for amusement, that there are other possibilities. Why cannot we manufacture these small computers somewhat like we manufacture the big ones? Why cannot we drill holes, cut things, solder things, stamp things out, mold different shapes all at an infinitesimal level? What are the limitations as to how small a thing has to be before you can no longer mold it? How many times when you are working on something frustratingly tiny like your wife's wrist watch, have you said to yourself, "If I could only train an ant to do this!" What I would like to suggest is the possibility of training an ant to train a mite to do this. What are the possibilities of small but movable machines? They may or may not be useful, but they surely would be fun to make.

Consider any machine—for example, an automobile—and ask about the problems of making an infinitesimal machine like it. Suppose, in the particular design of the automobile, we need a certain precision of the parts; we need an accuracy, let us suppose, of 4/10,000 of an inch. If things are more inaccurate than that in the shape of the cylinder and so on, it is not going to work very well. If I make the thing too small, I have to worry about the size of the atoms; I cannot make a circle out of "balls" so to speak, if the circle is too small. So, if I make the error corresponding to 4/10,000 of an inch, which correspond to an error of 10 atoms, it turns out that I can reduce the dimensions of an automobile 4000 times, approximately—so that it is 1 mm across. Obviously, if you redesign the car so that it would work with a much larger tolerance, which is not at all impossible, then you could make a much smaller device.

It is interesting to consider what the problems are in such small machines. Firstly, with parts stressed to the same degree, the forces go as the area you are reducing, so that things like weight and inertia are of relatively no importance. The strength of material, in other words, is very much greater in proportion. The stresses and expansion of the flywheel from centrifugal force, for example, would be the same proportion only if the rotational speed is increased in the same proportion as we decrease the size. On the other hand, the metals that we use have a grain structure, and this would be very annoying at small scale because the material is not homogeneous. Plastics and glass and things of this amorphous nature are very much more homogeneous, and so we would have to make our machines out of such materials.

There are problems associated with the electrical part of the system—with the copper wires and the magnetic parts. The magnetic properties on a very small scale are not the same as on a large scale; there is the "domain" problem involved. A big magnet made of millions of domains can only be made on a small scale with one domain. The electrical equipment would not simply be scaled down; it has to be redesigned. But I can see no reason why it cannot be redesigned to work again.

Problems of lubrication

Lubrication involves some interesting points. The effective viscosity of oil would be higher and higher in proportion as we went down (and if we increase the speed as much as we can). If we do not increase the speed so much, and change from oil to kerosene or some other fluid, the problem is not so bad. But actually we may not have to lubricate at all! We have a lot of extra force. Let the bearings run dry; they would not run hot because the heat escapes away from such a small device very, very rapidly.

This rapid heat loss would prevent the gasoline from exploding, so an internal combustion engine is impossible. Other chemical reactions, liberating energy when cold, can be used. Probably an external supply of electrical power would be most convenient for such small machines.

What would be the utility of such machines? Who knows? Of course, a small automobile would only be useful for the mites to drive around in, and I suppose our Christian interests do not go that far. However, we did note the possibility of the manufacture of small elements for computers in completely automatic factories, containing lathes and other machine tools at the very small level. The small lathe would not have to be exactly like our big lathe. I leave to your imagination the improvement of the design to take full advantage of the properties of things on a mall scale, and in such a way that the fully automatic aspect would be easiest o manage.

A friend of mine (Albert R. Hibbs) suggests a very interesting possibility for relavely small machines. He says that, although it is a very wild idea, it would be intersting in surgery if you could swallow the surgeon. You put the mechanical surgeon nside the blood vessel and it goes into the heart and "looks" around. (Of course the nformation has to be fed out.) It finds out which valve is the faulty one and takes a ttle knife and slices it out. Other small machines might be permanently incorpoated in the body to assist some inadequately functioning organ.

Now comes the interesting question: How do we make such a tiny mechanism? I eave that to you. However, let me suggest one weird possibility. You know, in the tomic energy plants they have materials and machines that they cannot handle irectly because they have become radioactive. To unscrew nuts and put on bolts nd so on, they have a set of master and slave hands, so that by operating a set of evers here, you control the "hands" there, and can turn them this way and that so ou can handle things quite nicely.

Most of these devices are actually made rather simply, in that there is a particular able, like a marionette string, that goes directly from the controls to the "hands." ut, of course, things also have been made using servo motors, so that the connecon between the one thing and the other is electrical rather than mechanical. When ou turn the levers, they turn a servo motor, and it changes the electrical currents in e wires, which repositions a motor at the other end.

Now, I want to build much the same device—a master-slave system which opertes electrically. But I want the slaves to be made especially carefully by modern rge-scale machinists so that they are one-fourth the scale of the "hands" that ou ordinarily maneuver. So you have a scheme by which you can do things at ne-quarter scale anyway—the little servo motors with little hands play with little uts and bolts; they drill little holes; they are four times smaller. Aha! So I manufacre a quarter-size lathe; I manufacture quarter-size tools; and I make, at the oneuarter scale, still another set of hands again relatively one-quarter size! This is ne-sixteenth size, from my point of view. And after I finish doing this I wire directly om my large-scale system, through transformers perhaps, to the one-sixteenth-size ervo motors. Thus, I can now manipulate the one-sixteenth-size hands.

Well, you get the principle from there on. It is rather a difficult program, but it is a ossibility. You might say that one can go much farther in one step than from one to our. Of course, this has all to be designed very carefully and it is not necessary simly to make it like hands. If you thought of it very carefully, you could probably rive at a much better system for doing such things.

If you work through a pantograph, even today, you can get much more than a ctor of four in even one step. But you cannot work directly through a pantograph hich makes a smaller pantograph which then makes a smaller pantograph— ecause of the looseness of the holes and the irregularities of construction. The nd of the pantograph wiggles with a relatively greater irregularity than the regularity with which you move your hands. In going down this scale, I would find the end of the pantograph shaking so badly that it was not doing anything sensible at all.

At each stage, it is necessary to improve the precision of the apparatus. If, for instance, having made a small lathe with a pantograph, we find its lead screw irregular—more irregular than the large-scale one—we could lap the lead screw against breakable nuts that you can reverse in the usual way back and forth until this lead screw is, at its scale, as accurate as our original lead screws, at our scale.

We can make flats by rubbing unflat surfaces in triplicates together—in three pairs—and the flats then become flatter than the thing you started with. Thus, it is not impossible to improve precision on a small scale by the correct operations. So, when we build this stuff, it is necessary at each step to improve the accuracy of the equipment by working for a while down there, making accurate lead screws, Johansen blocks, and all the other materials which we use in accurate machine work at the higher level. We have to stop at each level and manufacture all the stuff to go to the next level—a very long and very difficult program. Perhaps you can figure a better way than that to get down to small scale more rapidly.

Yet, after all this, you have just got one little baby lathe 4000 times smaller than usual. But we were thinking of making an enormous computer, which we were going to build by drilling holes on this lathe to make little washers for the computer. How many washers can you manufacture on this one lathe?

A hundred tiny hands

When I make my first set of slave "hands" at one-fourth scale, I am going to make 10 sets. I make 10 sets of "hands," and I wire them to my original levers so they each do exactly the same thing at the same time in parallel. Now, when I am making my new devices one-quarter again as small, I let each one manufacture 10 copies, so that I would have a hundred "hands" at the one-sixteenth size.

Where am I going to put the million lathes that I am going to have? Why, there is nothing to it; the volume is much less than that of even one full-scale lathe. For instance, if I made a billion little lathes, each 1/4000 of the scale of a regular lathe there are plenty of materials and space available because in the billion little ones there is less than 2% of the materials in one big lathe.

It does not cost anything for materials, you see. So I want to build a billion tiny factories, models of each other, which are manufacturing simultaneously, drilling holes, stamping parts, and so on.

As we go down in size, there are a number of interesting problems that arise. All things do not simply scale down in proportion. There is the problem that materials stick together by the molecular (van der Waals) attractions. It would be like this: After you have made a part and you unscrew the nut from a bolt, it is not going to fall down because the gravity is not appreciable; it would even be hard to get it off the bolt. It would be like those old movies of a man with his hands full of molasses trying to get rid of a glass of water. There will be several problems of this nature that we will have to be ready to design for.

Rearranging the atoms

But I am not afraid to consider the final question as to whether, ultimately—in the great future—we can arrange the atoms the way we want; the very *atoms*, all the way down! What would happen if we could arrange the atoms one by one the way we want them (within reason, of course; you cannot put them so that they are chemically unstable, for example).

Up to now, we have been content to dig in the ground to find minerals. We heat them and we do things on a large scale with them, and we hope to get a pure substance with just so much impurity, and so on. But we must always accept some atomic arrangement that nature gives us. We have not got anything, say, with a "checkerboard" arrangement, with the impurity atoms exactly arranged 1000 Å apart, or in some other particular pattern.

What could we do with layered structures with just the right layers? What would the properties of materials be if we could really arrange the atoms the way we want them? They would be very interesting to investigate theoretically. I cannot see exactly what would happen, but I can hardly doubt that when we have some *control* of the arrangement of things on a small scale we will get an enormously greater range of possible properties that substances can have, and of different things that we can do.

Consider, for example, a piece of material in which we make little coils and condensers (or their solid-state analogs) 1000 or 10,000 Å in a circuit, one right next to the other, over a large area, with little antennas sticking out at the other end—a whole series of circuits. Is it possible, for example, to emit light from a whole set of antennas, like we emit radio waves from an organized set of antennas to beam the radio programs to Europe? The same thing would be to *beam* the light out in a definite direction with very high intensity. (Perhaps such a beam is not very useful technically or economically.)

I have thought about some of the problems of building electric circuits on a small scale, and the problem of resistance is serious. If you build a corresponding circuit on a small scale, its natural frequency goes up, since the wave length goes down as the scale; but the skin depth only decreases with the square root of the scale ratio, and so resistive problems are of increasing difficulty. Possibly we can beat resistance through the use of superconductivity if the frequency is not too high, or by other tricks.

Atoms in a small world

When we get to the very, very small world—say circuits of seven atoms—we have a lot of new things that would happen that represent completely new opportunities for design. Atoms on a small scale behave like *nothing* on a large scale, for they satisfy the laws of quantum mechanics. So, as we go down and fiddle around with the atoms down there, we are working with different laws, and we can expect to do different things. We can manufacture in different ways. We can use, not just circuits, but some system involving the quantized energy levels, or the interactions of quantized spins, etc.

Another thing we will notice is that, if we go down far enough, all of our devices can be mass produced so that they are absolutely perfect copies of one another. We cannot build two large machines so that the dimensions are exactly the same. But if your machine is only 100 atoms high, you only have to get it correct to one-half of one percent to make sure the other machine is exactly the same size—namely, 100 atoms high!

At the atomic level, we have new kinds of forces, new kinds of possibilities, and new kinds of effects. The problems of manufacture and reproduction of materials will be quite different. I am, as I said, inspired by the biological phenomena in which chemical forces are used in a repetitious fashion to produce all kinds of weird effects (one of which is the author).

The principles of physics, as far as I can see, do not speak against the possibility of maneuvering things atom by atom. It is not an attempt to violate any laws; it is something, in principle, that can be done; but in practice, it has not been done because we are too big.

Ultimately, we can do chemical synthesis. A chemist comes to us and says, "Look, I want a molecule that has the atoms arranged thus and so; make me that molecule." The chemist does a mysterious thing when he wants to make a molecule. He sees that it has got that ring, so he mixes this and that, and he shakes it, and he fiddles around. And, at the end of a difficult process, he usually does succeed in synthesizing what he wants. By the time I get my devices working, so that we can do it by physics, he will have figured out how to synthesize absolutely anything, so that this will really be useless.

But it is interesting that it would be, in principle, possible (I think) for a physicist to synthesize any chemical substance that the chemist writes down. Give the orders and the physicist synthesizes it. How? Put the atoms down where the chemist says, and so you make the substance. The problems of chemistry and biology can be greatly helped if our ability to see what we are doing, and to do things on an atomic level, is ultimately developed—a development which I think cannot be avoided.

Now, you might say, "Who should do this and why should they do it?" Well, I pointed out a few of the economic applications, but I know that the reason that you would do it might be just for fun. But have some fun! Let us have a competition between laboratories. Let one laboratory make a tiny motor which it sends to another lab which sends it back with a thing that fits inside the shaft of the first motor.

High school competition

Just for the fun of it, and in order to get kids interested in this field, I would propose that someone who has some contact with the high schools thinks of making some kind of high school competition. After all, we have not even started in this field, and even the kids can write smaller than has ever been written before. They could have competition in high schools. The Los Angeles high school could send a pin to the Venice high school on which it says, "How's this?" They get the pin back, and in the dot of the "i" it says, "Not so hot."

Perhaps this does not excite you to do it, and only economics will do so. Then I vant to do something; but I cannot do it at the present moment, because I have not repared the ground. It is my intention to offer a prize of $1000 to the first guy who an take the information on the page of a book and put it on an area 1/25,000 smaller n linear scale in such manner that it can be read by an electron microscope.

And I want to offer another prize—if I can figure out how to phrase it so that I do ot get into a mess of arguments about definitions—of another $1000 to the first guy vho makes an operating electric motor—a rotating electric motor which can e controlled from the outside and, not counting the lead-in wires—which is only /64 inch cube.

I do not expect that such prizes will have to wait very long for claimants.

Periodic Table of Elements

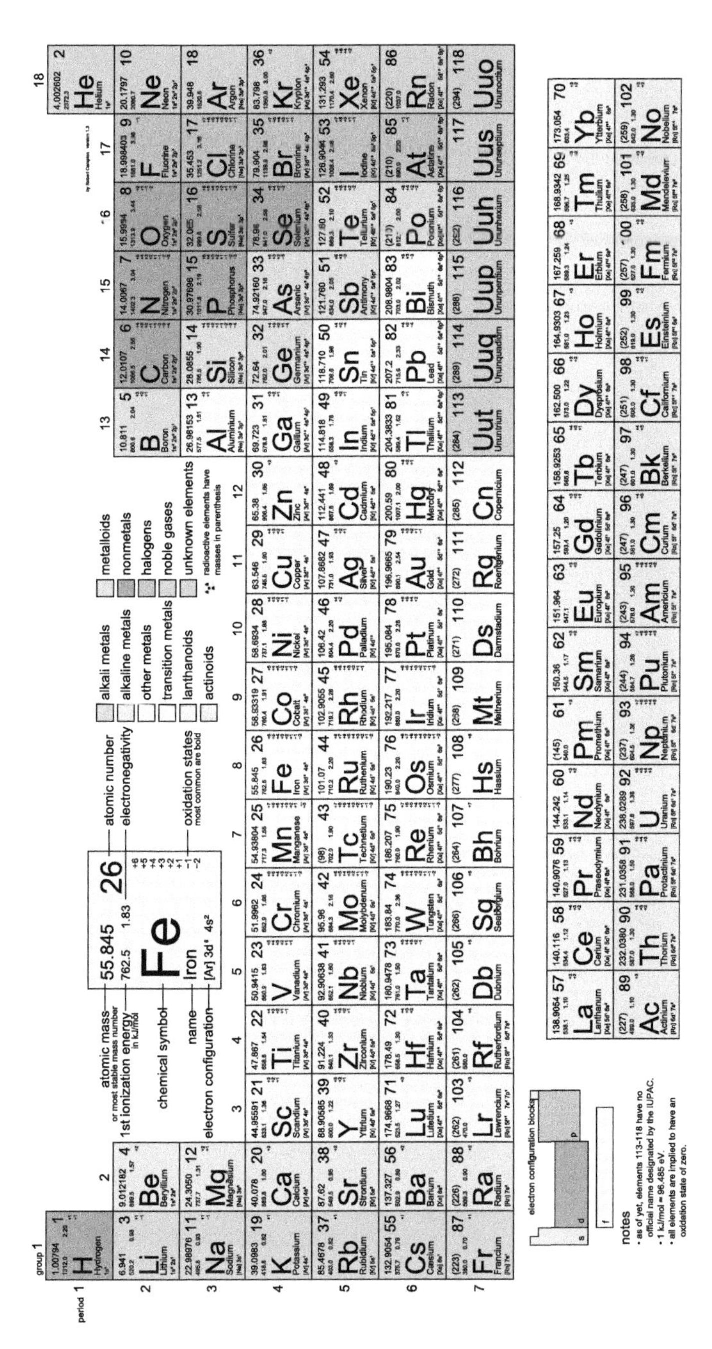

Graphene Sheet

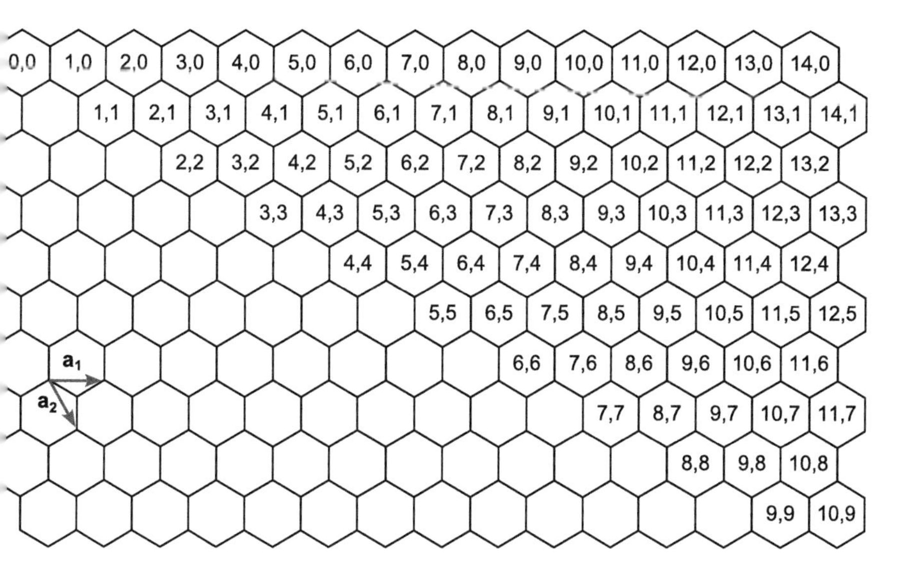

APPENDIX D

Simulations Using Matlab®

Code rule of mixtures

```
% Code Rule of Mixtures.m
% ©2013 Marcio R. Loos
% This routine calculates the Young's modulus of composites reinforced
with CNTs
% according to the rule of mixtures.
% There are three entries necessaries for the simulation:
% Vf: Volumetric fraction of CNTs
% Ef: Young's modulus of CNTs (in GPa)
% Em: Young's modulus of the matrix (in GPa)
% The output of the routine is:
% Ec: Young's modulus of the composite (in GPa)
% Message is printed on the screen:
disp('This code calculates the Young's modulus of composites reinforced
with CNTs.');
disp('Please enter the values below');
% The values of the variables are requested:
Vf = input('Type Vf:');
Ef = input('Type Ef:');
Em = input('Type Em:');
% The volumetric fraction of the matrix phase is calculated:
Vm = 1-Vf;
% Rule of mixtures:
Ec = Vf*Ef + Vm*Em;
% The result of the simulations is printed on the screen:
disp(['Young's modulus of the composite = 'num2str(Ec) 'Volumetric
fraction of CNTs = 'num2str(Vf)]);
% End of the code.
```

Code Halpin-Tsai

```
% Code Halpin-Tsai.m
% ©2013 Marcio R. Loos
% This routine calculates the Young's modulus of composites reinforced
with CNTs
```

```
% according to the Halpin-Tsai model.
% Description of the variables:
% l: average length of CNTs (in meters)
% d: average diameter of CNTs (in meters)
% eta: aspect ratio of CNTs
% Vf: volumetric fraction of CNTs
% Ef: Young's modulus of CNTs (in GPa)
% Em: Young's modulus of the matrix (in GPa)
% ksi: parameter related to the geometry of the filler
% etaL: term related to the longitudinal modulus of the composite
% etaT: term related to the transversal modulus of the composite
%
% The routine performs the following steps:
% i) calculates the Young's modulus in the range Vf = 0-1
% ii) exports the results to a file in the hard disc
% iii) prints the result of the simulation on the screen
% iv) plots a graph of Vf x Ef and print it on a new window
% vi) exports the graph to a file Graph-Result.tif on the hard disc
%
% Definition of the variables and terms:
l = 5d-6;
d = 15d-9;
eta = l/d;
Ef = 900;
Em = 3;
ksi = 2*(l/d);
etaL=((Ef/Em)-1)/((Ef/Em)+ksi);
etaT=((Ef/Em)-1)/((Ef/Em)+2);
% The file 'Result.txt', where the result of the simulation will be
printed is open:
fid = fopen('Result.txt','wt');
% Calculations start. A volumetric fraction of CNTs is varied from 0 to 1
% in increments of 0.01.
for Vf=0:0.01:1
% Halpin-Tsai model:
Ec = Em*(((3/8)*((1+(ksi*etaL*Vf))/(1-(etaL*Vf))))+((5/8)*(((1+
(2*etaT*Vf))/(1-(etaT*Vf))))));
% The result of the simulations is exported to the file 'Result.txt':
fprintf(fid, '%6.3f %12.8f\r\n',Vf,Ec);
end;
% The file 'Result.txt' is closed:
fclose(fid);
% The result of the simulations is printed on the screen:
type Result.txt
```

```
 The result of the simulation is imported through the command load:
oad Result.txt;
 A graph of Vf (column; 1) versus Ef (column; 2) is plotted on the
creen:
lot(Result(:,1), Result(:,2));
 The title 'Modulus of the composite' is added to the graph:
itle(' Modulus of the composite ');
 The x-axis is named:
label('Vf');
 The y-axis is named:
label('Ec (GPa)');
 A legend is added to the graph:
egend('Halpin-Tsai')
 The graph is exported to the file 'Graph-Result.tif ':
et(gcf, 'PaperPositionMode', 'auto');
print('-dtiff',' Graph-Result.tif ');
 End of the code.
```

Questions and Exercises

Chapter 1

1.1. Calculate the number of atoms contained in a copper cube having a volume of 5.2 cm^3. Consider the copper density equal to 8.94 g/cm^3.

1.2. A gold sphere (density of 19.30 g/cm^3) has a radius of 2.19 cm.
- **(a)** Calculate the number of atoms contained in the sphere.
- **(b)** Assuming that the price of gold is $ 53.00 per gram, calculate the price of this sphere.
- **(c)** Calculate the mass of a gold atom. Refer to the periodic table of elements in Appendix B to obtain the necessary data.

1.3. How many nanocubes with an edge of 10 nm are required to produce the same surface area of a cube with an edge of 1 m? What volume would be occupied by these cubes?

1.4. Element iron has a molar mass of 55.84 g/mol.
- **(a)** Calculate the volume of 1 mol of iron. Consider $\rho_{Fe} = 7.87$ g/cm^3.
- **(b)** Using the value found in (a) calculate the volume of an iron atom.
- **(c)** Now, calculate the volume of one Fe atom whereasthe atomic radius of iron is 126 pm. Compare this value with that obtained in (b) and explain the difference.
- **(d)** Repeat the above calculations for nickel.

Consider $\rho_{Ni} = 8.91$ g/cm^3 and atomic radius of 124 pm.

1.5. What does the term nano mean?

1.6. How many nanometers are contained in:
- **(a)** A micrometer?
- **(b)** A millimeter?
- **(c)** A centimeter?
- **(d)** A meter?

1.7. Who was the first to use the term nanotechnology and when?

1.8. What is the title of the famous presentation by Richard Feynman on December 29, 1959?

1.9. How many oxygen atoms should be aligned to occupy a space of 1 nm? Repeat the exercise for hydrogen, sodium, carbon (sp^3), and chlorine.

1.10. Calculate the specific surface area (m^2/kg) of a gold cube with an edge of:
- **(a)** 1.0×10^{-3} m.
- **(b)** 1.0×10^{-5} m.
- **(c)** 1.0×10^{-7} m.
- **(d)** 1.0×10^{-9} m.

Consider the density of the gold equal to 19.30 g/cm^3.

1.11. What is the main effect of reducing the dimensions of a material?

1.12. What is considered the starting point in triggering nanoscience and nanotechnology?

1.13. What is nanoscience and nanotechnology?

1.14. Consider a sphere with a diameter of 1 μm and a cube with a 1 μm edge (the sphere fits inside the cube). What is the ratio of the surface area of the cube and sphere? Now assume that the cube and sphere have the same volume. Which has the largest surface area? What is the implication of this result in nature?

1.15. Derive equations for the ratio surface area/volume (S_A/V) of the following geometric shapes:

(a) A cube with edge a.

(b) Sphere of diameter d.

(c) Disc with height h and diameter d ($h \ll d$).

(d) Cylinder of height h and diameter d ($d = h$).

(e) Cylinder of height h and diameter d ($h \gg d$).

1.16. What do you find most interesting or surprising about nanotechnology and nanoscience?

1.17. Calculate the diameter of a spherical particle of lithium with a density $\rho = 0.534$ g/cm^3 and a specific surface area of 200 m^2/g. What is the ratio surface area/volume (S_A/V) for this particle?

1.18. What is the ratio surface area/volume (S_A/V) and the specific surface area of osmium particles with a diameter of 1.33 nm and density $\rho = 22.6$ g/cm^3?

1.19. Name five current or envisaged applications of nanotechnology.

1.20. The following table shows the approximate amount estimated from different chemical elements present in a human body with an average of 70 kg. Based on these data, calculate the average number of atoms present in a human body.

Element	Amount (kg)	Amount (mol)
Oxygen	43	2700
Carbon	16	1300
Nitrogen	1.8	130
Calcium	1	25
Phosphorus	0.78	25
Sulfur	0.14	4.4
Potassium	0.14	3.6
Sodium	0.1	4.3
Chlorine	0.095	2.7
Magnesium	0.019	0.78
Silicon	0.018	0.64
Iron	0.0042	0.075
Fluorine	0.0026	0.14
Zinc	0.0023	0.035
Rubidium	0.00032	0.0037

—Cont'd

Element	Amount (kg)	Amount (mol)
Strontium	0.00032	0.0037
Bromine	0.0002	0.0025
Lead	0.00012	0.00058
Copper	0.000072	0.0011
Aluminum	0.000061	0.0023
Cadmium	0.00005	0.00044
Boron	<0.000048	0.0044
Barium	0.00022	0.00016
Tin	<0.000017	0.00014
Iodine	0.000013	0.0001
Manganese	0.000012	0.00022
Nickel	0.00001	0.00017
Gold	<0.000010	0.000051
Molybdenum	<0.0000093	0.000097
Chromium	<0.0000018	0.000035
Cesium	0.0000015	0.000011
Cobalt	0.0000015	0.000025
Uranium	0.00000009	0.00000038
Beryllium	0.000000036	0.000004
Radium	3.1×10^{-14}	1.4×10^{-13}

Chapter 2

2.1. What are the different types of materials? Classify them according to their density, stiffness, and specific strength.

2.2. What are polymers?

2.3. What is the difference between thermoplastics and thermosets polymers?

2.4. List some advantages and disadvantages of polymeric materials in relation to metals.

2.5. What are composites?

2.6. Describe the different phases of a composite material and its functions.

2.7. Cite examples of natural and synthetic composite materials.

2.8. What are the classifications of composite materials according to the matrix and reinforcement?

2.9. What are the main types of thermosetting resins used for the manufacture of structural composites? Discuss them briefly.

2.10. Why is the processing of thermoplastic matrix composites shorter than the processing of thermosets?

2.11. Describe the main properties of glass, carbon, and aramid fibers. Compare their properties with aluminum and steel.

2.12. What are the main advantages and disadvantages of composite materials?

2.13. Cite 10 different applications of composite materials.

Chapter 3

3.1. Discuss the importance of carbon atom to life.

3.2. What was the role of silicon in electronics evolution?

3.3. What are the main allotropes of carbon? Describe the main differences between these forms and the reason for such differences.

3.4. Draw a timeline of major events related to the history of CNTs. Also include the discovery of the fullerene.

3.5. Calculate the diameter of the following single-walled nanotubes:

(a) (6, 6)
(b) (11, 1)
(c) (10, 10)
(d) (11, 5)
(e) (12, 12)
(f) (6, 4)
(g) (12, 0)
(h) (9, 3)
(i) (10, 2)
(j) (11, 7)

3.6. What are the experimental methods that can be used to determine the diameter of nanotubes?

3.7. Why do SWCNTs agglomerate to form bundles?

3.8. How many C_{60} would be required to fill out an SWCNT with a 15-μm length and diameter of 1.3 nm?

3.9. What is the inner diameter of an SWCNT with an outer diameter of 1.68 nm?

3.10. The energy stored in a C=C bond is 614 kJ/mol and its length is 0.134 nm. The covalent radius of carbon is r_c 0.077 nm. Assuming that the bond can elongate 2.5 times before breaking, what is the ultimate strength of the bond? Repeat the calculation for a bond C≡C, where the stored energy is 839 kJ/mol and the length is 0.120 nm.

3.11. Based on the values obtained in exercise 3.10 and example 3.1, make a table comparing the stored energy, maximum strength, and bond length for bonds C–C, C=C, and C≡C. What is the relationship between tensile strength and the maximum force required to break each type of bond?

3.12. Calculate the specific surface area of the following CNTs:

(a) Nanotube armchair (10, 10) $\rho = 1.33$ g/cm^3.
(b) Nanotube zigzag (17, 0) $\rho = 1.34$ g/cm^3.
(c) Nanotube chiral (12, 6) $\rho = 1.40$ g/cm^3.

Consider length $l = 20$ μm.

3.13. Calculate the tensile strength of the following CNTs:

(a) (9, 0)
(b) (5, 5)
(c) (13, 0)
(d) (12, 12)

3.14. Consider the nanotubes from the previous exercise. How many CNTs of each type would be needed to support the weight of a 1500-kg car?

3.15. Plot a graph of tensile strength (in GPa) as a function of the nanotube diameter for the following CNTs:
(a) (5, 0)
(b) (6, 0)
(c) (7, 0)
(d) (8, 0)
(e) (9, 0)
(f) (10, 0)
(g) (11, 0)
Suggestion: Use Microsoft Excel® to facilitate the calculations and plot the graph.

3.16. Which of the following SWCNTs are conductors and semiconductors?
(a) (6, 4)
(b) (5, 3)
(c) (10, 1)
(d) (5, 3)
(e) (10, 4)
(f) (12, 3)
(g) (11, 0)
(h) (6, 1)
(i) (13, 1)
(j) (6, 4)
(k) (9, 0)

3.17. Find out why some nanotubes are insulators.

3.18. Build the following nanotubes using a graphene sheet printed on a transparency sheet (such as those used on overhead projectors) and use tape to keep the wall of the nanotube stable. Use the graphene sheet in Appendix C.
(a) (9, 0)
(b) (9, 9)
(c) (7, 0)
(d) (11, 11)
(e) (5, 3)

3.19. Build a model of a zigzag nanotube (10, 0) using Styrofoam spheres and toothpicks. Note that the C–C bonds have a defined angle.

Chapter 4

4.1. Describe and compare the different production methods of CNTs. Build a table with the main characteristics (advantages and/or disadvantages) of each method.

4.2. Compare the price of SWCNTs, DWCNTs, and MWCNTs. What are the more expensive types of CNTs and why?

4.3. Based on Section 4.3, prepare a list of good practice measures that should be followed when dealing with carbon nanotubes.

Chapter 5

5.1. Use Eqns (5.5) and (5.4) to deduce Eqn (5.6).

5.2. Calculate the density of the NTC samples with the following characteristics:

(a) MWCNTs, $d = 18$ nm and $d_i = 4$ nm.
(b) MWCNTs, $d = 50$ nm and $d_i = 4$ nm.
(c) SWCNTs, $d = 2.68$ nm and $d_i = 2$ nm.
(d) SWCNTs, $d = 1.55$ nm and $d_i = 0.87$ nm.
(e) MWCNTs, $d = 15$ nm and $d_i = 3$ nm.

5.3. The table shows the density of components from different polymeric composites containing varying concentrations of CNTs. Complete the blank columns with the concentration (V_f) and volume fraction (vol%) for each case.

ρ_{CNT} (g/cm^3)	ρ_m (g/cm^3)	*Wt%*	V_f	*Vol%*
1.50	1.25	0.1	0.00083	0.083
1.9	1.3	0.08		
1.3	1.19	0.5		
2.16	0.96	4		
0.99	1.01	0.75		

5.4. The mechanical properties of ultra-high-molecular weight polyethylene (UHMWPE) can be improved by the addition of CNTs. It is known that the Young's modulus of the UHMWPEs and MWCNTs are worth 640 MPa and 400 GPa, respectively. Compute the Young's modulus of a composite containing 0.1, 0.5, 1, 2, and 5 vol% of CNTs. Use the Voigt and Reuss models for the calculations.

5.5. Repeat the calculations from exercise 5.4 using the Cox model. Consider the case of nanotubes aligned unidirectionally and randomly distributed. Use $l_{CNT} = 10$ μm and $d = 25$ nm.

5.6. Estimate the maximum tensile modulus for a polyurethane (PU) matrix composite containing 10 vol% of CNTs. Assume moduli of 2.79 GPa and 1 TPa for the PU and CNTs, respectively.

5.7. Derive the Eqns (5.47) and (5.48) based on Eqns 5.45 and 5.23 and considering $\zeta = 2l/d$. Start from Eqn (5.18).

5.8. A phenolic resin, with elastic modulus of 5.13 GPa and $\rho = 1.03$ g/cm^3, is reinforced with 0.02, 0.06, 0.09, and 1.1 wt% of the CNTs. The CNTs have a length, diameter, density, and modulus of 2 μm, 40 nm, 900 GPa, and 1.3 g/cm^3, respectively. Calculate the modulus of the composite with CNTs randomly oriented according to the Halpin-Tsai model and modified Halpin-Tsai model. Compare the values obtained and discuss the observed difference.

5.9. A scientist must choose between two types of CNTs for manufacture of a composite with polyamide (PA) matrix. Data for the PA and the CNTs are shown in the following table. Given that a volume fraction of CNTs equal to 0.01 is used (i.e., 1 vol%), suggest what type of CNT is the most suitable to obtain an isotropic PA matrix composite with improved modulus of elasticity. Use the Halpin-Tsai model during calculations.

	E (GPa)	**ρ (g/cm³)**	**l (nm)**	**d (nm)**
PA	2.42	1.3	–	–
CNT 1	700	1.9	50	5
CNT 2	700	1.9	1	10

.10. An undergraduate student works on a project aimed to strengthen high-density polyethylene (HDPE) with MWCNTs. Given that: $E_m = 1.10$ GPa, $\rho_m = 0.96$ g/cm^3, $\alpha = 86$, $\rho_{\text{NTC}} = 2.16$ g/cm^3 and $E_{\text{NTC}} = 910\,GPa$, plot a graph of the elastic modulus as a function of the volumetric concentration of CNT (from 0 to 100 vol%). Adopt the Cox model and consider both cases of CNTs aligned and randomly distributed.

.11. Appendix D at the end of the book presents a Matlab code for calculation of the elastic modulus of composites reinforced with CNTs, according to the rule of mixtures. Write a code similar for the calculation of the modulus based on the series model and another based on the Cox model.

.12. A composite consists of a matrix of polymethylmethacrylate (PMMA) reinforced with 1 wt% of CNTs.

(a) Compute the modulus of the composite according to the Nielsen model assuming the CNTs are aligned. Consider $E_m = 1.55$ GPa, $\rho_m = 1.19$ g/cm^3, $\alpha = 100$, $\rho_{\text{NTC}} = 1.7$ g/cm^3 and $E_{\text{NTC}} = 800$ GPa.

(b) Repeat the calculation in (a) assuming that the CNTs are in a random distribution.

(c) What is the maximum resistance of the composite if $\sigma_{\text{NTC}} = 15$GPa and $l > \sim 10 l_c$?

.13. Carbon nanotubes from three different suppliers (A, B, and C) are available to improve the ultimate strength of an epoxy resin used in the manufacture of aircrafts. Using the data provided in the following table below and knowing that $\tau_{\text{NTC}} = 125$ MPa, $d = 20$ nm, $d_i = 5$ nm, and $V_{\text{NTC}} = 0.01$, find out which supplier offers the CNTs most suitable for the fabrication of composites.

	σ (MPa)	l (μm)
Epoxy	65	–
CNT A	45,000	28
CNT B	45,000	50
CNT C	45,000	1

5.14. The following table shows the effect of CNT X and CNT Y in the modulus of elasticity of polyvinyl acrylonitrile (PVA). Calculate the Young's modulus and ultimate strength of both the types of CNTs. Consider $\rho_m = 1.3$ g/cm^3 and $\rho_{NTC} = 2.15$ g/cm^3.

	Wt%	E	σ (MPa)
CNTs	0	4.34	80
CNT X	0.2	5.31	104
CNT Y	0.2	6.33	122

5.15. The effects of adding randomly dispersed CNTs in the modulus of elasticity and ultimate strength of an epoxy matrix are presented in the following table.

V_f	E (GPa)	Standard Deviation	σ (MPa)	Standard Deviation
0	3.11	0.0524	64.51	2.72
0.00163	3.23	0.0354	68.23	3.80
0.00326	3.48	0.0418	73.05	3.85
0.00652	3.67	0.1245	78.11	7.16
0.0098	3.89	0.0846	85.01	2.99
0.0131	4.20	0.0758	89.41	4.73
0.0197	4.45	0.0962	100.10	4.50
0.0264	4.70	0.0557	104.50	3.51
0.0398	4.98	0.0557	108.20	4.64
0.0535	5.22	0.0863	109.00	1.14
0.0674	5.41	0.0557	110.10	1.22

(a) Plot a graph $E_c \times V_f$ and $\sigma_c \times V_f$. Use a computer and appropriate software (Microsoft Excel, Matlab, Maple, etc.). Include the standard deviation in the experimental points.

(b) Try to fit the data for the modulus of elasticity using the Halpin-Tsai model. Assume $l = 1$ μm and $d = 40$ nm.

(c) Given $l > \sim 10 l_c$, fit the graph $\sigma_c \times V_f$ plotted in (a).

(d) Based on the graph plotted in (c), discuss the discrepancy between the experimental and theoretical values to high concentrations of nanotubes.

5.16. Appendix D at the end of the book presents a Matlab code for calculation of the elastic modulus of composites reinforced with CNTs, according to the Halpin-Tsai model. Enter this code into a computer and perform simulations varying the length, diameter, and modulus of the CNTs. Plot different graphs of E_c as function of V_{NTC}.

5.17. Calculate the thermal conductivity of a PP matrix composite ($k_m =$ 0.11 W/m K) containing 0.27 vol% CNTs ($k_{NTC} = 1000$ W/m K). Perform calculations using the series and parallel models and an arrangement of both in parallel.

5.18. A research project aims to increase the thermal conductivity of an epoxy resin having conductivity of 0.19 W/m K. To this end, CNTs will be used in three concentrations: 0.01, 1, and 5 vol%. Knowing that $k_{NTC} = 190$ W/m K, estimate the thermal conductivity of the composites based on the geometric and the Nan models.

5.19. An engineer must prepare an epoxy matrix composite with thermal conductivity of 1 W/m K. For the application envisioned, the composite must have the shape of a cube with a 10-cm edge. The price of the composite should be as low as possible and three different NTC available are shown in the following table.

	k (W/m K)	Price ($/g)
CNT A	800	300
CNT B	500	125
CNT C	80	13

(a) What would be the required volume fraction of each type of nanotube to increase the thermal conductivity of epoxy from 0.18 W/m K to 1 W/m K? Use the Nan model during calculations.

(b) Given that $\rho_m = 1.18$ g/cm^3 and $\rho_{NTC} = 1.9$ g/cm^3, calculate the mass required of CNT A, CNT B, CNT C, and epoxy for the manufacture of the composites. Consider the values obtained in (a) and remember that $\rho_c = \rho_{NTC} V_{NTC} + \rho_m V_m$.

(c) Knowing that the epoxy used costs $17/kg, calculate the final price of the composites fabricated using different nanotubes. What type of CNT is more feasible to obtain a composite conductivity 1 W/m K and the lowest possible cost?

5.20. Apply the Hatta model to estimate the thermal conductivity of a polystyrene matrix composite ($k_m = 0.14$ W/m K) containing 7 wt% of CNTs with aspect ratios of 15, 45, and 150. The density of the matrix and CNT are worth 1.75 and 5.1 g/cm^3, respectively. Consider that all CNTs have the same thermal conductivity: 250 W/m K.

5.21. The thermal conductivity of polycarbonate can be improved by the addition of CNTs. Consider CNTs with a length of 445 nm, diameter of 3 nm, and conductivity of 230 W/m K. Given $k_m = 0.20$ W/m K, plot a graph of the thermal conductivity of the PC matrix composite as a function of volume fraction of CNTs (0–1). Adopt the parallel, series, geometric, and Hatta models. Tip: Use Microsoft Excel® or similar program.

5.22. To increase the thermal conductivity of PP, a master student prepared composites with different concentrations of CNTs. The samples prepared were characterized and the results are shown in the following table:

Sample	CNTs (wt%)	CNTs (vol%)	*k* (W/m K)
PP	0.0	0.0	0.206 ± 0.02
PP + CNTs	1.5	0.68	0.307 ± 0.014
PP + CNTs	2.5	1.14	0.340 ± 0.043
PP + CNTs	4.0	1.84	0.390 ± 0.023
PP + CNTs	5.0	2.31	0.395 ± 0.007
PP + CNTs	6.0	2.79	0.424 ± 0.021
PP + CNTs	7.5	3.52	0.492 ± 0.011
PP + CNTs	10.0	4.76	0.619 ± 0.011
PP + CNTs	15.0	7.36	0.837 ± 0.068

The CNTs used in the study are characterized by the following properties: $d = 10$ nm, $l = 10$ μm and $k = 20$ W/m K.

(a) Using a computer plot a graph of the thermal conductivity of the composite as a function of volume concentration (vol%) of CNTs. Use the data in the table and include the standard deviation.

(b) Use the Nielsen model to try to fit the data plotted. Consider $A = 13.1$, $\psi = 0.20$ and vary the thermal conductivity of the CNTs: 200, 800, and 1500 W/m K. What is the best fit obtained? Why is the value used for the CNT conductivity in this case so inferior to the theoretical for CNTs (6600 W/m K)?

5.23. Select three available models for calculating the thermal conductivity of composites containing CNTs. Write a Matlab® code for calculating the thermal conductivity for each model selected. The program should perform the following steps:

(a) Calculate the thermal conductivity as a function of volume fraction of CNTs ($V_f = 0 - 1$).

(b) Export the results to a file on your hard disk.

(c) Print the simulation results on the screen.

(d) Plot a graph of V_{NTC} versus k and display it in a new window.

(e) Export the chart to the file. tif hard disk. Hint: Use the "Code Halpin-Tsai.m" presented in Appendix D as a starting point.

5.24. Show that Eqns (5.58)–(5.61) reduce to the rule of mixtures when $A \rightarrow \infty$ and $\varphi_{max} = 1$.

5.25. Show that Eqns (5.58)–(5.61) reduce to the inverse rule of mixtures when $A \rightarrow 0$.

5.26. Nylon matrix composites were prepared with different concentrations of CNTs. The electrical conductivity measured for the nylon and different samples prepared are presented in the following table:

CNTs (vol%)	σ (S/cm)
0	5.00×10^{-16}
0.086	7.60×10^{-16}
0.15	8.80×10^{-16}
0.38	4.14×10^{-10}
0.48	3.00×10^{-7}
0.98	3.99×10^{-5}
1.97	3.45×10^{-4}

(a) Using the experimental data presented in the table, plot a graph of electrical conductivity as a function of CNT volume fractions. Discuss the graph.

(b) Fit the experimental data and determine the parameters σ_0 and t to fit with $R^2 > 0.85$.

Chapter 6

6.1. Describe in detail the three main methods used for preparation of polymer matrix composites reinforced with CNTs.

6.2. Consider the processing method in solution, melt, and in situ. Which methods could be applied for preparing a composite of:

(a) Epoxy-reinforced with SWCNTs.

(b) Nylon 6,6 with MWCNTs.

(c) Polyethylene with CNTs.

(d) Vinyl ester with CNTs.

6.3. Compare the techniques of magnetic and mechanical stirring. Discuss the effectiveness of these techniques in preparing composites reinforced with CNTs.

6.4. What are the main characteristics of equipment such as a dual asymmetric centrifuge and high shear mixer?

6.5. Describe the process of sonification in detail.

6.6. What are the advantages of using a three roll mill (calender) for dispersion of CNTs?

6.7. As we saw in the previous chapter, CNTs with long lengths are desirable for obtaining composites with improved mechanical properties. Discuss how increasing CNT length may affect the processability of composites.

6.8. From an industrial point of view, what are the most attractive techniques for manufacturing nanocomposites? What techniques are already used in companies?

Answers for the questions and exercises

Chapter 1

1.1. $N = 4.41 \times 10^{23}$ atoms

1.2. **(a)** $N = 4.41 \times 10^{23}$ atoms
(b) $44.85
(c) $wt_{Au} = 3.27 \times 10^{-22}$ g

1.3. 1×10^{16} nanotubes
$V = 1 \times 10^{-8}$ m^3

1.4. **(a)** $V = 7.10$ cm^3
(b) $V = 1.18 \times 10^{-23}$ cm^3
(c) $V = 8.35 \times 10^{-30}$ cm^3
(d) $V = 6.59$ cm^3; $V = 1.09 \times 10^{-23}$ cm^3; $V = 7.98 \times 10^{-30}$ cm^3

1.6. **(a)** 10^3 nm
(b) 10^6 nm
(c) 10^7 nm
(d) 10^9 nm

1.9. O = 7.58 atoms; H = 16.13 atoms
Na = 3.01 atoms; C = 6.49 atoms
Cl = 5.05 atoms

1.10. **(a)** $S_{SA} = 0.31$ m^2/kg
(b) $S_{SA} = 31.09$ m^2/kg
(c) $S_{SA} = 3108.81$ m^2/kg
(d) $S_{SA} = 310880.80$ m^2/kg

1.14. $S_{A-\text{cube}}/S_{A-\text{sphere}} = 6/\pi$
The cube has a surface area 1.24 times larger than the sphere.

1.15. **(a)** $S_A/V = 6/a$
(b) $S_A/V = 6/d$
(c) $S_A/V = 2/h$
(d) $S_A/V = 3/d$
(e) $S_A/V = 4/d$

1.17. $d = 56.2\,nm$
$S_A/V = 1.07 \times 10^8\,m^{-1}$

1.18. $S_A/V = 4.52 \times 10^9\,m^{-1}$
$S_{SA} = 200\,m^2/g$

1.20. The human body consists of approximately 10^{27} atoms arranged in a physical structure. Forty-one kinds of chemical elements are most commonly found in body composition; however, C, H, O, and N atoms represent 99% of the atoms.

Chapter 3

3.5. **(a)** $d = 0.81$ nm
(b) $d = 0.90$ nm

(c) $d = 1.36$ nm
(d) $d = 1.11$ nm
(e) $d = 1.63$ nm
(f) $d = 0.68$ nm
(g) $d = 0.94$ nm
(h) $d = 0.85$ nm
(i) $d = 0.87$ nm
(j) $d = 1.23$ nm

3.8. 13,600 fullerenes

3.9. $d_i = 1$ nm

3.10. $\sigma_{C=C} = 273$ GPa
$\sigma_{C\equiv C} = 415$ GPa

3.11. The ratio "tensile strength"/"force required to break the bond" is the same for each type of bond: $\sim 5.38 \times 10^{19}$ m^{-2}.

3.12. **(a)** 2.21×10^6 m^2/kg
(b) 2.24×10^6 m^2/kg
(c) 2.30×10^6 m^2/kg

3.13. **(a)** $\sigma_{NTC} = 56.86$ GPa
(b) $\sigma_{NTC} = 68.87$ GPa
(c) $\sigma_{NTC} = 39.39$ GPa
(d) $\sigma_{NTC} = 28.77$ GPa

3.14. 6.54×10^{11} *NTC* (9, 0)
5.89×10^{11} *NTC* (5, 5)
4.54×10^{11} *NTC* (13, 0)
2.45×10^{11} *NTC* (12, 12)

3.15.

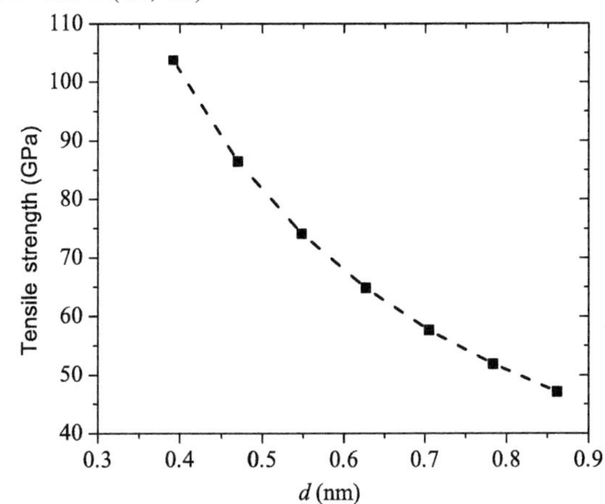

3.16. Conductors: c, e, f, i, k
Semiconductors: a, b, d, g, h, j

Chapter 5

5.2. **(a)** $\rho = 2.14\ \text{g/cm}^3$
(b) $\rho = 2.24\ \text{g/cm}^3$
(c) $\rho = 1\ \text{g/cm}^3$
(d) $\rho = 1.54\ \text{g/cm}^3$
(e) $\rho = 2.16\ \text{g/cm}^3$

5.3.

V_f	Vol%
0.00083	0.083
0.00055	0.054
0.0046	0.46
0.018	1.8
0.0077	0.77

5.4.

Vol%	Reuss (GPa)	Voigt (GPa)
0.1	0.6406	1.039
0.5	0.6432	2.637
1	0.6465	4.634
2	0.6530	8.627
5	0.6736	20.61

5.5.

Vol%	E_c
0.1	0.71
0.5	0.99
1	1.35
2	2.07
5	4.25

5.6. $E_c = 103$ GPa

5.8.

Vol%	H–T E_c (GPa)	Modified H–T E_c (GPa)
0.016	5.151000464	5.132392904
0.048	5.193019779	5.137179935
0.071	5.224550362	5.140771281
0.87	6.294196664	5.262219144

5.9. CNT 1 is the most appropriate. $E_{c1} = 5.0$ GPa, $E_{c2} = 3.5$ GPa

.10.

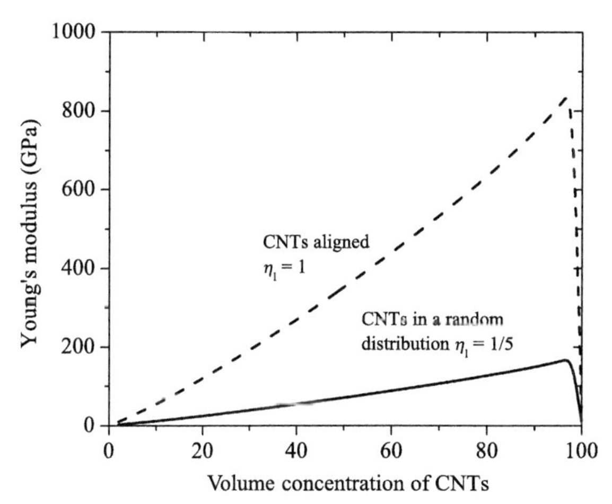

.12. **(a)** $E_c = 3.13$ GPa
(b) $E_c = 1.65$ GPa
(c) For $\sigma_m = 55$ MPa, $\sigma_c = 160$ MPa

.13. CNT A $\sigma_c = 487$ MPa
CNT B $\sigma_c = 514$ MPa
CNT C $\sigma_c = 127$ MPa
CNT B is the most appropriate.

.14. $E_{CNTs-X} = 806$ GPa
$E_{CNTs-Y} = 1.65$ TPa
$\sigma_{CNTs-X} = 20$ GPa
$\sigma_{CNTs-Y} = 35$ GPa

.17. Series $k_c = 0.11$ W/m K
Parallel $k_c = 2.81$ W/m K
Arrangement $k_c = 1.46$ W/m K

.18.

Vol%	Nan k_c (W/m K)	Geometric k_c (W/m K)
0.01	0.196	0.19
1	0.823	0.204
5	3.36	0.268

.19. **(a)** $V_{CNTs-A} = 0.0031$; $V_{NCNTs-B} = 0.0049$; $V_{CNTs-C} = 0.031$
(b) $wt_{CNTs-A} = 5.84g$; $wt_{epoxy} = 1.18$ kg;
$wt_{CNTs-B} = 9.35g$; $wt_{epoxy} = 1.17$ kg;
$wt_{CNTs-C} = 58.42$ g; $wt_{epoxy} = 1.14$ kg
(c) Cost A = \$21.75; cost B = \$21.13; cost C = \$20.20.

5.20.

α	k_c (W/m K)
15	0.13
45	0.49
150	0.19

5.26. **(a)**

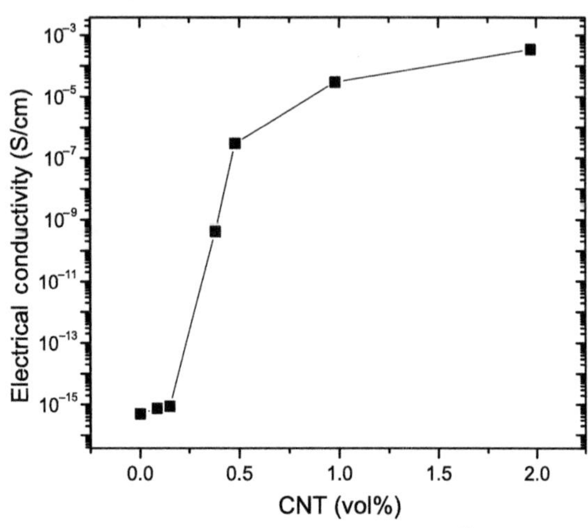

(b) Considering $V_c = 0.25$, we get $\sigma_0 = 6.26 \times 10^{-5}$ S/cm, $t = 4.96$, and, in this case, $R^2 = 0.88$. Alternatively, if $V_c = 0.30$, we get $\sigma_0 = 8.91 \times 10^{-5}$ S/cm, $t = 4.30$, and, in this case, $R^2 = 0.92$.

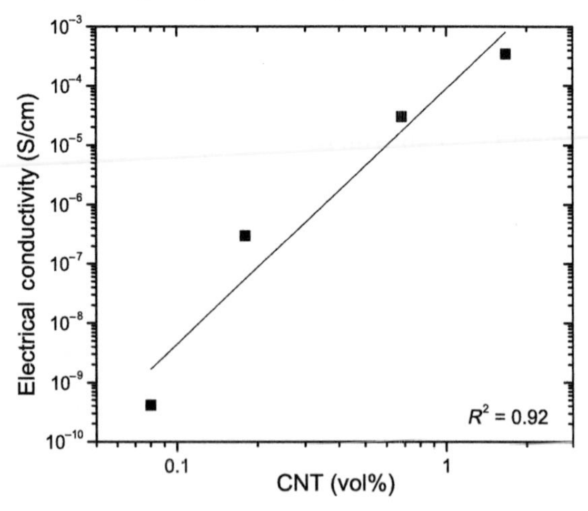

Index

Note: Page numbers followed by "b", "f", and "t" indicate boxes, figures, and tables respectively.

Printed and bound by CPI Group (UK) Ltd, Croydon, CR0 4YY
08/05/2025
01864838-0003